Praise for
Dangerous Doses

"A riveting account of a 2½-year investigation in South Florida . . . As Eban recounts, the scam was broken wide open by a 'ragtag' group of seasoned investigators who seem as if they were cast right out of an episode of *The Wire*."
—Bernadine Healy, *U.S. News & World Report*

"This is a book that comes along so rarely in nonfiction—brilliantly reported, written with the pace of a potboiler and harrowing in its societal repercussions. Katherine Eban takes us on a journey into the underbelly of the pharmaceutical industry so spooky and strange and sinister and deadly, you will have a hard time believing it is true. But it is, every word, which only makes *Dangerous Doses* shine even more."
—Buzz Bissinger, author of *Friday Night Lights*

"The book is the page-turner result of a painstaking investigation into what happens when a poorly regulated industry, enabled by a lack of price controls on drugs in the USA, is allowed to operate."
—*Lancet*

"A riveting tale. *Dangerous Doses* is part detective story, part pharmacological primer."
—*The New York Sun*

"Reads like a thriller."
—*Business Wire*

Dangerous Doses

Katherine Eban

Dangerous Doses

A True Story of Cops,
Counterfeiters, and the
Contamination of
America's Drug Supply

A HARVEST BOOK
HARCOURT, INC.
Orlando Austin New York
San Diego Toronto London

Requests for permission to make copies of any part of
the work should be mailed to the following address:
Permissions Department, Harcourt, Inc.,
6277 Sea Harbor Drive, Orlando, Florida 32887-6777.

www.HarcourtBooks.com

"A Counterfeit—a Plated Person" reprinted by the permission
of the publishers and the Trustees of Amherst College from *The
Poems of Emily Dickinson*, Thomas H. Johnson, ed., Cambridge,
Mass.: The Belknap Press of Harvard University Press,
Copyright © 1951, 1955, 1979, 1983 by the President and
Fellows of Harvard College.

The Library of Congress has cataloged the hardcover
edition as follows:
Eban, Katherine.
Dangerous doses: how counterfeiters are contaminating
America's drug supply/Katherine Eban.—1st ed.
p. cm.
1. Pharmaceutical policy—United States. 2. Drug adulteration.
3. Drugs—Safety measures. 4. Product counterfeiting.
[DNLM: 1. Fraud—United States. 2. Pharmaceutical
Preparations—supply & distribution—United States.
3. Consumer Advocacy—United States. 4. Government
Regulation—United States. 5. Legislation, Drug—United
States. QV 736 E15d 2005] I. Title.
RA401.A3E25 2005
363.19'4—dc22 2004025581
ISBN-13: 978-0-15-101050-9 ISBN-10: 0-15-101050-1
ISBN-13: 978-0-15-603085-4 (pbk.) ISBN-10: 0-15-603085-3 (pbk.)

Text set in Ehrhardt
Designed by Kaelin Chappell

Printed in the United States of America

First Harvest edition 2006
K J I H G F E D C B A

For my beloved friend Karen Avenoso (1967–1998)
Whom I admire and miss every day

A Counterfeit—a Plated Person—
I would not be—
Whatever strata of Iniquity
My Nature underlie—
Truth is good Health—and Safety, and the Sky.
How meagre, what an Exile—is a Lie,
And Vocal—when we die—

—EMILY DICKINSON

Contents

A Note to Readers

DANGEROUS DOSES IS BASED ON TWO-AND-A-HALF YEARS OF reporting in which I conducted more than 160 significant interviews with people involved in the problem of contaminated medicine in America. Those I interviewed include government investigators, pharmacists, doctors, patients, drug makers, lobbyists, politicians, regulators, lawyers, researchers, and pharmaceutical wholesalers, as well as those involved in drug counterfeiting and diversion. I also conducted dozens of background interviews with sources that asked not to be named but whose contributions helped to shape the book.

In the course of this reporting, I obtained more than 13,000 pages of documents. These include government records received in response to sixteen Freedom of Information requests I submitted to the Food and Drug Administration, the Florida health department, the Florida Department of Law Enforcement, and other state pharmacy boards. Wherever possible, I have relied on original material including internal memorandums, police reports, sales records, shipping records, correspondence, investigative records, inspection reports, search warrants, surveillance videotapes, photographs, and court documents. This material includes the investigative records from Operation Stone Cold of both the Florida Department of Law Enforcement and the Miami-Dade Police Department.

My reporting has benefited from the extensive help of federal and state law-enforcement officials, including members of the Operation Stone Cold task force. In the book's endnotes, the reader will find references to documents I used to re-create scenes and make factual assertions. Material that came from interviews and events that I reported firsthand is not cited in endnotes. Instead, after the endnotes, I have provided a list of substantive on-the-record interviews I conducted for the book. I have not included the names of the hundreds of people I interviewed whose information did not directly shape the narrative.

I made sixteen trips to Florida, as well as trips to Nevada, Missouri, Tennessee, Georgia, and Washington, D.C. Scenes I did not witness were re-created through interviews with those present as well as through investigative records, transcripts, audiotapes, and videotapes. I worked to resolve discrepancies through repeated interviews with as many of those present as possible. The few instances where discrepancies remain are documented in endnotes. In drawing on investigative reports, I did my best to independently confirm the information contained in them.

Each of those persons or companies described in this book as being involved in potentially criminal or other unlawful activity was given an opportunity to comment, respond to allegations, or correct the record, either firsthand or through their lawyers. Readers should keep in mind that, except as expressly stated otherwise in the text or endnotes: At the time this book went to press none had been determined to have violated any criminal or other laws; all have, or should be considered to have, denied any wrongdoing; and each of those indicted has pleaded not guilty. All should be deemed innocent of any accusations of unlawful activity unless and until proved otherwise in the final outcome of judicial proceedings. Readers are encouraged to consult the Who's Who immediately following and the endnotes.

Winter 2004/2005

Who's Who

FLORIDA

Operation Stone Cold

HORSEMEN OF THE APOCALYPSE—Five investigators who penetrated the corruption of America's drug supply.

Cesar Arias, Drug Inspector, Bureau of Statewide Pharmaceutical Services

Gene Odin, Drug Inspector, Bureau of Statewide Pharmaceutical Services

Randy Jones, former Detective, Miami-Dade Police Department

John Petri, Sergeant, Miami-Dade Police Department

Gary Venema, Special Agent, Florida Department of Law Enforcement

SUPPORTING INVESTIGATORS

Jack Calvar, Senior Investigator, Medicaid Fraud Control Unit, Attorney General's Office

Steve Zimmerman, Detective, Miami-Dade Police Department

CATALYSTS OF THE INVESTIGATION

Martin J. Bradley, CEO & Founding Partner, BioMed Plus, Miami

Sydney Dean Jones, pharmacy technician; cooperating witness

Annette Mantia, former secondary wholesaler; cooperating witness

Government

DEPARTMENT OF HEALTH

Dr. John O. Agwunobi, Secretary

Robert Daniti, General Counsel

Jerry Hill, Chief, Bureau of Statewide Pharmaceutical Services

Gregg Jones, Pharmaceutical Program Manager, Bureau of Statewide Pharmaceutical Services

Sandra Stovall, Compliance Officer, Bureau of Statewide Pharmaceutical Services

Robert Loudis, former Drug Inspector, Bureau of Statewide Pharmaceutical Services

ATTORNEY GENERAL'S OFFICE

Charlie Crist, Attorney General

Peter Williams, Statewide Prosecutor

Melanie Ann Hines, former Statewide Prosecutor

Oscar Gelpi, Assistant Statewide Prosecutor

Stephanie Feldman, former Assistant Statewide Prosecutor

Robert Penezic, former Assistant Statewide Prosecutor

FLORIDA DEPARTMENT OF LAW ENFORCEMENT

Michael Mann, Assistant Special Agent in Charge

Tim Moore, former Commissioner

HOUSE OF REPRESENTATIVES

Representative Ed Homan (R-Temple Terrace)

STATE SENATE

Senator Walter "Skip" Campbell Jr. (D-Tamarac)

FEDERAL

Food & Drug Administration

OFFICE OF THE COMMISSIONER

William Hubbard, Senior Associate Commissioner for Policy, Planning and Legislation

Nevada Board of Pharmacy

Louis Ling, General Counsel

Keith Macdonald, Executive Secretary

Texas Department of Health

John Gower, Director, Drugs, Devices & Cosmetics

Albert Hokins, Jr., former Investigator, Drugs, Devices & Cosmetics

MANUFACTURERS

J. Aaron Graham, Vice President, Corporate Security, Purdue Pharma; former special agent, FDA Office of Criminal Investigations

Jon Martino, Security Specialist, Amgen Inc.

PRIMARY DISTRIBUTORS

Chris Zimmerman, Senior Director, Security and Regulatory Affairs, AmerisourceBergen

Neil Spence, former Vice President, National Specialty Services, Cardinal Health Inc.

SECONDARY DISTRIBUTORS

Florida

Susan Cavalieri, former Principal, AmeRx Pharmaceutical; cooperating witness—Relinquished her wholesale license and paid $900,000 as part of her agreement with Florida prosecutors.

Brian Hill, former Chief Operating Officer, Jemco Medical International—Company license revoked.   .

José Castillo, former President, Jemco Medical International—Company license revoked.

Nevada

Lance Packer, former Principal, Legend Pharmaceuticals—Company license revoked.

Paul DeBree, former President, Dutchess Business Services—Company license revoked.

Robb Miller, President, Caladon Health Solutions

Missouri

Doug Albers, Owner, Albers Medical Distributors

Noah Salcedo, Buyer, Albers Medical Distributors

WHOLESALER TRADE GROUPS

Healthcare Distribution Management Association (HDMA)—Represents primary wholesalers.

Bonnie Basham, Lobbyist

Pharmaceutical Distributors Association (PDA)—Represents secondary wholesalers.

Sal Ricciardi, President

Bruce Krichmar, Legislative/Regulatory Affairs Director(s)

Ross McSwain, Lobbyist and Lawyer, Blank, Meenan & Smith

Anthony Young, Lawyer, Kleinfeld, Kaplan & Becker

MICHAEL CARLOW'S WORLD

Michael Carlow, ex-convict charged with trafficking in bad medicine—Indicted in July 2003 on charges of racketeering, conspiracy to commit racketeering, organized scheme to defraud, grand theft, dealing in stolen property, sale or delivery of a controlled substance, possession with intent to sell prescription drugs, sale and delivery of a controlled substance and purchase or receipt of a prescription drug from an unauthorized person. Pleaded not guilty and is in jail awaiting trial.

Alleged Shell Companies

BTC Wholesale, Florida

G&K Pharma, Maryland

Complete Wholesale, Maryland

Accucare, Florida

JB Pharmaceutical, Texas

Pormis Wholesale Distributors, Texas

Alleged Accomplices

Candace Carlow, wife—Indicted in July 2003 on charges of racketeering, conspiracy to commit racketeering, and organized scheme to defraud. Pleaded not guilty and is awaiting trial.

Thomas Atkins, brother-in-law—Indicted in July 2003 on charges of racketeering, conspiracy to commit racketeering, organized scheme to defraud, and possession with intent to sell prescription drugs. Pleaded not guilty and is awaiting trial.

Marilyn Atkins, mother-in-law—Indicted in July 2003 on charges of racketeering, conspiracy to commit racketeering, and organized scheme to defraud. Pleaded not guilty and is awaiting trial.

José L. Benitez—Indicted in July 2003 on charges of racketeering, conspiracy to commit racketeering, and organized scheme to defraud. A fugitive with a warrant out for his arrest.

Fabian Díaz—Indicted in July 2003 on charges of racketeering, conspiracy to commit racketeering, organized scheme to defraud, grand theft, and dealing in stolen property. A fugitive with a warrant out for his arrest.

David Ebanks—Indicted in July 2003 on charges of racketeering, conspiracy to commit racketeering, organized scheme to defraud, and possession with intent to sell prescription drugs. Pleaded not guilty and is awaiting trial.

Henry García—Indicted in July 2003 on charges of racketeering, conspiracy to commit racketeering, organized scheme to defraud, and possession with intent to sell prescription drugs. Pleaded not guilty and is awaiting trial.

Others

> Gina Catapano—Did secretarial work for Carlow.
>
> Jean McIntyre—Carlow's banker and prospective business partner.
>
> Mark Novosel, former Carlow deputy; cooperating defendant. Pleaded guilty in June 2003 to organized scheme to defraud and is awaiting sentencing.

JOSÉ GRILLO'S WORLD

> José Grillo, charged with counterfeiting medicine— Indicted in July 2003 on charges of organized scheme to defraud, purchase or receipt of a prescription drug from an unauthorized person, and possession with intent to distribute prescription drugs. Indicted in May 2004 on charges of racketeering, conspiracy to commit racketeering, organized scheme to defraud, product tampering, sale of prescription drugs to an unauthorized person, selling or offering for sale counterfeit goods, and forging or counterfeiting private labels. Pleaded not guilty and is in jail awaiting trial.

Alleged Accomplices

> Nicholas Just, Co-owner, Playpen South—Indicted in May 2004 on charges of racketeering, conspiracy to commit racketeering, organized scheme to defraud, product tampering, purchase or receipt of a prescription drug from an unauthorized person, sale of a prescription drug to an unauthorized person, and selling or offering for sale counterfeit goods. Pleaded not guilty and is awaiting trial.
>
> Dr. Paul Perito, Co-owner, Playpen South—Indicted in May 2004 on charges of racketeering, conspiracy to commit racketeering, organized scheme to defraud, money laundering, purchase or receipt of a prescription drug from an unauthorized person, sale of a prescription drug to an unauthorized person, and selling or offering for sale counterfeit goods. Pleaded not guilty and is awaiting trial.

Carlos Luis, Owner, Medix International—Indicted in
May 2004 on charges of racketeering, conspiracy to
commit racketeering, organized scheme to defraud,
Medicaid fraud, possession with intent to sell
prescription drugs, possession with intent to sell or
deliver a controlled substance, money laundering,
product tampering, purchase or receipt of a
prescription drug from an unauthorized person, and
selling or offering for sale counterfeit goods. Pleaded
not guilty and is awaiting trial.

Eddie Mor, Owner, Express Rx—Indicted in May 2004
on charges of racketeering, conspiracy to commit
racketeering, organized scheme to defraud, purchase
or receipt of a prescription drug from an unauthorized
person, possession with intent to sell prescription
drugs, product tampering, and selling or offering for
sale counterfeit goods. Pleaded not guilty and is
awaiting trial.

Others

Maria Castro and Jesús Benitez—Ran pharmacy alleged to
have diverted medicine. Indicted in October 2003 on
charges of organized scheme to defraud, possession
with intent to sell prescription drugs, and sale of a
prescription drug to an unauthorized person. In
February 2005 Benitez pleaded guilty to the
unauthorized sale of prescription drugs, received
probation, and agreed to cooperate. The charges
against Castro were dropped, and she agreed to
cooperate.

Armando Rodriguez—Admitted that he bought vials for
Grillo. Cooperating witness.

Silvino Cristobal Morales—Admitted that he helped
Grillo relabel vials. Cooperating witness.

Prologue

EVERY YEAR AMERICANS FILL MORE THAN THREE BILLION PRE-
scriptions through reputable pharmacies and assume their medicine will be pure and effective—precisely what their doctors
ordered. Since 2000, an increasing number of Americans who
went to their pharmacies got counterfeit medicine instead. The
medicine looked the same; the packaging appeared identical; the
pharmacists could not tell the difference. But the medicine was
different. Counterfeiters seeking profits had diluted it, replaced
it with cheaper ingredients, or relabeled it to appear stronger.

The costly, brand-name medicine affected by this ranged
from treatments for cancer and renal failure to drugs for
AIDS and high cholesterol. The patients who wound up with
the counterfeit medicine were not risk takers who sought discounts across borders or over the Internet. They had bought
legitimate medicine from the heart of America's drug supply.
When these patients suffered terrible side effects like rashes
or stinging or when their drugs abruptly stopped working,
counterfeits were often the last possibility they considered.

DANGEROUS DOSES IS A BOOK ABOUT THE SYSTEMIC CONTAMINA-
tion of the very medicine we trust the most—the medicine

that is prescribed by our doctors and distributed within our borders. By 2002, federal and state investigators faced a dramatic increase in cases of domestic pharmaceutical counterfeiting. Each incident seemed to grow in scale. One involved enough cancer medicine to treat 25,000 patients for a month; another involved a best-selling cholesterol drug that may have reached 600,000 patients.

The counterfeits were not a one-time fluke but rather a consequence of America's distribution system. Our medicine moves through a gray market of middlemen who trade the drugs as they would any other commodity, seeking profits from ever-fluctuating prices. These sales can obscure the medicine's origin and make its purity impossible to guarantee. They also open the door to counterfeits.

Investigators realized by 2002 that felons including narcotics traffickers and those with ties to organized crime had infiltrated America's drug supply. After becoming licensed wholesalers, they poured compromised medicine into our distribution system. Their profits rivaled those in the narcotics trade. As investigations proceeded across the country, one of the most intense ones was in South Florida, where many of the counterfeits originated.

By the fall of 2002, after receiving a tip from a longtime source in the federal government, I was reporting on the issue of domestic counterfeits for *Self* magazine. In eight years as an investigative medical reporter, I had never encountered a story like this one. Counterfeits had reached patients across the country. People had been hurt. Yet the public knew almost nothing about the problem, which had been growing quietly for years.

On my first reporting trip to South Florida, several experiences led to the development of this book. Driving out of Fort Lauderdale, I pulled off a busy road to take a call from the public relations director of a major drug company. I asked him to comment on what I'd learned: that numerous lots of his company's lifesaving drug had been relabeled to appear twenty times the actual strength; that investigators from the Food and Drug Administration and his own company were on their way to Florida to assist with the investigation; that a licensed distributor whose home I was driving to was suspected of trafficking in counterfeit versions of that very medicine.

"I'd hate to have you short the stock [try to profit from a decline in the share price]" over local and contained incidents, he responded. The comment surprised me. The drug company had been a victim of the counterfeiting and was confronting a public health issue. Why would it deny or minimize the problem? The man's remark inadvertently revealed that at least some in the drug industry associated exposure of the counterfeiting problem with financial loss. If drug makers had downplayed the problem, who else had? And what could patients do about it?

On that same trip, I met with a group of Florida investigators who were trying to track the sources of the tainted medicine and bring the problem to wider attention. Their work began haphazardly with a burglary at a medicine warehouse and had morphed into an enormous case; their suspects included one-third of Florida's 455 licensed drug wholesalers. The problem was so big that they likened their efforts to "shoveling sand in a hurricane."

Their task force, called Operation Stone Cold, consisted principally of three cops, Gary Venema, John Petri, and Randy Jones, and two pharmacists, Cesar Arias and Gene Odin. Four were over the age of fifty. All relied to varying degrees on prescription medicine they no longer trusted. Three had been ready to retire before this case came along. Now they worked seven days a week, often around the clock, and called their indissoluble team the Horsemen of the Apocalypse. "The public is so snookered," Cesar Arias told me. "If they really knew, they'd be outraged."

The public needed to know. Yet everywhere I looked, dangerous public health implications had been veiled in secrecy. Drug makers are not required to report counterfeit medicine to the FDA. While the FDA oversees drug manufacturers, the regulation of drug wholesalers is left to each state and a patchwork of regulations. Our drugs sometimes pass through a dozen hands on their way to our pharmacies. And a law enacted almost two decades ago that might have strengthened our supply chain had been gutted under pressure by various business interests.

I spent the next two-and-a-half years following the investigators who uncovered the counterfeits and the patients who received them. I met with drug makers, distributors, and regulators, as well as those accused of criminal misconduct. Most of the events in this book I reported as they unfolded. I allowed the haphazard path of the counterfeits to define the unfolding narrative, which weaves together the stories of strangers in disparate parts of the country whose lives become inextricably and fatefully linked by bad medicine. Patients who understand the hidden dangers can demand the safest drugs possible. I hope this book will help begin that process.

Part One

1. A Victim of Success

January 2002
Miami, Florida

ABOARD A CRUISE SHIP TO COZUMEL, THE WEATHER SPARKLING and an Absolut and soda in hand, Marty Bradley stared at the Gulf of Mexico from inside a locked suite, his silent misery fed by thoughts of betrayal, financial ruin, and even physical danger.

He had brought sixty employees on the gleaming white ship for his company's annual blowout of dancing, gambling, and sunbathing, a reward for meeting their sales targets. But all Bradley could think of now was armed guards, locksmiths, and lie detector tests—and which of the employees on board had sold him out and gotten away with the score of their lives.

Just twenty-four hours ago, while he had been packing for the cruise, a white van had backed into an alleyway behind his Miami warehouse. Some men climbed from the van and managed to twist the dead bolt, tear off the rear metal door, and enter the warehouse around 8 P.M. Once inside, they knew exactly what to look for.

Shortly afterward, the distribution manager, René Perez, returned from his errand at Dadeland Mall and swung past the back alley as was his habit, partly because he was responsible for keeping it clear and partly because Bradley was his brother-in-law. Perez felt protective and viewed the business as a family affair.

Usually the narrow, dimly lit alley was empty. But tonight Perez saw the white van parked outside the warehouse, its motor running and its side door open. He turned and drove toward it along the dusty street. As he angled to park, a man in a long-sleeved T-shirt holding something in his hand ran from the warehouse and leapt into the van as it pulled out, tires screeching.

Perez followed, peeling down the alley behind the warehouse, almost crashing into a garbage dumpster as he struggled, unsuccessfully, to make out the license plate through the uneven street light. After about fifty yards the van barreled north onto another side street and disappeared into Miami's traffic-clogged arteries.

Stunned, Perez stopped and called Bradley at home to tell him about the burglary.

BRADLEY'S COMPANY, BIOMED PLUS, ONE OF THE LARGEST PRI-vate wholesale distributors of blood products, sells fragile plasma derivatives and other specialty medicines to doctors' offices, hospitals, and even competing wholesale companies.

The thieves had headed directly for a freezer that contained blood products destined for patients with compromised immune systems, hemophilia, and other rare disorders. All told, they had taken 344 vials of clear liquids that for many patients meant the difference between life and death. Some of the vials cost almost $4,000 apiece. The heist was worth about $335,000.

What bothered Bradley most was not what they had taken but *when*. The break-in occurred just hours after the delivery of a shipment that included a rare formula called NovoSeven to help hemophiliacs form blood clots. The thieves had taken all of it. Bradley spent the next day—in the hours before boarding the cruise—hiring an armed security guard for $8,000 a month, repairing the damaged door, and installing new locks and metal gates for $150,000. He also arranged for several of his employees to take lie detector tests.

Then he reported the theft to the Bureau of Statewide Pharmaceutical Services, a regulatory requirement that was sure to solve nothing. The inspector he knew there, Cesar Arias, a tousled Cuban-American with a chipped front tooth whose heart was certainly in his job, had no juice whatsoever. One glance at the man's car, a dilapidated blue Buick that looked like it had been pulled off a junk heap, told the story of his agency's budget woes.

The local cops took a report, but they were too busy chasing cocaine and other street drugs to care much about a theft of clotting factor.

But Bradley knew the stolen vials posed a serious danger. The medicine inside was for critical care. It had to remain motionless at a constant temperature and could only be transported with careful planning. At best, it would become useless to a patient and at worst, might do harm. Bradley also worried that the men in the van would return. Or that next time they would come back armed when his employees were there.

Bradley shared these fears with two of his managers, who also stayed in their cabins during the cruise, drinking and avoiding the festivities. On the last day during the group photograph, the three men looked uneasy amid the smiling, well-tanned sales force.

Bradley was in the ship's cocktail lounge waiting to disembark when his cell phone rang. His purchasing manager, Marlene Caceres, was calling about a deal that had been offered in Bradley's absence. A small pharmaceutical wholesale company, the Stone Group, had called and sent faxes offering to sell some plasma derivatives. Bradley had done business with the fledgling company before.

The pharmaceutical wholesale market operates as an all-hours auction, with deals and discounts materializing suddenly and medicine passing through many hands. And while few patients know that these middlemen exist, much of the nation's medicine passes through companies like Bradley's BioMed Plus and the Stone Group. On this deal, the Stone Group had made a typical pitch.

But as Caceres read off the details of the offer, Bradley said, "I don't believe it." Everything she mentioned—including twenty-two vials of NovoSeven at 1.2 milligrams, another twenty-nine at 4.8 milligrams, along with specific amounts of Gammimune, Gammagard, and Iveegam, all for the steeply discounted price of $229,241—was identical to his list of stolen goods, right down to the specific quantities. The medicine was too rare and was almost never traded freely. Bradley knew the medicine was his.

2. Flamingos in Missouri

Spring 2002
Harvester, Missouri

AT THE DOOR OF HIS MODEST TRACT HOUSE, ED BLOUNT STOOPED to pick up the morning paper when he caught sight of the shiny pink shapes on his lawn. They were staked into the dewy grass right next to his pickup. They had not been there last night. A slow smile warmed his face, weathered from years of outdoor work. "Maxine," he called into the house, which sat amid one-story homes with unguarded bicycles and open garage doors in St. Louis's oldest suburb. "Come here. I want to show you something."

His sixty-year-old wife came slowly behind him wrapped in a bathrobe and tried to focus on the lawn. In addition to everything else, Maxine thought her vision must be going. But the shapes out front were unmistakable: A small flock of pink flamingos, shiny plastic birds on spikes, had been stuck into the ground.

She laughed out loud, delighted by the mystery, the pink birds just the latest gift from a secret group of friends. Since she had gotten sick, they had brought flowers, cards, and candies, but had never revealed their names. Maxine, in turn, had printed up posters with a picture of a rat, offering a reward to anyone who could help identify her friends. But the secret persisted and the gifts continued. Maxine loved a good conspiracy.

When her breast cancer returned in the spring of 2001, after a mastectomy that had allowed her several years of good health, Maxine underwent chemotherapy. Her son, Bill, who had left the house at eighteen, moved back home to be with her.

Her friends did everything they could to boost her spirits. Maxine had visitors all day long, first at the Mailboxes Etc. store she ran in nearby O'Fallon, and then later at home when her health deteriorated. Her friend Patti Silvey, a bank branch manager, drove her to Illinois to buy lottery tickets for a $206 million jackpot that she didn't win. She enjoyed the trip nonetheless. Maxine liked to gamble and still went to the casinos, which made her feel lucky.

On good days when her energy came back, she again commanded the big rough-hewn kitchen, and her touches were everywhere, from the family photos taped to the refrigerator to the cheerful posters that brightened the walls. For years Maxine fed her husband and their six children here, cooking two meals every day from scratch, one before and one after work. Three children came from Ed's first marriage, one from her first, and two from their twenty-eight-year marriage. They had met when she walked into a bowling alley one night. Ed, thirteen years her senior, a lanky contractor with a delicate manner and big, rough hands, was immediately drawn to her sparkle and wit. It took three years to bring their families together.

She had such abundant energy that everyone around her came to rely on it. She not only ran the household, prepared the meals, paid all the bills, worked as a notary public, and held a succession of full-time jobs, but also was active in local Republican politics.

A box of plaques attested to the impression she made: KI-WANIS OF THE YEAR from the local Kiwanis club; FRANCHISEE OF THE YEAR from Mailboxes Etc.; BEST SALESPERSON from the Harry Lloyd Toy Company, a job that had sent her traveling all over the world.

When the O'Fallon Chamber of Commerce named her PERSON OF THE YEAR, the state Senate had passed a resolution recognizing Maxine as an outstanding citizen and community leader. Her "infectious smile, twinkle in her eye, and wicked sense of humor continue to make those she meets feel special," the resolution declared and was signed by John Ashcroft, then a Missouri senator. She had already been elected the next president of the O'Fallon Chamber of Commerce when her illness interfered.

Largely self-educated, she also was an avid reader, particularly of legal thrillers. At one time she had wanted to work as an investigator for a law firm and fancied herself a minor sleuth. She wanted to know all about the world—how it worked, who pulled the strings, and where the hidden interests lay.

IN 2002 MAXINE FELL INTO THE EXCRUCIATING TORPOR OF ANE-mia, the result of her red blood cell count plummeting under the onslaught of toxic chemotherapy.

In March her doctor prescribed Procrit, a medicine that fights anemia by stimulating red blood cell growth. She picked up the vials of clear solution in a nearby pharmacy in St. Peter's and went to a local clinic for injections. The weekly shots of Procrit, at the highest dose of 40,000 units per milliliter (U/mL), had a dramatic energizing effect.

To Maxine, the medicine was like liquid gold—and at almost $2,000 for four tiny vials, it cost far more. The family health insurance, which Ed got through the automotive union by doing maintenance at a car dealership, covered most of it. The co-payment alone was $100.

Also sold under the brand name of Epogen, the medicine is a triumph of genetic engineering. And though Maxine did not pretend to understand the science, she was grateful for the result. "This is a wonder drug," she told Ed after one weekly injection. "It makes you feel great." She looked forward every Wednesday to getting her Procrit.

At the clinic, the medicine generally burned as it entered her body, so the nurse injected it slowly. Then one day, the injection didn't burn anymore. The change was so marked that she said to the nurse, "I must be getting used to it."

But her anemia returned. She grew tired again. Her world contracted. She struggled to get out of bed. Sometimes she made it to the kitchen she once dominated, ate, and returned to bed. Other times she was unable to get up at all.

Fighting to maintain her routine, she slipped farther behind, as though life was a treadmill going too fast and she was falling off the back end. She felt almost as if she hadn't taken the Procrit. She was sicker the day after her injections than she was the day before. Her doctor considered upping the amount she took.

She viewed this decline as her fault, as though somehow she had failed to get better and to engage the world as she had always done, and this feeling became painful to her, separate and apart from her disease.

It was in June, months after she first started taking the medicine, that an astute nurse at her clinic held one of Maxine's vials to the light and studied the label.

Any nurse who administers powerful medicine fears deadly accidents, the possibility of zeros added or decimal points moved by a tired pharmacist or lab technician hurriedly translating a doctor's scrawled order. This kind of mistake can kill a patient, sometimes almost as quickly as if you took out a gun and pulled the trigger. It was the kind of nightmare that made you square the dose on the label against the dose recorded in the chart.

But the nurse was now scrutinizing Maxine's vials for a different reason. Days before, the clinic had received a letter from Ortho Biotech, the drug company that sold her medicine, warning that a counterfeit product labeled as Procrit 40,000 U/mL with the lot number P002384 had been circulating and was actually one-twentieth of its labeled strength. To the naked eye, the two products were indistinguishable.

This notice came on the heels of a first warning from the company about a different lot, P002641. The letter did not say how the company had detected the counterfeits. But it stated:

> *It appears that counterfeiters may be acquiring PROCRIT 2,000 U/mL vials and relabeling the product with counterfeit 40,000 U/mL labels. Because of the lower than labeled strength of the counterfeit vials, it is possible that patients could be under-dosed. In addition, other potential health risks cannot be ruled out at this time.*

As the nurse studied the vial no bigger than a thumbnail, she noticed negligible imperfections: a blurry letter here, text a little squished there, and almost invisible lines running diagonally through the zeros in the printed lot number, P002384.

Suddenly Maxine's situation took on a strange new shape: Her medicine was counterfeit and her body had become a crime scene. How much tainted medicine had she received? Most of the evidence had been dispersed inside her, the empty vials discarded, the ultimate effect on her health unknown.

At the time, Missourians were more familiar than most with the notion of adulterated medicine. Ten months earlier a Kansas City pharmacist named Robert Courtney had been charged with diluting the prescriptions of cancer patients and selling the extra medicine for profit. Courtney had locked eyes with his patients, taken their prescriptions, befriended them, and exploited their trust. He had also very likely shortened their lives.

Eventually he admitted diluting seventy-two different drugs over a decade—affecting some 4,000 patients.

The public viewed Courtney as an anomalous evil in an otherwise safe system, in which the American drug supply was inviolable, guarded by regulators and laws carrying severe penalties for those who tampered with it.

A Missouri judge, in sentencing Courtney to thirty years in prison, reinforced the notion that his crimes were unique. Calling his conduct a shock to the conscience of the country, he said, "You alone have changed the way a nation thinks. The way a nation thinks about pharmacists. The way the nation thinks about prescription medication. The way a nation thinks about those institutions that we trusted blindly."

Even a year after Courtney was caught, the *New York Times Magazine* described him as "consigned to freakdom in the annals of pharmacology." In other words, his conduct was rare and so profoundly demented that it seemed unlikely anyone might repeat it. When Ed Blount learned about the diluted Procrit, he thought immediately of Courtney—of "some guy in a back room, cutting it," as he said.

But Courtney was not the explanation now. A partial explanation lay 1,200 miles away in South Florida, where a dangerous gray market in stolen and counterfeit drugs made Courtney's activities seem minor by contrast.

Maxine, who knew none of this, approached the problem in her typical fact-finding way. She called Schnucks pharmacy in St. Peter's, where she had purchased the medicine, and the staff pointed her not to the drugmaker, but to a company that she had never heard of, Cardinal Health Inc. That company, a big pharmaceutical distributor in Ohio, had sold the medicine to Schnucks. The pharmacy had no idea what had gone wrong.

The clinic that had detected the counterfeits contacted the U.S. Food and Drug Administration. Three FDA investigators accompanied by the local sheriff came to the Blount home and stayed for hours, asking dozens of questions and documenting everything. They took several of the suspect vials for testing. The Blounts stored the rest in their refrigerator.

Their daughter, Tina, a hairdresser in a Kansas City salon, found herself telling a regular customer what had happened to her mother. The man was an attorney at a prominent Missouri law firm. Lawyers from the firm's St. Louis branch ended up taking Maxine's case. In September 2002, they filed a lawsuit alleging negligence against Amgen Inc., the drugmaker; Ortho Biotech Inc., the company that labeled and sold the Procrit under an agreement with Amgen; and Cardinal Health Inc., the distributor.

Before this, Maxine had never once worried about the purity of her medicine. She assumed that it traveled to her pharmacy along a safe and controlled route. Schnucks was a well-established chain with more than one hundred branches in seven states. There were no warning flags or perceptible risks in filling her prescription. Now, all her certainty had been replaced by unanswered questions.

Who had counterfeited her drugs? Where had the tainted medicine entered the supply chain? Should the manufacturer have recalled the Procrit, like carmakers do when defects come

to light? How come no one, including government investigators, could tell her where her drugs had been? How did she
know if her other medicine was legitimate? Why go through
agonizing treatments if she couldn't be sure?

   In his work as a contractor, Ed had always been good at
seeing how things fit together and the structure that lay beneath them. While the infrastructure that had supported the
counterfeiting remained invisible to him, he knew from experience that it must be deep. Seated at the kitchen table with
his hands around a coffee mug, he asked, "What greater conspiracy could there be than people counterfeiting medicine?"

3. Is Anything Okay?

November 2001
South Florida

THREE MONTHS BEFORE THE BURGLARY AT BIOMED PLUS, CESAR Arias was in his Buick Century amid documents, boxes, old coffee cups, and medicine bottles when his cell phone rang. A detective with the Miami-Dade Police Department was on the line. "You're the pharmaceutical guru in South Florida, right? We got a kid here caught stealing medicine from Jackson [Miami's largest public hospital] and reselling it to a guy in Broward. He says he wants to talk. Can you come down?"

Arias got this type of call all the time. The cops were always gunning for "drug crimes," which to them meant cocaine, heroin, or marijuana. They had little training and no separate department to handle crimes involving prescription medicine. In effect Arias *was* the department. He worked for the understaffed Bureau of Statewide Pharmaceutical Services (BSPS), a small backwater within the sprawling health department responsible for protecting Floridians from adulterated, expired, or mishandled medicine.

In his frayed guayabera shirt and worn khakis, gray sprinkling his once-dark hair, the forty-seven-year-old inspector spent most of his days driving up and down South Florida's highways to regulate drug manufacturers, oxygen distributors, and pharmaceutical wholesale companies. As a drug agent supervisor, Arias also oversaw three inspectors whom he largely left alone. He was rarely at his desk. He had no mind for management and no gift for office politics. A pharmacist by training, he was a medicine detective and could glance at a label or a pill and see depredations that few others could spot. Invisible to the public and overlooked by his supervisors, Arias had worked this job for the last fourteen years.

Now as he steered his Buick toward the police department in the crawl of rush-hour traffic, he returned to one of his usual preoccupations. Had the time to quit his job arrived? He wondered if there was a way to resign that would make a lasting political statement and embarrass his bosses in Tallahassee. He had entertained this thought before, usually when the drug supply and the atmosphere within his own bureau grew especially toxic.

As one of only nine drug inspectors to regulate 2,699 businesses statewide, he was thwarted by weak, outdated laws and apathy from Tallahassee, where the streets rolled up and the bureaucrats set their phones on redirect by 4:30 P.M. each day. Over the years he had watched pharmacists become drug inspectors hoping to better protect the public, only to give up in the face of entrenched political interests. His former colleague, Robert Loudis, had left to become a pharmacy manager at Walgreens, complaining that he often felt like a "potted plant," ordered not to interfere with the drugstore chains that had political pull, even when they flouted the regulations.

Years earlier Loudis and another inspector had gone to an Eckerd's pharmacy that had just installed an automated pill-counting machine. The pharmacists there had failed to label each bin with the medicine's expiration date and lot number, which drugmakers assign uniquely to each specific batch. This left no way for the pills in the bins to be identified in case a drug needed to be recalled.

Loudis and the other inspector immediately quarantined the sorted pills, imposing what is known as a "stop sale" on hundreds of thousands of dollars of inventory. Eckerd's, a political powerhouse in the state, began making phone calls. Within weeks, the men's bosses in Tallahassee ordered them to lift the stop sale for the most expensive drugs, but allowed them to maintain it on the inexpensive generics.

The inspectors never forgot this. During the four years Loudis worked for the department, he had kept a special file at the ready marked "Tallahassee" for directives sent to him by headquarters. It remained empty. There were few directives from Tallahassee except the unspoken one: do as little as possible. Unsurprisingly, turnover within the bureau was high and morale low.

Why Arias had not yet quit remained a mystery to those who loved him. His wife, an accountant, could not understand why he tolerated the archaic bureaucracy and petty insults of state employment. His mother did not see the logic in settling for a $30,000 pay cut when he could make six figures at Eckerd's or CVS as a pharmacist. His father agreed that his son appeared to be treading water, having sacrificed money and freedom for a job in which he was unappreciated and underpaid.

Even the wholesalers he regulated, like Marty Bradley, regarded Arias with curiosity and pity. "Did you see his car?" Bradley asked. "They should give that guy more muscle." But the Buick—with more than 100,000 miles on it, rust outside and stains inside, and a passenger door that wouldn't shut properly—actually suited Arias's purpose. It let him drive into Miami's worst neighborhoods and visit storefront pharmacies and clinics without arousing any suspicion.

Arias had a stubborn, oxlike temperament that he had inherited from his parents. In Cuba, his mother had been a

pediatrician and his father the country's director of engineering for the department of public works until they left Castro
and all their possessions behind in 1961. They arrived in
America fiercely committed to upward mobility and cleaned
hotel rooms and picked tomatoes until they were able to resume their professions.

But Arias didn't care about money. Being a pharmacist at
a chain drugstore, a cog in someone else's corporate wheel,
held no appeal. He had a different obsession, and evidence of
it lay scattered around the Buick: letters to politicians, drafts
of opinion pieces to newspapers, and e-mails to his superiors,
documenting the dangers to the drug supply that he witnessed every day. He wrote them late at night after his wife
and three sons had gone to sleep.

He had no idea if anyone read his missives, but he suspected not. His most recent treatise—"Pharmaceutical Drug
Diversion, the Medicaid Monster!"—was somewhere in the
pile. The three-page letter to his bureau chief, Jerry Hill, described in sometimes hard-to-follow detail one of the schemes
that he saw every day.

Pass any Medicaid clinic in South Florida and you'd see
Hummers and BMWs idling outside, their occupants just
waiting for patients to exit with their government-paid medicine, which the patients often were willing to sell for ready
cash. In his letter, Arias explained that the buyers or "runners"
would bring this street medicine to their bosses, corrupt
wholesalers who would remove the patient dispensing labels
with lighter fluid or paint remover, and resell the medicine as
new, despite a telltale sticky trace on the box where the original labels had been.

The drugs would circle in this revolving door—distributed, sold, and resold, Medicaid paying for it each time—
until a retail pharmacy dispensed the drugs to an unwitting

patient (who would end up taking medicine now degraded by the handling) or until the box became too beat up to be saleable.

Arias's letter was dramatic in tone as it warned of skyrocketing Medicaid costs and grave public health dangers. He had copied this one to various authorities and reiterated the main points at every opportunity. The response, as usual, was silence. At such moments he would ask of the drug supply, "Is anything okay?" And of his agency's performance in securing it: "We have met the enemy and they is us."

In fairness, his bosses were not the only ones who disregarded him. For years, he had tried to get other law enforcement agencies to pursue the cases he developed. The Florida Department of Law Enforcement, the Miami-Dade Police Department, the Food and Drug Administration, and the FBI had ignored him at various times too.

ON BISCAYNE BOULEVARD, ARIAS CUT ACROSS FOUR LANES OF wild traffic and pulled into the familiar parking lot of the police operations bureau. This small branch office responded to crime at Jackson Memorial Hospital and at Miami's cacophonous port along the city's southeast coast.

Arias headed up the stairs of the honeycomb-shaped building and into the air-conditioned offices. In a small back room several detectives stood watching a tall, cocoa-complexioned young man handcuffed to a chair.

Sydney Dean Jones, twenty-five and a Jamaican native, looked impassively at them, his handsome face framed by small hoop earrings. He had accepted their offer of a Coca-Cola and asked for his lawyer. Within weeks he would agree to cooperate fully, divulging with some pride the scheme he had developed and the lifestyle it allowed.

Things of value walked out of Jackson Memorial all the time, as he explained. The monolithic public facility was bigger than some towns, with 1,700 beds for patients, 11,000 employees, and more than 500 exterior doors among its adjacent buildings. The hospital not only cared for the sick and indigent; it also fed a thriving black market for durable medical equipment, from glucometers and defibrillators to anesthesia machines. While the stream of theft strained the hospital's coffers, it also helped to prop up the underground economies of Little Havana and other cash-starved Miami barrios, allowing the small medical businesses that lined the streets to resell the equipment.

Eloquent and in training to become a pharmacist, Jones did not work for Jackson Memorial. He had been employed briefly by Cardinal Health Inc., a multi-billion-dollar, publicly traded company headquartered in Dublin, Ohio. Cardinal, one of the nation's three biggest pharmaceutical wholesalers, is a medicine middleman, buying from the nation's drugmakers and reselling to major pharmacy chains and hospitals. It was the same company that had sold Maxine Blount's Procrit to Schnucks pharmacy. In other branches of its business, Cardinal also served as an all-purpose helpmate—staffing the nation's pharmacies and hospitals through a roster of temp workers and helping them to automate services and improve their medicine packaging. Jones, one of the part-time pharmacy workers in the company's army, was called in to do an odd shift here and there at Jackson Memorial and pharmacies in the area.

Since age sixteen, Jones had worked at a neighborhood pharmacy, helping behind the counter and delivering medicine to elderly customers. The pharmacist there became like a second father to him. His own father was a respected police officer in nearby Hollywood. Jones went on to complete a college degree in biology and enroll in pharmacy school. The flexibility of his job with Cardinal allowed him time to study and play evening basketball. It also gave him ready access to expensive prescription medicine.

Technically, Jones had only worked one shift as a Cardinal employee at the Jackson Memorial pharmacy. But with his lab coat, the security code to enter the pharmacy, a temporary ID card, and an air of legitimacy, he entered easily and once inside tried to look busy. On five separate occasions he had reached into a small refrigerator and put tiny vials of medicine into his pockets. Neupogen, which helps cancer patients to fight infection, was among the drugs he took. A single box, which contained ten vials of the clear liquid, sold for about $2,000. Unlike narcotic painkillers such as Oxycontin and Dilaudid that were known to have a high street value and were kept under lock and key, the far more expensive and fragile drugs to treat cancer and AIDS were left unlocked and within reach.

Just weeks earlier, Jones had been arrested and charged with similar crimes at two branches of a Humana pharmacy chain in Boca Raton and in Plantation, where his company had sent him to work. Without authorization he had ordered medicine from Cardinal and charged it to the Humana chain. His haul included three injectable medicines, each genetically engineered to help strengthen patients undergoing chemotherapy, kidney, and AIDS treatments.

Jones generally brought the medicine to his contact, a man he called "Fred" who worked at Memorial West Hospital up in Pembroke Pines. Each week for the last seven months, Jones had handed Fred, an Iranian whose real name was Fariborz Sheie, $10,000 to $30,000 worth of medicine in the hospital parking lot. Sheie then sold it to a woman up north and later paid Jones his cut in cash—about a third of the value of the drugs.

Jones saw this as a low-risk way to make extra money, which he used for pharmacy school and helping friends with rent and other expenses. He thought of it as a Robin Hood scheme, though admittedly it also helped him pay for a leased Mercedes-Benz and gifts for his girlfriend of the moment, a Russian model who would later become a *Playboy* cover girl. "We live in a society where everyone wants to have some kind of pleasure," Jones later explained. "The average cop is looking for the hard-core drugs. Criminals just figure, 'We can get real medication for real money that no one will ever know.'"

But they did know. In September, three months after Jones began pilfering from the Jackson Memorial freezer, an alert technician caught him stuffing medicine into his pants pocket. A video surveillance camera also captured the grainy but unmistakable image of him reaching into the freezer.

Two months before Jones's arrest, officers had responded to Jackson Memorial's artificial kidney department, where a door had been shattered and a small refrigerator stolen from an office that was usually left open. The refrigerator contained 210 vials of Epogen, the injectable liquid used to boost the red blood cell count of dialysis patients. The medicine was worth about $24,000.

In September and later in November, a chain of dialysis clinics around South Florida suffered similar thefts. Some of the cops who responded wondered why anyone would steal Epogen, since you couldn't get high off it. But Arias and the detectives who arrested Sydney Jones knew the motive: The thieves could resell the medicine for big money.

Once the drugs were stolen they were technically adulterated, since their origin and quality could not be ensured. As stolen medicine passed from hand to hand in this street-level trade, it could reenter the supply chain and even wind up sold to the very hospitals and pharmacies from which it had been taken.

Despite this threat to public health, Jones's scheme presented Arias with an all-too-familiar enforcement problem. Crimes of medicine were usually prosecuted under the state's flimsy administrative health code, Statute 499, its penalties amounting to a slap on the wrist. Under this statute, Jones's conduct, taking illegal possession of medicine with intent to sell, would at most result in probation and a $250 fine. That is why Arias spent so much time trying to make the police care about stolen medicine—so that Jones and those like him would face felony criminal charges that carried the prospect of prison. In short, he needed someone with a badge and a gun to help him secure the drug supply.

The Miami-Dade detectives also faced difficulties. Sydney Jones's case presented jurisdictional issues, because he had resold the medicine in a different county from the one where he had stolen it. They needed someone with jurisdiction over it all to take Jones's case.

AT THE FLORIDA DEPARTMENT OF LAW ENFORCEMENT'S SMALL branch office in Fort Lauderdale, Gary Venema looked up wordlessly as his supervisor stepped into his cramped, windowless office. It was a rare day that Venema, a special agent at FDLE, had little to say.

Even at his most subdued, Venema was impossible not to notice. A big, sandy-haired man of fifty with a wise-ass grin, he had a magnetic, even manic, presence that drew every eye in a room. Typically he greeted his supervisor, Michael Mann, with a burst of verbiage—random facts, ribald jokes, new evidence from the big case that they had been working on for two years. He talked as though patching the listener in to an ongoing conversation with himself.

Today his down mood was evident. Though dressed with his usual exuberance in a bright, rumpled Hawaiian shirt and a gold sailboat charm around his neck, he said nothing. Instead, Mann started the conversation: "You're doing a one-day sting for Miami-Dade," he said, explaining that the matter of Sydney Jones was a nothing case, a "hit-and-run," as they called it. Mann offered that maybe it would connect a dot or two in their own investigation of drugstore thefts. Venema—two decades older than most of his other agents—let out a skeptical grunt.

Mann, who wore his usual gray suit with a gun strapped to his ankle, was skeptical too. The Sydney Jones case as described by the Miami-Dade detective sounded like a dog. The multi-county problem was the usual excuse for dumping cases on FDLE, the police agency that had statewide jurisdiction.

All Mann and Venema wanted to do was work on their big case, which involved the theft of over-the-counter goods from pharmacy chains. Thieves were stealing millions of dollars in diabetes test strips, razor blades, aspirin, and electric tooth-brush tips. They shoveled these items into beach bags lined with aluminum foil, which defeated security buzzers, and walked out the front door. The major chains, CVS and Eckerd's among them, were the victims of this organized scheme, in which stolen goods were shuttled to fences or middlemen and then sold across state lines.

Three months ago, Mann had reached out to a federal prosecutor and criminal investigators from the FDA so they could hunt down the out-of-state criminals. But the feds had remained noncommittal and the prosecutor never called back, effectively sidelining their big case.

The feds' lack of interest had left Venema despondent and angry. And now he figured that he was being dragged into a time-waster involving a Miami kid who pilfered cancer medicine from a hospital freezer. At the time, he had no way to imagine where the case of Sydney Jones might lead.

Venema drove home that night to his one-story ranch house in Davie, a community west of Fort Lauderdale, angrily thinking about the Feds. A former cop with twenty-four years' experience, fifteen as a homicide and narcotics detective in Hialeah, Venema's days of adrenaline-pumping shootouts and marathon shifts of pulling dead bodies out of social clubs were long gone. In 1991, after a bad political clash, the Hialeah police chief demoted him to a road sergeant. Venema spent the next six years supervising cops who responded to car wrecks and petty mischief complaints and helped to secure major crime scenes until his former colleagues, the detectives, showed up.

He had jumped at the chance to join FDLE, an agency with power and panache and notoriously low pay. Despite all his experience, his starting salary was $42,000. He was the oldest guy in his academy class by at least a decade. Many of his new colleagues were just a few years older than his three sons. But he enjoyed the training and liked shooting his gun lavishly during target practice, since the agency didn't ration the ammo. Almost immediately he embarked on the drugstore theft case and felt sure he had a winner—one that would reverse his long run of bad luck.

At home, he parked his pickup truck in the driveway beneath his bright front lights and the mahogany tree growing at a tilt and headed up the brick path, past the American flags sprouting from the lawn. The tidy houses in this neighborhood heavily concentrated with cops sat elbow-to-elbow on

less-than-quarter-acre plots, each adorned with American
flags since the terror attacks of September 11.

Inside, his wife, Sandy, had made a healthy dinner, since
Venema's blood pressure had been climbing again. She noticed that he seemed low. After twenty-four years of marriage,
she was used to his down moods after work setbacks. After
eating, Venema sat on the back porch, which overlooked a
small pool and, farther down by the lake, the gazebo and dock
that he built years ago. The sound of the wind chimes that
Sandy had hung soothed him and helped quiet the tape loop
in his head, which these days inevitably started with the words
"Fuck the Feds."

4. The R Word

WHENEVER DRUG INSPECTOR ARIAS CAME TO THE BRINK OF quitting, he tended to fall back on the hope that change lay right around the corner. On November 13, he felt hopeful as he drove to a meeting at the Statewide Prosecutor's Office to discuss the matter of Sydney Jones. He viewed any sit-down with a prosecutor or law enforcement agent as a golden opportunity to make them care about dangerous medicine.

For the other inspectors and investigators headed to the prosecutor's office, Sydney Jones was a minor felon amid a blur of more urgent cases. The FDLE agent Gary Venema arrived tired and unshaven and wanted the meeting to be short. He knew nothing about prescription medicine and had never met Arias or some of the other investigators. He was surprised as they filed in to see the Healthcare Fraud Priority Leader. She turned out to be Assistant Statewide Prosecutor Stephanie Feldman, a petite twenty-eight-year-old with five years' trial experience who stood about five feet one inch in heels.

Dwarfed by her desk, Feldman listened silently as those assembled described how Jones had been caught stealing medicine from Jackson Memorial in Miami-Dade County, had been planning to resell it to someone in Broward County, and was willing to cooperate. That is, if she was interested.

Feldman usually remained steely at work, her expression dull and slightly menacing as though to warn those who failed to take her seriously. Bridging her hands together and gazing through silver-rimmed glasses, Feldman said coolly, "Let's set up a wire on Jones and see who he sells to."

Venema, who had been brooding silently, became instantly alert, wondering whether she might be interested in *his* unprosecuted case of over-the-counter thefts at drugstores. Similar wheels turned in the heads of the other investigators, as they mentally catalogued their stalled and overlooked cases and thought, *here was a prosecutor who seemed ready to work.*

Feldman, all business, handed out assignments: Arias was to procure the medicine that would be used in the sting operation. Venema was to get his agency to charter a surveillance plane so the group could track Jones's medicine as it moved (tailing a vehicle in the traffic-clogged streets and highways of South Florida was dangerous and often impossible). A Broward detective was assigned the task of wiring Jones.

With one phone call, Feldman had Jones's case transferred from Miami, where a local prosecutor was all too eager to unload what appeared to be a humdrum case of theft from a hospital freezer.

CESAR ARIAS WENT HOME THAT NIGHT WOWED—SIMPLY BLOWN away—by the idea of a surveillance plane for medicine. He never dreamed that so big and high-tech an instrument could be deployed in the service of what he did every day. For years, he had fought a losing battle with his bureau to equip the inspectors with thermometers so they could measure the temperature at which drugs were stored, as required by state statute. The health department's lawyer had determined that thermometers could be unreliable and their results challenged in court and therefore should not be used. Now Feldman was talking about an airplane.

Arias had first met the young prosecutor a few weeks earlier. He and a federal agent had gone to talk to her about a prescription drug wholesaler who they suspected was selling stolen medicine. After the U.S. Justice Department had turned down the agent's request to run a wiretap, the men hoped to interest a state prosecutor instead. To that meeting Arias also had brought along his perpetual sidekick, a seventy-two-year-old drug inspector named Gene Odin.

Odin, a registered pharmacist with a Ph.D. in medicinal chemistry, had an impish wit and two hearing aids that often conked out. He had no intention of retiring anytime soon and regularly gave Arias hell if excluded from anything, no matter how dull or unpromising. Because of the health department's depleted resources, Odin was solely responsible for regulating the drug businesses in twelve counties.

The two men had worked together for thirteen years and, if not exactly soul mates, were inseparable allies. Each relied on the other's strengths and knew the other's weaknesses.

The two inspectors had uncovered crack houses stocked like pharmacies with pills, vials, and brand-name medicines bound for patients through clandestine back-room sales. They had found pills packaged with cotton that was tattered and yellow with dirt, and intravenous containers with mold inside. All too often, they came across boxes and bottles of delicate medicine that had been "cleaned" with lighter fluid or torched with heat guns to remove or affix labels so that the drugs could be resold.

That day at Feldman's office, they told her that for years they had tried to get someone, anyone, interested in prosecuting the theft and illegal resale of pharmaceuticals, but they had no takers. As health inspectors, they had jurisdiction over almost nothing and had only "Mickey Mouse" weapons at their disposal.

To Feldman, the cause of their problem was obvious: The penalties for diverting or misbranding medicine—misdemeanors and insignificant fines—were far too low to interest federal prosecutors.

After listening to their description of the case involving the wholesaler, she told the FDA agent, "Bring me all your files." For once, Odin was silent. Arias, astounded by her interest, stood up and actually took her hand, shedding his habitual reticence. "I just want to touch you because I want to know if you're real," he said, almost but not really joking.

Odin, similarly stunned, asked, "Do you *like* these kinds of cases?"

FELDMAN DID LIKE THESE CASES AND WANTED TO PROSECUTE them. In part she saw a career opportunity because no one else would touch them. She also was the clearing house for all the state's health-care cases; any potentially criminal matter with a polysyllabic medical term got thrown to her. But there was also a personal reason for her interest: Her life depended on the integrity of her medicine.

Just weeks before Arias and Odin's visit, she had been admitted to the hospital in a coma, the latest chapter in her ongoing battle with juvenile diabetes, the disease's most severe form. She had been managing diabetes since age fourteen. She gave herself insulin shots five times a day and regularly took Neurontin, a medicine that helped restore sensation in her feet. She also had received home infusions of an injectable antibiotic, Rocephin, for treatment of a kidney infection that had resulted from her diabetes.

She rode these waves of disease, snatching control wherever she could, persisting through iron discipline and the syringe of insulin that she always carried with her.

In her family, too, medicine had been everything. Her father had owned nursing homes. Now eighty-three, he took Coumadin to help prevent heart attacks. Her mother, a geriatric nurse, had been killed in a car accident when Feldman was only seven. She believed that her mother's death, and the inadequate investigation that followed it, had sparked her later interest in being a prosecutor. Her entire upbringing and experience made her the person to care about what Arias and Odin were saying.

ON THE STORMY, OVERCAST MORNING OF NOVEMBER 15, SEVENteen officers from five agencies, including Arias and Venema, teamed up in unmarked cars near the Memorial West Hospital in Pembroke Pines. Sydney Jones, who had been wired with a recording device in a nearby Holiday Inn, was dispatched to the hospital parking lot with $50,000 worth of Neupogen. From a surveillance van, the investigators videotaped and listened in as Jones handed off his loot to Fariborz Sheie, an older man with gloomy hazel eyes who worked as a maintenance engineer at the hospital. As Sheie drove to the other side of the parking lot, the investigators noticed a blue minivan following close behind. Sheie stopped and handed off the medicine to its driver, a man with a mullet haircut and wraparound sunglasses.

The man in sunglasses turned out to be far more wary than Sheie. With the investigators on his tail, he kept looking in his rearview mirror as he drove north to his home in Shenandoah. He obviously detected or "made" the surveillance, as Venema would note in a later report. He began maneuvers to shake off the cops, making wide turns, zooming through shopping malls, doubling back along side streets. The cops responded with maneuvers of their own but ended up losing him. The surveillance plane, socked in by unusually heavy clouds, also lost him. An hour later, the team picked up his van as it returned to the hospital parking lot. Feeling spooked, the man in sunglasses had decided to return the medicine to Sheie and thereby rid himself of the evidence.

The cops moved in on both men separately and threatened them with arrest and prosecution unless they agreed to cooperate. Sheie and Melvin Otto, the man in sunglasses, quickly agreed to help the cops.

MELVIN OTTO EXPLAINED THAT HIS NEXT STOP WAS TO HAVE been in Palm Beach County, where he planned to deliver the medicine to a woman named Annette Mantia.

Mantia was a licensed drug wholesaler known to Arias's partner Gene Odin. Her business operated in one of Odin's counties, and he had long suspected she used it as a front for illegal activities. Her office was empty whenever Odin stopped by. One time an "assistant" materialized and claimed that she had no records to show him because Mantia had done no business in prescription medicine.

Odin was powerless to revoke her license but got her fined for a minor procedural violation: failing to be available for inspection during regular business hours. While the two had never met face-to-face, she had told him recently over the phone that she planned to give up her license in December.

On the day after the Jones sting, Odin watched as Melvin Otto handed Mantia the Neupogen and she handed him a package that contained $20,000 in cash. The investigators followed her car as she and her daughter went about some errands, the fragile medicine baking in the car trunk. After Mantia emerged from an hour-long appointment at an eye doctor, Odin and Michael Mann from FDLE approached her.

Reminding her of their earlier phone conversation, Odin asked, "Are you still planning on giving up your permit in December?"

"Is there a problem?" She asked from behind Jackie O sunglasses.

As Mann explained the rather sizeable problem and her daughter started to cry, Mantia said coolly, "It's okay, dear."

STEPHANIE FELDMAN FOUND HERSELF THINKING BROADLY ABOUT the Jones case. She wondered if the theft of medicine could be viewed as something larger, not merely as an issue of stolen property but of lost confidence in the drug supply. She figured that any theft raised the possibility that medicine of uncertain origin, transported in unknown conditions, could reach patients.

Could one argue, she wondered as though back in some law school exercise, that Jones had participated in a scheme that defrauded an immutable class of victims, those with diseases such as cancer, AIDS, and diabetes, that required lifelong medication and management? As a discrete class, these people deserved, and were entitled to, legal protection.

Excited, she floated her theory past a senior prosecutor, who laughed dismissively. Her colleague had recently recovered from a heart attack and was not about to tilt after fanciful prosecutions; it took all his energy just to get through the day.

But Feldman, feeling that she was on to something, continued to think through the facts and the law. If there was some criminal enterprise profiting from stolen medicine, who sat at the top of it? Not Sydney Jones, and almost certainly not his accomplices. Among the four suspects, Annette Mantia was certainly their biggest fish, caught dead to rights in a parking lot handing off cash for stolen medicine. Venema and the others had been eager to arrest her right then.

But Feldman had urged them to wait, explaining that to build a big case they would need a cooperating witness at each level who could lead them to even bigger fish. Since Mantia was not buying and selling that much medicine, logic dictated that someone, somewhere above her, was selling more.

SEVERAL DAYS AFTER THE PARKING-LOT STING, FELDMAN SAT opposite Mantia, her daughter, and her lawyer. "You have to promise her immunity," the lawyer commanded.

"Then you can leave my office right now," Feldman said, as though not the least bit concerned. The lawyer stayed put and, as Venema and Arias looked on, they hammered out the formal cooperation agreement that would allow Mantia to help with the investigation.

As soon as they left, Feldman began to expand on her view of the underlying case: If there was a runner stealing from a hospital and a middle person selling it to someone else, that meant there already was a chain of command and an enterprise committing acts of fraud—all with the goal of moving cheap medicine into a high-priced supply chain.

"Racketeering," she announced to Venema and Arias, who had sat in on the meeting with Mantia.

They were surprised. The "R word," as they called it, was one of the government's strongest weapons. Most prosecutors, not to mention most detectives, ran in the opposite direction from the crazy amount of work needed to build a racketeering case. Arias knew only that the federal government had used racketeering as a tool to prosecute the mob. After years of begging more prosecutors than he could count to take cases, here was a young woman who had offered the R word with no prompting.

IN THE FOLLOWING WEEKS, MANTIA TOLD THE INVESTIGATORS about a gray market where buyers and sellers, whether licensed or not, hunted for discounted brand-name drugs and asked few questions about their origin. Though a "legitimate" wholesaler with a state license, she told them that 95 percent of the medicine she sold had been stolen first. Her customers, mostly other wholesale companies, bought even the medicine lacking a veneer of legitimacy, she explained.

By law, medicine was supposed to come with a sales history or "pedigree," like a car title, that tracked the previous owners back to the manufacturer. Mantia's customers referred to these essential records as "paper." Some did not mind if paper was invented to conceal the truth. Others didn't even care about this artifice and would buy medicine without any paper at all.

FARIBORZ "FRED" SHEIE, THE FIRST LINK IN SYDNEY JONES'S ILlicit distribution chain, was an Iranian-born biomedical engineer who made repairs to the physical plants of hospitals around South Florida. He had been well liked at Memorial

West until a Broward detective called the hospital's security director to describe the sting and recommend that Sheie no longer be allowed keys to the hospital's locked areas. He lost his job as a result.

Even before being fired, Sheie struggled with a perpetual cash crunch. He had an ex-wife and a young daughter to support and was addicted to narcotic painkillers. His job gave him some access to the medicine he craved, but he had also come to rely on Jones's deliveries as a way to finance his addiction. Though his wife had left him three years earlier because of this problem, he remained devoted to his daughter and had custody of her every other weekend. He picked her up punctually, never missing a date.

Long before Sheie's arrest, his addiction had made him a hazard on the roads. The police stopped him on three occasions, once for hitting a cyclist and another time for hitting a car stopped at an intersection. In February of 1999, a Broward County police officer pulled him over and found the painkillers Percocet and Vicodin and the muscle relaxant Soma in his car. Unable to walk a straight line, Sheie was arrested for driving under the influence. After losing his job at Memorial West Hospital, he sank into a deep depression.

On January 13, 2002, less than three months after the sting operation, Sheie did not show up as usual to pick up his daughter. His ex-wife found him sitting upright and naked in a living room chair with EKG pads attached to his upper arms and lower abdomen, as though he planned to measure his heart rhythm as it stopped. The medical examiner who performed the autopsy found that his blood was infused with opiates. Sheie died from an overdose.

Devastated, Stephanie Feldman was certain that crucial information had died with him.

———

THOUGH VENEMA HAD WORKED INTERMITTENTLY ON THE IN-formation that came from his new informant Annette Mantia, he had remained preoccupied with his case of organized theft from drugstore chains, which seemed larger and more important. But the phone call about Sheie's suicide grabbed his attention and made him wonder if the case of stolen cancer medicine was bigger than he had initially thought. He burst into his supervisor Michael Mann's office and declared, "Fred is dead!"

5. Medicine in the Laundry Room

January 2002
Miami, Florida

WAITING IMPATIENTLY TO DISEMBARK FROM THE JUMBO S.S. *Sovereign of the Seas* after four terrible nights, Marty Bradley cursed cruise lines and thievery equally. Then he sought advice from the one person whose number he had with him: Cesar Arias at the state's pharmaceutical bureau. He explained to Arias that he believed the Stone Group wanted to sell him back his own medicine.

"Stand by," said Arias excitedly, viewing this unexpected event as a rare opportunity to unravel the theft. He called his new colleague in law enforcement, Gary Venema, with whom he'd worked the Sydney Jones case. He then called back Bradley to relay Venema's advice at top volume. "Buy it back! Buy it back!" Arias yelled as Bradley was herded off the boat.

THE STONE GROUP HAD APPARENTLY CALLED WHOLESALE COMpanies around the country offering to sell the medicine. Even in a market where few buyers inquired about a drug's origin, Bradley's competitors were suspicious: rare blood derivatives, especially NovoSeven, were almost never sold between wholesale companies.

Bradley was the only one to bite, playing perfectly his role of the chump. He negotiated a price of $229,241, one-third off the typical wholesale cost of $330,000, the dramatic discount suggesting that the medicine was both figuratively and literally a *steal*.

While Florida authorities expressed surprise at the brazenness of the Stone Group's offer, Bradley thought of it as stupidity. They obviously knew nothing about the medicine they were selling. The drug makers rarely marked it down, and the prices they offered to wholesalers differed by fractions of points, if at all, unlike other medicines with lots of play in the prices. Bradley, as one of the nation's biggest distributors of these blood derivatives, already enjoyed one of the best discounts.

Even if burglars had not just stolen the same drugs from him, how could some "bullshit company out of the blue," as he put it, have a better price than his own? He knew the answer and it was everything that was wrong with today's medicine business: The wholesale market was dominated by *don't-ask-don't-tell* transactions in which everybody sought the lowest price, regardless of the medicine's origin. And if the price of the drugs didn't tell the whole story, the kid climbing out of the sports car helped clarify it.

That afternoon at Bradley's warehouse, the young Stone Group salesman, Sean Dana, arrived in a souped-up Trans Am. He was wearing shorts and a T-shirt and carried an ice cooler full of the medicine stolen from Bradley just six days earlier.

After Dana dropped off the medicine in the warehouse, Bradley brought him to the conference room, presumably to complete the deal. But as the door swung open, Dana took in the five people arrayed around the conference table: Cesar Arias, Bradley's lawyer, and three Miami-Dade detectives with badges on their belts and guns in holsters.

In turn, each of them took in Dana, his bewildered pale face, one eye opening wide in shock and the other, a lazy eye, drifting off as though trying to leave the room on its own.

"Why don't you take a seat," one of the detectives suggested. And so Dana shakily sat down as Bradley suppressed an urge to strangle the kid.

When asked if he knew that the medicine had been stolen, Dana stammered, "I don't know anything about that." He explained that he was just a salesman and knew little about where the medicine had come from, having just seen it for the first time that morning.

Watching silently, Arias felt inclined to believe him. Dana seemed clueless, just a sales rep who probably spent every day, all day, on the phone brokering sales. Struggling through a few more questions, Dana then offered that he would like to help but wanted to consult a lawyer first. At that magical word *lawyer*, the questions had to end. The detectives acquiesced.

After the meeting, Bradley felt little relief. The people who stole from him were still at large. And though he had his medicine back, he would have to destroy it all because he couldn't guarantee that it had been kept at a constant freezing temperature. He also had little faith that the Florida authorities could actually help him, and was dismayed that the investigators had let Dana walk off after the meeting.

THE NEXT DAY, AS VENEMA AND ARIAS HEADED BACK TO Bradley's warehouse to photograph the returned medicine, one of the Miami detectives from the interview with Dana reached them. She said the president of the Stone Group, Adam Runsdorf, had called and wanted to cooperate. Runsdorf told her that his firm had purchased the drugs from a company up north in Kissimmee called BTC Wholesale. When pressed, he elaborated that his contact there was a man named Michael Carlow, whom he believed to be the owner. On paper, the owner of BTC was Michael Carlow's brother-in-law, a former mattress salesman named Thomas E. Atkins, Jr.

BTC was a licensed prescription drug wholesaler in Osceola County near Disney World. In South Florida, the drug inspector Gene Odin had been to two other companies that BTC did business with and had seen BTC records detailing millions of dollars in drug sales. While BTC had passed on the required pedigree records that showed the drugs' previous buyers, Odin told Arias that he suspected the paperwork was bogus.

Carlow, however, was another matter. As much as anyone, his face could have been on the post office wall depicting what was wrong with Florida's medicine business. In the distant past, Carlow had served time in prison for armed robbery and got probation for grand theft. Afterwards, he had owned AIDS clinics and substance abuse treatment programs. In 1998, the state gave his wife, Candace, a prescription drug wholesale license for a company that Carlow ran as president. In June 2000, Carlow was arrested for buying $90,000 of stolen Neupogen in the parking lot of a Miami restaurant, much as Sydney Jones had done on a smaller scale.

Leaving aside his felony record, one would have thought Carlow's arrest in 2000 would have closed the door on his prescription medicine career. He pled guilty, paid a nominal fine, was sentenced to eighteen months probation, and he and his wife voluntarily surrendered their state license—the harshest justice that could be meted out under Florida's weak statute. His alleged involvement with BTC suggested that he might be making a comeback.

THIS NEW INFORMATION CONFIRMED VENEMA'S FEELING THAT the case had more to it than one warehouse rip-off. The aspects of the break-in—the crowbar, the faceless perpetrators, the getaway van, the exorbitant cost of the medicine—had seized his attention in a way that Jones's pilfering could not.

Stephanie Feldman also leaped on the new information about Carlow, a line drawn straight from the robbery to a career criminal. She directed Venema and Arias to go to BTC first thing next morning.

THE NEXT MORNING AT 10:00, VENEMA'S RED TRUCK RATTLED UP to a fading little office building in Kissimmee and he and Arias climbed out. They would have missed the company altogether were it not for the letters "BTC" on a door on the second floor. A round-faced man with owl spectacles and dark hair opened the door. It was the registered owner, Thomas Atkins, Jr., whom they had instructed to be there. He stood in a room furnished with a desk, bare shelves, and a device in one corner that packed Styrofoam peanuts into boxes. The only medicine was four damaged boxes of nitroglycerine patches to treat chest pain in a cabinet marked "Quarantine" above an empty refrigerator. Atkins had a phone pressed to his ear and was talking to his lawyer in Miami.

He showed them a few pieces of paper, including invoices for medicine and the accompanying pedigree records. He said that no company checkbook, delivery receipts, bills of lading, or any other documents were kept on the premises, although the investigators noticed that a manual of BTC policies and procedures included a highlighted copy of the state statute, which specified that all records must be kept on site.

Atkins declined to answer most of their questions, including queries about Carlow, whom he acknowledged was his brother-in-law. He wouldn't say how he had gone from selling mattresses to selling medicine. As Venema noted in his report on the visit, "He also refused to state who put up the money to get him started, what he does all day, who he ships to, how the shipments are made and how the payments are made. . . . He also refused to answer how other shipments to other companies were handled, whether any product was ever stored at his business, or who his contacts were."

Atkins did say that he knew nothing about the drugs he was selling, except that some of them needed to be refrigerated, and that his business hours were once a week on Wednesdays.

"Are you basically a front for someone else in this business?" Venema asked.

Atkins refused to answer this question too. But he signed a sworn statement, saying that the invoices he had shown them represented the entirety of BTC's business dealings. These included records of transactions with the Stone Group and a Texas company called El Paso Pharmaceutical Brokers Inc.—which he said sold him Bradley's medicine. He said he had no idea it was stolen. Also included was a record of one transaction with a company in Boca Raton called Omnimed.

THE DRUG INSPECTOR AND THE INVESTIGATOR EMERGED FROM the meeting convinced that Atkins was lying and that BTC was nothing more than a shell company, its true nature unclear. On the ride back, Venema started banging out a report on his laptop while Arias drove. Exultant, Arias buzzed Gene Odin on his walkie-talkie phone and shared the details of their visit.

Odin's thin, bitter voice came back over the speaker: "You're in Kissimmee? And how come I'm not there with you?"

Venema stopped typing and looked up.

Again Odin's voice: "Why is it that the routine shit goes to me but whenever anything is a big case it goes to you? You've got Dade County, Cesar. You can't have the rest of the world."

Cesar wore a grave and chastened expression. "Anyway," Odin continued angrily from his home office in Boynton Beach, "if you want the nineteen other invoices between BTC and Omnimed, you'd better stop in here on your drive down."

After some silence, Venema, who had never met Odin, offered, "God, he sounds pissed."

ARIAS AND ODIN INTERACTED LIKE AN OLD MARRIED COUPLE. Odin hated being sidelined and was quick to suspect that if he wasn't vigilant, Arias would leave him out of everything. He resented any time that Arias closed his office door, held a sotto voce conversation that excluded him, or spoke in Spanish to those whose businesses they inspected. During their first year working together, they had tracked a man smuggling unapproved medicine from the Bahamas. In the teeming area near the port, it took days to find the man's cargo van. Their team waited near the vehicle for hours until Odin finally announced, "I'm going to take a leak," and headed into a nearby business. Seconds later, the suspect appeared and drove off in his van—and Arias and the others followed him. Odin emerged to find his colleagues gone and, in the days before cell phones, had no way to reach them.

That incident was one touchstone for their work relationship, with Odin's fear of being left out cropping up now and then. Another was their shared commitment to pure medicine and to ending the status quo within their agency. To some degree, they were perfect foils, Jewish and Cuban halves of the same stubborn coin. Both were prone to look on the dark side and see corruption and mismanagement. Both tried to fix such problems noisily and with little political grace.

As he drove, Arias now explained to Venema the frustration that he and Odin had faced over the years, trying to protect the public from adulterated medicine without guns or meaningful subpoena power. They had repeatedly urged their bureau to crack down on corrupt and dangerous manufacturers and wholesalers and to revoke licenses or even to refer some cases to other law enforcement agencies. In response, their bosses often accused them of "impeding commerce."

Whenever they seized bad medicine, they inevitably angered someone with means to hire a lawyer. Instead of lending support, their bosses often expressed doubt about their inspections. "We got all these drugs that are bad," Arias explained to Venema, "and they're asking me, 'Why are you there? How come?'"

Tensions between Tallahassee and the inspectors recently flared when management demanded that inspectors document their work days in fifteen-minute intervals. The stated goal was to measure their productivity. The inspectors, however, saw an ulterior motive. They feared that management would use the reports to "triangulate" them—to compare their daily activities with their gas receipts and highway tolls to find discrepancies of even a few minutes that would allow supervisors to impose discipline. Never mind that such reviews could burn hours of state time.

Arias and the other inspectors had proposed, without success, that management measure productivity by a simple count: the volume of businesses inspected, investigations completed, and pharmaceuticals seized. They figured this made more sense, since the whole point of their work was to find and destroy bad medicine before it reached patients.

The inspectors, some of whom likened their work to a crusade, believed that Tallahassee had the productivity problem, not them. They complained that the reports they filed often languished for months and even years without being entered into the bureau's computer system, leading the bureau's functionaries to renew the licenses of companies they were actively investigating and even to grant licenses to convicted felons, narcotics traffickers among them.

The dispute over daily activity reports, part of a larger struggle over the bureau's core mission, became a standoff. Arias and the others flatly refused to fill out the reports. As one inspector, who later quit in disgust, wrote in her resignation letter, "Management has refused to take any responsibility for the widespread discontent in this Department. Instead it has responded with intimidation tactics and micro management where more time is spent documenting the work than doing it."

"They want to monitor our activities," said Arias, "but they don't want us to do anything. Isn't that ironic?"

Venema, amazed by this stream of grievances, wondered how Arias could stand his job, which amounted to watching crime unfold and doing nothing. "Why didn't you quit?" he asked, his laptop idle and his report half-finished, his attention focused squarely on Arias.

As though it were obvious, Arias explained, "I love my job. It's really the perfect job, if it weren't for all the bullshit."

BY NOW THEY HAD ARRIVED AT BOYNTON BEACH AND THE GATED community of retirees where Odin refused to retire. When Odin opened the door, Venema saw a silver-haired man with glasses and rubber-soled shoes who walked with a slight shuffle. He had a hearing aid embedded in each ear and a string of Kabbala beads—reflecting a mystical Jewish belief that they would protect against evil spirits—around his wrist.

"Don't ever do that to me again," Odin said to Arias in greeting before ushering the men into his kitchen, where he had a pot of coffee and plate of cookies waiting on the table. He also had set out a stack of papers: the real business records of sales between BTC and the company called Omnimed. On twenty occasions, the two companies had bought and sold medicine to one another, the lowest sale being $33,000 and the highest nearing $100,000. Atkins had failed to give them nineteen of these records, despite his sworn statement that he had given them everything.

Odin often took a concerned, grandfatherly approach on his regulatory visits. He counseled those with new licenses on how to stay within the law's guidelines and continued to be affable even as they drifted into criminal activity. He could appear utterly fooled until he would walk outside and say, "You know they're lying because their lips are moving." Lulled into a belief in his senility, those he regulated inevitably divulged more information than they would have otherwise.

Given the records that Odin had produced, a perfect hammer to hold over Atkins's head, Venema was all too happy to include him. Also, Odin knew Carlow and Atkins's mother,

Marilyn, who had functioned as Carlow's bookkeeper for years.

"Can I make copies of these?" he asked Odin, who responded, "You can have them."

TEN DAYS LATER, THE THREE MEN SET OUT FOR BOCA RATON TO interview the employees of the Stone Group.

For years, Arias and Odin had been treated with casual disdain during regulatory visits: They were the flies that wouldn't stop buzzing. A joke. An irritant. But this time they arrived on the heels of the break-in at Bradley's and the subsequent police sting in the conference room, accompanied by a badge-wearing FDLE agent.

Since salesman Sean Dana's encounter with law enforcement, the employees of the Stone Group had been waiting for whatever the next step might be, fearing and suspecting they had been implicated in a criminal scheme. Venema did little to assuage their fear.

"We're going to put people in jail," he said by way of introduction.

It was five months after the September 11 terror attacks. Florida's flight schools had emerged as integral to the deadly planning. The highways had been cut up with countless road closures, emergency operations, terror alerts, and high-speed police pursuits and searches. No one wanted to be on the wrong side of law enforcement. And Venema mirthfully played upon that fear. As he said within earshot of the men, "Inside every Al Qaeda member is an FDLE informant just waiting to emerge," a phrase that made no particular sense but had the desired effect.

The Stone Group employees appeared terrified. Four of

five, including Dana, were there, and all seemed inclined to cooperate.

Arias and Odin began a site inspection. Odin was well known to the employees because he had given them one of his long-winded lectures about the do's and don'ts of the prescription drug business when they first got their license six months ago. They could only buy medicine from companies licensed by Florida. They needed to document the purchase of medicine from previous buyers, keep drugs in a refrigerated and secure warehouse that maintained regular office hours, and open their records to regular state inspections.

It took just minutes for Odin and Arias to turn up a number of *don'ts*. They found a drug called Marinol, a controlled substance and cannabis derivative that the Stone Group had purchased from BTC. Neither company had been licensed by the U.S. Drug Enforcement Administration to handle controlled substances. Doing so without a license was a felony.

The inspectors also found that several boxes of the cancer medicine Procrit were sticky to the touch, suggesting that they had already been dispensed to patients, then either stolen or resold and their pharmacy dispensing labels removed, leaving a gummy residue. "You guys must have thought you'd died and gone to heaven, getting such a good deal," Odin said with disgust. "This stuff came off the streets."

Sean Dana listened awkwardly. Meanwhile, Venema had taken Steve Gorn, the company's national sales manager, into another room to get his statement. Gorn, followed by the other Stone Group employees, explained that the company's growth had been aided by the relationship with Michael Carlow—a story that Odin and Arias could have guessed at from the new Lexuses and BMWs in the parking area outside. Each employee identified a mug shot of Carlow—suntanned and

smiling with a diamond stud in one ear—from his arrest in
2000.

For about seven years, the Stone Group had been a real
estate company before becoming a wholesaler of prescription
medicine six months ago. BTC quickly became one of the
group's suppliers, selling the fledgling wholesaler a stream of
remarkably discounted medicines, particularly injectable biotechnology drugs to treat cancer, AIDS, anemia, and other
conditions. So far Stone Group employees had purchased
nearly $2 million in medicine from BTC, dealing almost al-
ways with the jocular and helpful Michael Carlow and taking
his calls day and night whenever he had "product" to sell.

Marty Bradley's drugs were no exception. They came into
the Stone Group's possession on Sunday, January 20, 2002,
five days after they were stolen from Bradley's Miami warehouse. That afternoon, the Stone Group's salesman Doug
Brilliant, who three months earlier had worked as a stock-
broker, went to Carlow's home in Weston near Fort Lauderdale
to pick up the medicine.

When dealing with BTC, the pickup almost always involved a trip to Carlow's house, a lavish new construction
worth $1.3 million in the gated community of Windmill
Ranch Estates. The Stone Group employees would get waved
past the guardhouse at the community's edge, drive along the
quaint lanes to Carlow's house, and enter through the garage,
which housed a Ferrari convertible, a yellow Dodge Viper,
and a number of other vehicles. They would make their way
to the adjacent laundry room, where Carlow kept medicine in
a refrigerator, and he would pack up boxes of drugs on one of
several large tables. Usually someone was there to help him.

Gorn told Venema that they called Carlow after the sting
at BioMed Plus. Carlow acted surprised and apologetic and
wanted to know if they could still do business. "Basically, he

said he's going to have some more products with a paper trail and everything next week," said Gorn.

Venema asked when that discussion took place.

"I know we had a conversation this morning about some products," said Gorn. "But we haven't done anything since."

THE INSPECTORS SEIZED SEVERAL BOXES OF MEDICINE FROM THE Stone Group. Outside, they loaded it into the truck of Miami-Dade police sergeant John Petri, who had come for surveillance and protection.

Petri, short and muscular with a well-groomed mustache and a lively, sensitive face, was a decorated marksman with a wall of trophies. But his real gift was for surveillance, for waiting invisibly in his truck, either tailing a suspect or idling in a cul-de-sac or parking lot.

His attention rarely flagged during these hours-long stretches, in part because his truck allowed for a world of activity. It contained two police radio systems, photography and video surveillance equipment, a global positioning satellite device, a laptop that plugged into the cigarette lighter and had its own motionless stand, and a specially designed sun shade that allowed him to videotape right through it.

Now Petri gathered with the others in the parking lot and listened as Arias explained that the case against Carlow couldn't be much simpler: "You can't have a pharmacy in your house," he said.

6. The Cheshire Cat

January 20, 2002
Windmill Ranch Estates
Weston, Florida

THE GLISTENING "SUPERBURB" OF GOLF COURSES, LAKES, SHOP-ping malls, and gated communities of Italianate mansions that Michael Allyn Carlow called home was a developer's dream, dredged out of a swamp two decades ago. Planned down to each mailbox and hibiscus plant, Weston had become a destination address, described by the *Sun Sentinel* as a planned community "where hyperbole meets reality."

Once rare birds, snakes, and alligators occupied the 10,000 acres of wetlands west of Fort Lauderdale that bordered the Everglades and helped to regulate and drain its fragile ecosystem. Environmentalists lost their battle to preserve it and soon Starbucks, the Gap, and Florida's largest mall, Sawgrass Mills, replaced the rich marshland, with many of Weston's new residents driving gas-devouring Hummers.

Windmill Ranch Estates—lane after lane of palazzos on a grid of sparkling lakes and costly palm trees, each one trucked in and replanted—was among Weston's costliest gated communities. Carlow lived here on expansive landscaped grounds. Some thought he was a retired dot-commer. Others believed he had a furniture business. Regardless, his neighbors knew him as a gregarious family man.

Shortly after sunrise, on the Sunday the Stone Group was to pick up Bradley's medicine, Carlow opened the door of his mansion to confront an entirely separate problem from his past. Strangers had parked a giant moving truck in his driveway. The group included a lawyer, a paralegal, a police officer, four United States marshals, several movers, and a locksmith. The men offered him a court order, which Carlow perused. Appearing relaxed as ever and wearing only boxer shorts, he invited them in. They had come to seize everything that he owned.

At fifty, Carlow had a sunny, joking manner that little seemed to disrupt. Tan, with unkempt gray hair, slouching posture, and a paunch, he was casual about his appearance, if not indifferent to it. He appeared affable and low-key, his diamond stud earring the only hint of wildness. On occasion his features held a predatory glimmer—the low-slung jaw, the downturn in his narrow lips, the slight jutting of the bottom teeth, and his brown eyes small and intense below a broad brow.

The officers who now trooped through his mansion had nothing to do with the theft of medicine from Bradley's warehouse four days earlier. They had arrived on the matter of *Medical Distribution Inc. v. Quest Healthcare Inc. and Michael Carlow,* one of the most persistent in a trail of lawsuits in which Carlow had been named a defendant. So many people had fought to recover money from him it was actually hard to keep them all straight. He owed the Internal Revenue Service almost $500,000. A jewelry store in Fort Lauderdale had demanded $54,000 for watches and diamond rings that he bought on credit. Even his old homeowner's association had lodged a judgment against him for $2,006 plus lawyer's fees.

A Federal court in Louisville, Kentucky, had ordered him to pay $98,630 to Medical Distribution Inc., a company that had shipped pharmaceuticals to his corporations but had not gotten paid for them. When he failed to pay, the court authorized MDI to seize his belongings.

As the team sent by MDI moved from one enormous room to the next, Carlow followed calmly. He was not about to lose his head over some possessions. He knew from experience that even if they took everything today, he could start buying more tomorrow. He had done it four years ago after he filed for bankruptcy. He gave up his share of a beachfront condo to his third wife, from whom he was getting divorced. The bank had repossessed his $108,000 Bentley and his $675,000 Sea Ray Yacht.

The yacht, which he bought in 1997, had been a thrill. As he wrote to the man who arranged the financing, "I am very excited about this purchase and I will sing your praises for some time to come." He had named the boat the *Cheshire Cat*, which proved to be quite apt. The animal from *Alice in Wonderland* had been able to vanish when in danger, leaving only its famous grin behind to mark the spot where it had been.

After his bankruptcy, Carlow also vanished in his particular way. He put most of his new possessions in the name of his fourth wife, Candace. He also formed companies that appeared to belong to others, a scheme that MDI's lawyers spent the better part of a year untangling. Once they realized that Carlow, not the listed owners, owed the money, they began the legal proceedings that had brought them to his door today.

Inside Carlow's mansion, the Miami lawyer for MDI, Alejandro Hoyos, was amazed by the lavish décor: the marble bath rising from the center of the master bathroom like a throne; the Hadrian's Villa effect of the outdoor rooms, with several wet bars, overlooking the enormous swimming pool; and the master bedroom, with its four-poster bed and separate sitting room, that was as big as Hoyos's entire apartment.

Throughout their exploration, Carlow's pregnant, twenty-nine-year-old wife Candace hovered in the background. Barefoot and wearing shorts and a T-shirt, she made continual cutting remarks, even as her husband chatted cordially. While the big blonde struck Hoyos as strictly "trailer park," Carlow came across as smooth and knowledgeable, "the type of guy who could sell ice to Eskimos," Hoyos said later.

In the bedroom closet where Carlow kept a row of custom-made suits that he rarely wore, the Weston sheriff paused over a bag of sex toys.

"They're just for the benefit of my wife," said Carlow.

"Leave them," Hoyos told the sheriff. "I don't want to touch them."

The men spent all day inside the mansion. They seized jewelry, paintings, Carlow's collection of sixteen Gibson guitars, designer clothing, camera equipment, enormous bronze statues including one of four jockeys on horseback that took three people to carry out, flat-screen television sets, fitness equipment, fishing rods, a coin collection, and even Carlow's electric shaver with the charger too. They also took envelopes of cash—two containing $10,000 each and another with $5,000 inside.

By day's end, they had filled the eighteen-wheeler outside with 129 pieces of property. On their way out, one of the men took the Rolex off Carlow's wrist and plucked the diamond stud from his ear. The truck then disappeared down the palm-tree-shaded lane.

As Carlow stood in his denuded mansion, he was hardly vanquished. The men who had just seized his possessions had failed to take the most valuable items in the house, and they were now his to sell: 344 vials of rare blood derivatives, including NovoSeven and Gammimune, which he had stashed in the refrigerator in his laundry room for the Stone Group salesman to pick up that very afternoon.

BORN IN CONNECTICUT AND RAISED IN HOLLYWOOD, FLORIDA, Michael Carlow began his vanishing tricks in high school. He often sneaked into a local theme park, Pirate's World, which had rides and concerts, and sometimes he got caught when he became tangled in the barbed wire. His classmates remembered Carlow as the class clown, their best friend, and a bubbling personality who could lift their spirits with a single phone call. He played guitar in a popular rock band.

For all his sociability, his peers also recalled something private about him beneath the buoyant surface. His yearbook photo showed him with a wide grin and a sidelong glance, as though enjoying a secret joke.

After graduating in 1970, he drifted through a series of jobs, working in the claims department of an insurance company, as a printing-press operator, and in advertising for a car-wash equipment company.

He also embarked on a series of crimes. In 1973, at age twenty, he was convicted of armed robbery and served three years in prison. In 1984, he was arrested for dealing in stolen property. While the case against him was dismissed, he was ordered to complete a substance abuse treatment program. While enrolled in the program, he was arrested in Alabama for selling cocaine and this time fled. A warrant was issued for his arrest. In 1986, he turned himself in and resumed his drug rehab.

Perhaps more than anything, his time at the Village South Drug Rehabilitation Program changed his life. Not only did he quit the drinking and drug use that had fueled his petty crimes and revolving-door trips through the court system, but he also glimpsed a business that would earn him a fortune.

By the time MDI extended him credit in 1999, Carlow had shed any semblance of the drug-addled hood in thick glasses and a bushy mustache in his old mug shots. He now had a story that seemed believable.

He claimed to have entered the health-care field, particularly substance abuse treatment. And indeed he had—as a patient in court-ordered rehab. In 1991, he formed what he called a consulting company that later became Quest Healthcare Inc. As he explained to those from whom he wanted money or business, Quest oversaw more than a dozen mental health, male impotency, and HIV infusion clinics in six different states. "I trust this will give you a background sketch of what we do," he wrote to his yacht broker in 1997.

After his 1998 bankruptcy, Quest and his other spin-off companies, most of them not in his name, began branching out into pharmaceuticals, which allowed for the purchase of more luxury goods, including another yacht. This time the boat's name, *Tenacity*, reflected his talent for coming back. Just a few years after his bankruptcy, he owned the Weston mansion, five cars, a new boat, and numerous men's watches including an $110,000 Audemars Piguet with 458 diamonds. His net worth exceeded $5 million.

In addition to angry creditors, he left behind a trail of colleagues, whose names and credit ratings he had all but destroyed. Those who went into business with him were rarely blame-free: Some had criminal records, personal bankruptcies, or at least a self-serving view of the law. But most saw Carlow as being in a league of his own. "He would do anything that he could do that would make him money," said Michael Rosen, who helped manage Quest Healthcare and had a prior conviction for bank fraud. "He wouldn't care whose life was destroyed in the balance."

Crushed by debt after Carlow failed to pay Quest's bills,

Rosen went to Carlow's mansion and asked to borrow a few thousand dollars. Surrounded by $60,000 Gibson guitars and wearing one of his extravagant watches, Carlow responded, "I don't have that kind of money. Can't help you."

Jeffrey Schultz, a Coral Springs businessman who had been convicted of bank fraud alongside Rosen and was recovering from his own personal bankruptcy, had a similar experience. He had been trying to set his life right when Carlow hired him to manage another of his companies that raised money for investment in health clinics. He found Carlow to be "personable, very likeable," and said he did not think otherwise until he learned that MDI was suing him because Carlow had not paid for more than $260,000 in pharmaceuticals. Schultz said Carlow asked him to order the drugs.

Still imagining some misunderstanding, Schultz went to Carlow's mansion and pleaded with him, "This is not my debt." Carlow responded blithely that it was not his problem. "We're all big boys," he said. Schultz was forced to file for bankruptcy a second time.

Of all the lawsuits that Carlow faced, the one brought by MDI went the farthest in penetrating his business interests, in part by simply finding him behind the front people, shell corporations, and decoy mailing addresses that he used. The lawsuit alleged that he had been the silent operator of sham corporations in Ohio, Michigan, Kentucky, and elsewhere, and that he had conspired with his wife, Candace, to buy pharmaceuticals on credit and not pay for them. According to the lawsuit, the Carlows set Schultz and others up in business, told them what to do, presented management of the companies to them as a business opportunity, and left them "on the hook" for the unpaid accounts.

The principal lawyer for MDI, Dennis Murrell in Louisville, managed to depose most everyone involved. And as he did, a common allegation emerged: Carlow was the silent partner who misused those naïve enough to allow their names on his businesses, such as Norman J. Embree, who once ran a profitable AIDS clinic in Fort Lauderdale.

After a business slowdown, Embree sold a portion of his clinic to Carlow and so began a nightmare. Embree recalled that one day, Carlow marched into a treatment room and grabbed boxes of an experimental HIV vaccine that the clinic had gotten through a federal grant.

Though Embree protested, he said that Carlow walked out and resold the vaccine for a profit. On another night, Carlow brought a truck and removed medical equipment, selling a refracting lens back to the company from which Embree bought it. Shortly afterward, Embree said that Carlow changed the locks and expelled him from his own clinic.

Within a year, the clinic was closed and Embree's finances were in tatters. He later testified in a deposition that Carlow and his wife, Candace, "just disappeared and left my name on everybody's loans, so they never registered with the State as being involved in any way, shape or form."

When Murrell asked Embree what the result of that had been, he responded, "Um, I get invited to depositions like this one."

For more than a year, MDI attorney Dennis Murrell had tried unsuccessfully to question Carlow, who ignored court orders, dodged subpoenas, and failed to show up for scheduled depositions. In September 2001, when Murrell finally got him under oath in a Miami law office, he appeared as composed and jaunty as ever.

Carlow denied having any relationship with the companies in question and said that he was unemployed and survived on the salary of his wife, who had her own medical consulting company. He claimed to own almost nothing, as his various houses, cars, and credit cards were all in his wife's name. And he insisted that he was not responsible for the actions of those who had ordered medicine and failed to pay their bills.

He said of Jeffrey Schultz's complaint that he had ordered medicine on Carlow's behalf: "I think it makes fair sense to me that Mr. Schultz is trying to shift the weight, as it were." He added, "Jeffrey is a big boy. He has a couple of years on me, as a matter of fact."

It was four months after this deposition that MDI's representatives drove out to Carlow's mansion and seized his property. Three weeks after that, on February 12, 2002, Murrell deposed Carlow again. Despite his financial and legal woes, Carlow remained chipper and, in his jocular way, continued to deny any involvement in procuring pharmaceuticals without paying for them.

"All right. Back on the air," he said, resuming his testimony after a lengthy off-the-record debate. Offering to spell a complicated last name for the stenographer, he said, "I'll take it for a descriptive drive."

No longer "unemployed" as he claimed to be at the last deposition, he said he was dividing his time between two start-up companies: one, Navigator PC, was developing global positioning satellite devices; the other was a wholesale apparel company. He also acknowledged consulting for a new company, G&K Pharma, about which he declined to answer questions. It was licensed in Baltimore, Maryland, and on paper it was run by Carlow's brother-in-law Atkins. The initials stood for Atkins's two favorite cats, Gus and Kitty.

It seemed obvious from Carlow's irreverent and almost giddy manner throughout the deposition that he felt well insulated. He was expanding his businesses with obvious zeal, while thumbing his nose at the state's notoriously weak health laws.

FEW WHO TOOK ON CARLOW ACTUALLY WON THEIR BATTLES. JOAN Bardzick, the tough-as-nails owner of three check-cashing stores, was an exception. Her customers included migrant farm workers, loan sharks, and anyone else who needed instant cash. At her store in Davie, she worked behind two sets of locked bulletproof doors, her inner office seemingly frozen in the 1970s with its well-stocked backlit bar and black leather sofas.

A salty veteran of every imaginable scam attempted on her, Bardzick, with her leathery tan and velour track suits, took pride in never being played for a fool. But in the summer of 2000, Carlow had fooled her. She thought he was a legitimate businessman who ran AIDS treatment clinics because her daughter Wendy, who worked for a major pharmaceutical company as a sales rep, counted Carlow among her best customers. His doctors were some of the busiest prescribers in the area; her daughter knew this from the data that her company collected. She had also been inside several of his bustling clinics and vouched for him when he started bringing checks to her mother's store. That, and his Rolex watch and Mercedes convertible, convinced Bardzick that he was credible.

At Carlow's clinic in Fort Pierce, checks were "bouncing all over the place," recalled Hazel Hoylman, the clinic's community outreach coordinator. Even though the clinic received generous federal reimbursement for treating AIDS patients, doctors and nurses went unpaid and stopped coming to work. Disposable needles piled up after the company that collected them didn't get paid, said Hoylman.

At Bardzick's A-1 check-cashing store, Carlow soon bounced $93,120 in checks. Afterward, he showed up acting sheepish and promised to make things right. Bardzick, now on full alert, immediately hired a lawyer to put liens on his property.

Unabashed, Carlow proposed to her a way that she could get her money back and they could both wind up ahead: Her daughter could open an AIDS clinic and he would become a silent partner.

As Bardzick told her daughter, "The man's off his rocker."

Wendy felt responsible for what had happened to her mother and determined to set it right. First, she went to lunch with Carlow and tried, face to face, to convince him to repay the debt. He spent most of the meal on his cell phone, conducting other business.

Frustrated and angry, she called him at home several times, the last time threatening to take out an ad in the newspaper near where his kids went to school. He responded, "Don't you ever call me again," and something in his tone made her nervous, as though she had crossed some invisible boundary into a danger zone.

So Wendy decided to exact revenge the best way she knew how: leverage her knowledge and drive him out of business by steering his employees, patients, and federal grant money away from his clinics to other competitors in the area.

She succeeded. As his clinics faltered, Carlow was forced to sell one of his houses. Bardzick, who had already placed a lien on his property, was waiting for him. She succeeded in recouping her money plus lawyers fees. Though still furious months later, she remained one of the few who had tangled with Carlow and not lost a dime.

7. One Man's Trash, Another Man's Treasure

February 2002
Windmill Ranch Estates
Weston, Florida

WELL AFTER MIDNIGHT GARY VENEMA WAS AWAKE, STARING AT THE ceiling. He didn't even need to set the alarm. He couldn't sleep because tonight was Wednesday. And on Wednesdays, the city of Weston collected trash at the Windmill Ranch Estates.

Donning dark jeans, a T-shirt, and sneakers, Venema glided toward the door, forgetting his wife's admonition that he should wear gloves. Venema actually liked working at night when the world slowed down and he could focus his attention on whatever his goal might be. In Hialeah, he had made a lot of overtime working nights. But money had never been his motive. He believed that the harder he worked, the luckier he got. And he enjoyed thinking of himself as a thief in the night (albeit one on the right side of the law).

The drive to Weston took only five minutes with so few cars on the road. Outside the gated community he flashed his badge, and a sleepy guard let him in. Down the silent lanes, he noted all the garbage cans spotting the front lawns of the houses until he approached 3465 Windmill Ranch Road, which he had driven past once before. It was huge, a stucco mansion in the Mediterranean style with pillars and archways shaded by palm trees. A pool glimmered out back.

He drove past slowly, looking to see if anyone registered

his presence, but the house remained dark, a green van parked in the driveway and the winter moon bright. He doubled back, left the car running, and hopped out, within seconds closing the distance to Carlow's front lawn. He wrestled briefly with the trash bag, which was wedged inside the can, and winced at the rattling sound. Holding the bag between two fingers, he loped back to his truck, threw it in the flatbed, and kept motoring along the lane.

SANDY VENEMA AWOKE THE NEXT MORNING TO FIND HER HUSband still asleep and a familiar acrid smell wafting into their bedroom. She passed through the living room and peeked into the garage. There lay a disemboweled garbage bag and drying papers sprinkled with coffee grounds and other debris.

Venema returned twice the next week. No one could have asked for a more reliable garbage man. He sifted through the bags until he found the documents he was seeking nestled amid the cigarettes and Diet Coke cans that, despite the state's prohibition, he would come to associate with Carlow's late-night office work. Each time, Carlow's trash disgorged clear evidence that he was into pharmaceuticals, attempting to expand his various businesses, and indulging his taste for luxury items such as yachts, sports cars, and new homes.

For starters, the records showed that BTC, the company that supposedly operated out of the empty Kissimmee office, actually received mail at Carlow's home. The company's business also crossed state lines: BTC had bought the controlled substance Marinol from a Kansas City, Missouri, company and had sold other drugs to a company in Henderson, Nevada. Another company that Carlow appeared to control, Accucare LLC, the one listed as belonging to his wife, was buying and selling medicine to various companies in Florida and around the country including in Kansas City, Missouri; Henderson, Nevada; and Nashua, New Hampshire.

Venema had also unscrunched several faxes supposedly sent by Atkins, Carlow's brother-in-law. One, on BTC letterhead, was to the Arizona department of health, seeking a license to operate in the state. The other was to the Texas health department, applying to open a branch of G&K Pharmaceuticals there. It showed the office address for G&K at 300 Thunderbird Drive in El Paso, right near El Paso Pharmaceuticals, the company that had allegedly sold Bradley's stolen medicine to BTC.

The most interesting document was a memo from a woman named Lori Marvel at a company called Two M Enterprise with a 219 area code—somewhere in Indiana—and no listed address. Dated January 31, 2002, it was written in the lingo of the pharmaceutical trade and was a pitch letter of sorts, offering an array of discounted drugs:

> *Hi Michael,*
>
> *It was a delight talking to you today. I put together a list of product that could be possibly available next week. I talked to the gentleman that would have the products and he is willing to work with [us] on this. I put down what the WAC is and what 25% off that would be. I don't have anything added for myself, but I have to figure out where you would like to have the pricing on this so that I can start negotiating as soon as possible.*
>
> *He did state that there would be approx. $40,000 to $50,000 worth of product per week. He also told me that all the product is clean and that some of the patient inserts on the bottles may be missing. He stated that he does not get to "pick and choose" what he gets so he has to take what is available. However, he can exchange items if*

you don't like the way they look when you get them. All items will have at least a year expiration date. He will not sell short date product and if he has any he will let us know in advance.

The discount that Marvel's source was offering was one-quarter off the "WAC" or wholesale acquisition cost of the drugs. The investigators knew that only medicine first stolen or purchased for almost nothing off the streets would be that discounted. Marvel wrote in closing:

I also have a lot of paint that has to be exported. It is about 250,000 gallons. I don't know if you deal with this type of item but let me know and I will fax over what they sent me. They also have paint that can be sold in the US for less than any department or home improvement store can buy it for. Just let me know.

Clearly, whether knowingly or not, Marvel was a broker of whatever pilfered goods came her way, from paint to Procrit.

Venema also found a sales brochure for a luxury yacht, the Hatteras 6300, and records for the two boats that Carlow already owned, *Tenacity* and *Delicious*. He had just traded in his old Ferrari for a new one worth $249,000. And he had closed on another house nearby, in the Sunshine Ranches development, for $1.9 million.

The trash even offered clues to Carlow's personality, which appeared both smug and flirtatious. One document from a prospective seller listed medicine, its quantity and price, and a cheerful note to "Please indicate based on the qty listed what would be the best price." The seller Beth signed off with, "Thx Baby!"

Carlow wrote back by hand, "Beth, I see all these items. . . . Not in such large quantities though! Would probably take me a month to fill an order this size. (Size matters!)"

Venema, exuberant after his first uncensored look into Carlow's life, wrote and sent his own memo:

To: Stephanie Feldman / Statewide
From: Gary Venema / Starfleet Command

Steph—couldn't sleep last nite—and you know what I do when I can't sleep—I DO TRASH PULLS! Carlow had a w-9 or whatever showing misc. income for 2001 at $700,000 from some investment firm. Also a letter from Hatteras Yachts—FDLE agents don' motor around the waterways in Hatteras Yachts—[our] whole office couldn't buy gasoline for one!

He continued:

My strategy would be to:

1) Call Mr. Carlow real nicely for a little friendly chat . . .

2) Have the warrant ready for when his attorney, David Mandel tells me to pound salt

3) Hit his house like the weapon of mass destruction that I intend to be on this guy. . . .

And at the bottom he wrote:

Remember—INSIDE EVERY TALIBAN AND AL QAEDA WARRIOR IS AN FDLE INFORMANT OR COOPERATING DEFENDANT STRIVING TO EMERGE!!!!!

Venema's enthusiasm—and his manic endeavors—took some understanding. In part, they showed his gratitude for just being given the chance to work hard. But detective work also came naturally to him because he saw the world in black and white. There were *good people*, particularly those who worked hard. He had already decided that Cesar Arias was *good people*, since he had been taking the long way around every obstacle for fourteen years.

Then there were bad people, a category to which Carlow belonged, since he appeared to be stealing from everybody and giving to himself. Through his trash alone, Venema soon formed an intimate dislike of the man, and he began declaring to most anyone who would listen that he was *coming downtown, Charlie Brown*, to get Michael Carlow.

IN THE DAYS SURROUNDING THE BIOMED PLUS BURGLARY, CARlow had received a steady stream of visitors, as the records from the guardhouse of his gated community showed. The investigators were particularly interested in repeated visits from a man named Fabian Díaz, who had driven his sporty Mazda back and forth to see Carlow on the day of and the day after the Miami warehouse break-in.

Díaz, whose car was registered to a local family counseling center, was listed as an agent for El Paso Pharmaceuticals, the Texas company that Atkins claimed had sold him Bradley's drugs. He was also a convicted heroin seller who had spent years in the Florida prison system.

Phone records deepened the intrigue. The day after the break-in, Díaz had received a call from the Stone Group office and the next day a call from the home phone of Steve Gorn, the company's sales manager. More calls followed.

Meanwhile, the calls between Carlow and Díaz flowed back and forth. Suddenly Díaz, a man whom Gorn had never mentioned, seemed central to the theft at Bradley's warehouse.

NOTHING WAS WHAT IT SEEMED. MEDICINE THAT ON PAPER HAD traveled across the country in fact had never left South Florida. The Texas authorities had checked out El Paso Pharmaceuticals in Houston and found an unoccupied back room in an office building where a secretary said that in seven months, she had seen no one go through the door. Investigators found that the pedigree papers that the Stone Group had provided with Bradley's medicine listing its previous owners were a complete fiction. A call to the drugmaker that manufactured NovoSeven revealed that it had never sold the medicine to any of the companies listed on the pedigree.

Despite falsehoods, from a distance the sales appeared legitimate, made by legal and licensed companies operating within the margins of a legitimate industry.

On February 11, more than three weeks after their first visit, Odin, Arias, and Venema returned to the Stone Group. This time the company's thirty-year-old president, Adam Runsdorf, was in. He wore a Rolex watch and fine preppy clothes and had arrived at work in his Mercedes 500 SL. His first declaration was a mind-blower.

He told the investigators that he was suing BioMed Plus for nonpayment of the check that Bradley had handed over to his salesman that day in the conference room.

"You are kidding me," Venema said. "How can you expect them to pay for property that was stolen from them?"

"We are still entitled to a profit," Runsdorf persisted.

It was the kind of attitude that made Venema want to kill

slowly and painfully, another reason why he never carried his gun but kept it locked in his truck.

Runsdorf went on to explain that he had first met Carlow at his home in January after their companies had already done $2 million in business together. His own family had been in the pharmaceutical business for twenty-six years, he added.

"Then why were you purchasing over $2 million in pharmaceuticals from a residence, which is a felony?" Odin asked.

"It was an error in judgment," said Runsdorf, explaining that it was easiest for his employee to pick up the pharmaceuticals from Carlow's house because he lived nearby. "The Stone Group is tightening up its policies."

Odin scowled his disapproval. Venema then showed Runsdorf and Gorn a list of people, including Fabian Díaz, who had visited Carlow at his home and asked if they knew any of them. The men stared silently at the list. "I spoke with Fabian Díaz five months earlier," Gorn offered, saying nothing about the two calls from his home to Díaz less than a month before. He claimed to have done no business with him and said he could not remember the name of his company. "I can research that and get back to you," he offered.

As they left, Runsdorf extended a fig leaf: "Maybe we won't sue BioMed Plus after all."

Once outside in the parking lot, Venema exclaimed, "We're going to kill those assholes! We are going to rain death and destruction down on them!"

Arias also felt inspired. Because of Venema's subpoena power, they had been able to reach into visitor logs and phone records to catch the Stone Group in a matrix of lies. Maybe together, the three men could inflict some real damage. Like the Horsemen of the Apocalypse, thought Arias, they could exact a revenge of biblical proportion on those who sold bad medicine.

BEFORE HE HAD MET ARIAS AND ODIN, VENEMA HAD ASSUMED that his medicine traveled directly from the drugmaker to the pharmacy, through pristine laboratories and warehouses staffed by men of science in lab coats. But Arias and Odin worried about the medicine's transport, its temperature, where it originated, the path it took, the documentation of all this, and the patient who might wind up with it, dependent on its powers to heal but with no inkling of these concerns.

Venema began to think of himself as a student. He apprenticed himself to the two drug inspectors. The middle-aged detective who had thought the theft of toothbrush tips was a big deal set out to learn everything he could about the nation's medicine. He began to call Arias at all hours of the day and night, on weekends, whenever a question popped into his mind, oblivious to the possibility that Arias might be doing something else.

Arias was always ready to work. He began going over to Venema's house, where the two men would sit for hours at a table in the lush backyard beneath the palm trees. There, Arias would describe everything he knew about how medicine is distributed.

Arias's wife, who tended to keep her distance from his work, finally asked her husband, "Who is this guy?"

Arias, who was in heaven, told her about Venema, adding, "Usually I have to keep calling them. But he keeps calling me."

WITH ITS STRIP CLUBS, GLOOMY INDUSTRIAL PARKS, AND TASTE-less wealth, the city of Hialeah, northwest of Miami, was often thought of as a place best left behind. The Cuban immigrants who made it up and out to more elegant areas like Coral Gables referred to anything garish or downmarket as "muy

Hialeah." It was Florida's fifth-largest city, but many thought of it as the northernmost city of Latin America. In the 1980s, when Venema worked there, it was almost 90 percent Latino.

Hialeah was a dangerous place, awash in drugs and guns. It had attracted some of Florida's worst cocaine traffickers, the type that Al Pacino made famous in the movie *Scarface*. In 1982, murders in Dade County climbed to 508, twenty-five of which were in Hialeah. Though known as "Sparkle City," Hialeah was anything but. The place was a dump, Venema recalled. And he had loved working there.

The shifts of duty were relentless, a blur of car wrecks, armed robberies, murders, and burglaries. The cops were not permitted to get food even at a drive-through without "checking out their signal" and having the minutes deducted from their time off. Venema, like the other cops, learned to bolt his food and eat without chewing, a skill he retained.

Whenever the phone rang late at night, Venema would wake up instantly and get out of bed: Homicides at that hour were a small victory for his bank account, enabling him to collect time-and-a-half. "They never killed each other between 8 A.M. and 4 P.M.," he would say, not without gratitude for the late-night slaughter that helped boost his income.

He found cock-fighting rings in the middle of nowhere and abandoned cars with dead bodies (one car had three in the trunk). Two of Venema's colleagues were executed on the job: one investigating an armed robbery, the other responding to a noise complaint. The danger and stress energized Venema, and his personnel file grew thick with commendations.

His superiors praised his investigative ability, his hard work, and his bravery. He leaped from one balcony to another of a high-rise building to save a woman threatening suicide. He rescued a baby from a stolen car, tracked down the killers of deliverymen and taxi drivers, aided rape victims, and stung currency counterfeiters.

He even found humor amid the violence. One tiny bar in particular was the site of numerous murders. When a victim was killed in the bar, the witnesses would all claim to have been in the bathroom, a stall barely big enough to turn around in. On another occasion, Venema pulled a body from the bathroom, and the witnesses all claimed to have been in the bar and seen nothing.

But the highlight was the narcotics unit, not just because he loved to dress like a bum and grow a beard and go out and make deals undercover. He loved the guys he worked with. Together, their busts became so lucrative that they were practically minting money for his department. In January of 1990 alone, his unit arrested twenty-nine people, seized twenty-two kilos of cocaine and 126 pounds of marijuana, and confiscated eleven cars and $29,736 in cash. The detectives sported the best equipment—all of it seized from the narcotics dealers. They had nice cars, those big early-model cell phones, and fancy guns; Venema carried a repossessed $600 Sig Sauer.

He had no patience for those not willing to work hard. His parents had worked hard all their lives, his father as a mechanic for Eastern Airlines, his mother as a teacher. Those who didn't share his ethic invited his undisguised scorn, his warm and excited manner turning instantly to ice. He gave his own sister a cold shoulder because he felt she had sponged off his parents. Even when the two bumped into each other at a local diner, Venema would not get up or say hello. Sandy, as always, played the goodwill ambassador and approached her cordially.

Sandy, who Venema viewed as his best friend, was his opposite: reed thin compared to his big, sturdy build, quiet while he was voluble, brunette and pale instead of bronzed by the sun.

On superficial meeting, some might mistake the pretty former nurse for meek. But in her quiet way she was forceful and steadying. She had given Venema the sailboat charm when they owned a little Hobie Cat. She reminded him to take a sip of water or even to breathe when he got overexcited. And by taming his wildest emotions, she had probably saved his life on any number of shifts when rational thinking meant survival.

Unlike Sandy, Venema had little to say about his personal or inner life. He had no time for introspection and didn't even seem to understand the impulse. But he reacted strongly and passionately to the world around him. And for these reasons, even as his successes in Hialeah continued through 1991, his relationships with his superiors broke down.

Hialeah was run by what many viewed as a corrupt political machine. Mayor Raul Martinez, who was subsequently convicted of extortion and racketeering, handily won elections in which even the dead voted for him, sometimes more than once. He handpicked Police Chief Rolando D. Bolanos, who some in the department felt allowed nepotism, cronyism, and reverse racism to flourish, creating a morale crisis.

Even under the best circumstances, Venema was "always on the verge of being in hot water," said Dwight Snyder, a former colleague who now lives across the street from him. Venema bristled particularly at directions to arrest street-level drug dealers, who made the neighborhood look bad, instead of the kingpins and traffickers. This strategy, *to pursue rocks and not kilos*, as Venema said, sickened him. Never one to conceal his opinion, he piped up scornfully at one roll call, "We're breaking up the Medellín cartel, rock by rock."

He created an underground newspaper mocking the leadership and called it the *Starfleet Enquirer*. He had only to print a copy or two and, within hours, copies were everywhere. He posted a picture on his truck of the mayor with a "just say no" line drawn through his face and drove around town. Sandy recalled, "It's a miracle they didn't just fire him."

On December 6, 1991, after years of commendations, he received a three-sentence memo from Chief Bolanos with the subject: "*Reassignment.*" He was being demoted from the narcotics special investigations unit back to uniformed patrol, where he had begun his career. Bolanos went on to disband the narcotics and homicide units and redistribute those plum jobs to a loyal cadre of Latino officers.

Venema became depressed. His hatred of Bolanos gnawed away at him. "It became like an obsession," Sandy later recalled. "Don't bring your job home? Forget that." Six other Caucasian sergeants from narcotics and homicide also were demoted. In 1996, they filed a reverse discrimination lawsuit claiming that Bolanos favored Hispanics over Caucasians when handing out desirable jobs. Backed by the Equal Employment Opportunity Commission, the lawsuit crept through federal court. Meanwhile, Venema spent six years on uniformed patrol, responding to bus accidents and watching speeding motorists that he should have pulled over pass by him instead. Not wanting to concede anything, he refused to resign or retire just yet.

Venema never sought comfort in alcohol, explaining that he could "act like an asshole completely sober." Instead, during his darkest days in Hialeah, he stayed sane by doing woodworking, producing a remarkable flow of perfect bookcases, bowls of rare wood, and other gleaming objects from shelves to abstract sculptures. But the last time he had undertaken carpentry in a bad frame of mind, he severed the tip of his left thumb with a table saw. He had to get one of his three sons, Kevin, home on leave from the National Guard, to help locate the missing digit. It turned up in the driveway where he had hopped around hollering.

Meanwhile, he applied to other police departments in South Florida. One sent him a letter all but stating that he was too old. Some didn't even write back. He would remember this bad treatment and vow to reciprocate if given the chance. The job with FDLE fell into his life like a miracle.

From his new perch he watched the disintegration in Hialeah. Bolanos's two sons on the force were tried on police brutality charges. Bolanos also was investigated for concealing one of his sons' previous criminal convictions, which would have disqualified him from police work. Bolanos was later suspended for tampering with the campaign mail of a mayoral candidate. After Mayor Martinez was convicted of racketeering, his disappointed loyalists waged a Santeria campaign against members of the new city administration, mailing them severed animal parts. A muzzled goat head surfaced in the police department parking lot. The events there became a statewide embarrassment, even filling a column in the *New York Times*.

In 2000, Venema and the six other police sergeants won their reverse discrimination lawsuit against Bolanos. A federal jury awarded them $1.4 million—$200,000 each. While Venema told his new colleagues about the Hialeah of car chases, shootouts, and camaraderie, and even about the demotion he suffered, he almost never spoke of the context: the terrible politics, the agonizing lawsuit, and the depressing six years of being back in uniform. Instead, he said simply of that chapter in his life, "Hialeah was a great place to have in your rearview mirror."

VENEMA RETURNED TO CARLOW'S MANSION ON FEBRUARY 15, 2002. This time he was accompanied by Arias, Odin, and five other investigators including Randy Jones, a bear of a detec-

tive from the Miami-Dade Police Department slung with
cameras and video equipment. He held a search warrant
positing that the crimes of racketeering, conspiracy to racke-
teer, grand theft, dealing in stolen property, and prescription
drug fraud might have occurred on the premises.

Venema rang the doorbell and a startled maid opened the
door. Carlow wasn't home.

As the men trooped inside, they were stunned by what
they saw. By now Carlow had repaid his debt to MDI and recovered most of his lavish furnishings. To the investigators
who had toiled all their lives and whose wives cut coupons, it
was like walking into Versailles. Venema whooped in amaze-
ment while Detective Jones, the group documentarian, tried
to hush him as he began to videotape.

It was one thing to find evidence of Michael Carlow's Ferrari in his trash. It was another to see the car in his garage, to
smell the new, butter-leather interior and see the snazzy
bright-red horse figures on the plush floor mats. A zippy yellow Dodge Viper with black racing stripes sat just outside the
garage. In the house were gleaming antiques, a wraparound
indoor/outdoor shower off the master bedroom, flat-screen
televisions and computer monitors, a Sub-Zero refrigerator,
copper pots, and other accoutrements of major money, including several maids standing around looking nervous.

Carlow, who was away at a boat show, expressed no interest in returning when one of the maids called him. Instead, he
sent his lawyer, who stayed around only long enough to get a
voucher for the property the investigators took—records,
medicine, computers, and more.

Carlow's file cabinets turned up neatly indexed folders for
shell companies, financial records, and personal pursuits. One
folder was devoted to his yacht and another to a "smoking ces-
sation" program. A box of business cards listed Carlow as the

"principal" for BTC Wholesale. Paperwork showed the sports cars registered to Carlow's wife, Candace, including the Ferrari and the Viper, which was worth $87,000. A credenza in Carlow's office contained many bottles of the antibiotic Doxycycline. Arias suspected that Carlow was hoarding the medicine to sell at inflated prices during the ongoing anthrax crisis.

They also found two invoices from a corporation called Joskay in Homestead, south of Miami, for construction equipment costing $39,594 and $68,999 respectively. Alongside the invoices was a note from Lori Marvel in Indiana with two lists of blood derivatives and pharmaceuticals for the exact same prices and an explanation that Joskay would send Carlow invoices itemizing the products as "construction equipment."

For Arias and Odin, the trip through Carlow's files was like old-home week, revealing a common root system and intertwined business interests among several familiar bad guys. They would learn that Joskay had been newly formed by José L. Benitez, the half-brother of a diverter of medicine, named José A. Benitez, whose license to distribute prescription medicine was about to be revoked by the state. José A. was also on federal probation after serving time for marijuana smuggling. The investigators emerged with dozens of names, companies, bank account records, and other leads to mine. They were surprised, not just by the evidence, but also by the opulence they encountered. As Jones later reflected, "All of the *stuff* was just overwhelming, the amount of wealth he had acquired over such a short time."

FIVE DAYS LATER, PROSECUTOR STEPHANIE FELDMAN SUMMONED Gary Venema, Cesar Arias, Gene Odin, and Miami-Dade investigators John Petri and Randy Jones to her office to create a special task force. At her direction, the five men would investigate how stolen, diverted, and possibly counterfeit medicine was moving throughout Florida and into the nation's supply. She would call their work Operation Stone Cold. (The men assumed this was because they had caught the Stone Group *cold* selling Bradley back his medicine.)

Their goal, she said, would be to build racketeering cases against Michael Carlow, José L. Benitez, and their accomplices. Venema, who had spent the pre-dawn hours in Carlow's trash yet again and looked exhausted, would be their lead investigator. She expected indictments within six months.

The men were not a natural dream team. They had scant practice in complex investigations. None had worked such a high-stakes case. As a group, their work lives had fallen short of what they had hoped to accomplish. They didn't follow orders well. Though they were the types to stand on principle, they had usually stood alone or fought battles that would annoy more people than they would impress.

At least three of them had strong reasons to retire or quit: Arias to make more money, Odin to finally retire and collect the $36,000 the state would give him if he did it sooner rather than later, and Randy Jones to finally pursue his hobby of photography, which he had mostly practiced by taking stealth photographs of suspects entering and exiting South Florida condominiums. Except for Arias, they were all over fifty. They all followed some version of the Atkins diet, took medicine for high blood pressure, and had to read documents and restaurant menus at a distance. At forty-eight, Arias was fond of saying, "The memory is the second thing that goes."

But the men also shared other characteristics not lost on Feldman. They were tenacious, if not obsessive. They were old-school investigators, more comfortable sifting through garbage or documents and following their suspects silently than using fancy computer programs or cutting-edge forensics. They also were the ones who came early and stayed late. And none liked spending time behind a desk.

Arias and Odin would supply the essential knowledge of medicine. Venema, as lead investigator, would supply the adrenaline. Petri and Jones, who had worked together thirteen years and had known each other longer, would do the surveillance.

Though Petri supervised Jones at the Miami-Dade Police Department's career criminal unit, the two men had a friendship that knew no rank. Petri, extroverted with a polished manner and a taste for the limelight, loved giving toasts and speeches and ordering plaques to honor his friends. Jones, an intensely private man with a rapier wit and a shy smile, was content to let others do the talking. Though he was seemingly aloof, little escaped his notice.

In 1984, Jones noticed the couture seamstress who had been tailoring his shirts and sent Petri to pick up the garments. The seamstress, Gloria, became Petri's second wife and the two men continued their work together—watching, waiting, and living from their trucks. They loved the work with a strange intensity. "Every human being has a pattern," said Petri, "and if you follow someone for a week, you'll get all of that."

THE FIVE MEN BEGAN CALLING THEMSELVES THE HORSEMEN OF the Apocalypse. Arias had originated the name, and it fit their sense about the case. To some degree, they all envisioned bad medicine as a scourge on the drug supply and they planned to cut a wide swath as though wielding scythes on horseback. Because they had mostly worked alone, at odds with their own organizations, never part of the winning team or any team really, they secretly hoped that this time would be different.

Feldman instructed Venema to document everything they did. He would be the one to chronicle the investigation, so the men did not create conflicting reports. She also insisted that he collect the reports, chronologically, in a binder.

It was hard enough for Venema to sit still in a meeting, let alone stay seated to write it all down. She already knew that from when he tried to sell her on prosecuting his over-the-counter theft case. He had flown in pharmacy investigators from Atlanta and Tampa and had stayed on his feet the entire meeting, talking almost ceaselessly while pacing the length of the small, windowless conference room—more than sixty times, she estimated.

The report writing, she hoped, would get him to slow down—to think about what racketeering meant, to think spatially, to build a circle of evidence and then tighten it.

She also insisted on reviewing each search warrant affidavit before they brought it to a judge. She was particularly keen that they didn't run haphazardly after a suspect, especially since Venema, in his excitement, would say things like, "We know he's doing it!" as justification for a search warrant. That line of reasoning would get them laughed out of court.

Venema also memorized the trash pick-up schedules of their widening circle of suspects. He regularly spent the pre-dawn hours wading through dumpsters and stealing trash bags across South Florida, tangling with raccoons and chasing can lids that rolled down the streets. Pungent odors now wafted from his office, where he kept his best finds. Colleagues went the long way round to avoid walking past his office.

The self-named Horsemen began to show their findings to others. At one of their first presentations to a security director from a Walgreens pharmacy, Feldman recognized that

Venema was actually shy beneath his animated demeanor and dreaded public speaking.

When the Miami-Dade Police Department began to challenge the overtime accrued by Petri and Jones, who were spending long hours tailing suspects, the task force members held a PowerPoint presentation for a department chief. They showed him pictures of the suspects' fancy homes and cars and their bank-account statements, the significance of which was not lost on the man. Here were assets they could potentially seize, which might help the department weather a tough budget year. The chief became a convert.

Once the investigators got over their initial shock at Feldman's youth and diminutive size, their respect for her toughness and her vision grew. She took their calls day and night, gave orders, and laid the ground rules for their team work. She demanded cooperation and the pooling of information at all times. They wondered how she knew so much about racketeering and about teamwork after only five years as a prosecutor.

They did not know that her life depended on the syringe of insulin that she carried, or that she had dubbed the task force Operation Stone Cold because she viewed those trafficking in adulterated medicine as stone-cold killers. From all the time she spent in hospitals watching IV solutions drip into her own veins, she knew that patients' lives were threatened if they did not get exactly the right medicine, maintained in the right way. "What was happening was nothing short of murder by inches," she concluded early on.

8. A Cold Chain Gets Hot

February 2002

THE MEDICINE THAT HAD INITIALLY PROVED MIRACULOUS FOR Maxine Blount slipped off a conveyor belt in crystalline bottles, the liquid inside a living protein, its composition almost as delicate as the patients who would use it.

The clear solution, which had resulted from scientists' ability to penetrate the human genetic code, replicated a human protein, erythropoietin, that boosted red blood cell production. In 1983, scientists at a California biotechnology company, Amgen, had figured out how to capture the DNA, or the exact amino acid sequence, of the protein and then grow it in cells derived from the ovaries of Chinese hamsters. In 1989, the Food and Drug Administration approved the resulting formula, Epoetin Alfa, for use in humans. While Amgen continued to make and sell the drug under the name Epogen, it licensed Ortho Biotech, a subsidiary of Johnson & Johnson, to sell the formula in different concentrations under the name Procrit.

Immediately, the two medicines transformed the lives of patients like Maxine who suffered from anemia after cancer treatment, organ transplants, or kidney disease. The drug turned Amgen, once a tiny, unknown company, into the world's largest independent biotechnology company. And Epo, as it is

known in the trade, became the best-selling medicine of biotechnology with sales of over $6 billion in the United States
in 2004.

Epo requires careful handling from manufacture until it is
administered to patients. The medicine has to be maintained
at a constant temperature of 2 to 8 degrees Celsius and requires protection from moisture, frost, excessive heat, and
even light. Exposure to any of these can change the drug's
consistency, cause disintegration of its ingredients, and reduce its potency. The drug actually begins to degrade as soon
as it hits room temperature.

If shaken, the preservative-free medicine can change composition and become ineffective. Exposure to any impurity
can lead to devastating reactions in patients already in poor
health, from bloodstream infections to serum sickness, a reaction in which the body seizes up after the medicine is injected.

In 1999, fifteen patients at a Colorado dialysis center suffered bloodstream infections after being injected with Epogen
that had accidentally been exposed to bacteria after the staff,
attempting to preserve every drop, had reinserted clean syringes into the tiny used vials to extract the leftover medicine.
This practice had contaminated the medicine, a study by the
Centers for Disease Control and Prevention found.

For all these reasons, Amgen and Ortho Biotech—as well
as other drugmakers that produced delicate injectable medicine including vaccines, insulin, and chemotherapy treatments—
tried to design almost perfect cold chains, an unbroken set of
optimal and controlled conditions throughout the manufacturing process.

Transporting Epo is no simple matter. Amgen manufac-
tures the medicine in Puerto Rico. From there, it is transported by sea in air-conditioned cargo containers, and then in
refrigerated trucks, the medicine's temperature constantly

monitored by computer and the data retained to be checked stateside by Amgen and Ortho Biotech specialists.

Their packaging and compliance specialists consider the impact of such factors as temperature changes, shock, vibration, and magnetic fields on the drug's integrity. They study how different packing mediums might affect its overall temperature. In transit, some companies use designated chill rooms as places to hand off the medicine, thereby minimizing the chance of accidental exposure.

Those in charge of preserving this cold chain for Procrit and other medicine view their jobs as a way of preserving life. At a national conference, a compliance manager for one drugmaker exhorted his colleagues, "Develop and project a *passion* for temperature control in your organization." Amgen and other companies try to foresee every possible breach of this cold chain from the moment the medicine is manufactured until it leaves their loading docks.

FROM THERE, RESPONSIBILITY FOR PRESERVING THIS COLD CHAIN shifts to pharmaceutical middlemen who buy, sell, sort, repackage, and distribute 98 percent of the nation's medicine. William Hubbard, the FDA's associate commissioner for policy and planning, described the system as a "rigorous" one in which the medicine "moved through licensed individuals." The companies, 6,500 in all, range from publicly traded giants with pristine warehouses to small, obscure companies that operate from back rooms. These middlemen sit between the drugmakers' loading docks and the nation's pharmacies and hospitals. With rare exception, everyone's medicine moves through them.

The largest middlemen, McKesson Corp., Amerisource-Bergen, and Cardinal Health Inc.—multi-billion-dollar publicly traded entities known as the Big Three—control 90 percent of this market. Because they buy most of their drugs in huge volume directly from the manufacturers, they are known as "authorized distributors of record." Below them sit some fifteen regional wholesalers, also authorized distributors that do billions in business. And below them sit the smaller secondary wholesalers, a group to which Carlow's Quest and BTC claimed membership. The companies all buy from, and sell to, one another.

The drugmakers and the pharmacy chains benefit from this system, which allows them to centralize their selling and purchasing and save billions in distribution costs. The huge purchases by the largest wholesalers let the drugmakers show Wall Street big sales numbers, which helps boost their stock prices. In turn, the smaller wholesalers serve a stopgap purpose: They can supply small clinics and doctor's offices; and bigger wholesalers can turn to them to buy medicine in the event of a shortage or to unload excess drugs.

All the middlemen, regardless of size, aim to buy medicine as cheaply as possible and resell it for a profit, a system of arbitrage made possible by widely varying drug prices. Unlike in Europe and Canada, where governments largely regulate pharmaceutical prices, drugmakers in the United States fought off price controls, choosing instead to offer an array of targeted discounts that allow them to reach more patients and increase their market shares. As a result, they often sell the exact same drug for any number of prices.

While pharmacies pay "direct" prices, wholesalers get a small reduction, the WAC or "wholesale acquisition cost." Others receive even steeper discounts as long as they agree not to resell the medicine. Hospitals and so-called "closed-door" pharmacies, which solely supply facilities such as nursing homes, sometimes pay less than half the direct price. Foreign countries and charities receive even greater reductions, while doctor's offices receive free samples. Further variation exists within each of these categories, the ultimate discounts cloaked in secrecy and guarded as proprietary information.

These bargains sparked the frenetic trading that Marty Bradley and other wholesalers knew so well. The Big Three have trading divisions that scout the secondary wholesale market for discounted medicine. Cardinal, for example, maintained a spreadsheet showing how much the company saved by purchasing heavily discounted medicine from obscure wholesalers in South Florida and elsewhere.

The secondary wholesalers contend that aggressive trading, in which wholesalers buy and sell to one another, restores competition in the marketplace: It helps them reduce prices for mom-and-pop pharmacies and local hospitals that lack the buying power of the big chains. William Hubbard of the FDA explained that this trading could be perfectly legal and sometimes helped to even out surpluses in the supply chain.

But the bargains also drive a parallel and illegal practice called diversion, in which some middlemen resort to fraud or misrepresentation to obtain discounted medicine. Corrupt wholesalers often solicit closed-door pharmacies and others who qualify for discounts to buy more medicine than they need and sell the rest out the back door for kickbacks. In 2000, a task force for the National Association of Boards of Pharmacy estimated that up to four-fifths of the closed-door pharmacies that received discounted medicine resold at least a portion illegally to outside buyers.

By 2002, the FDA's criminal investigators faced a problem that they could not clearly measure or solve: A huge volume of the nation's medicine no longer flowed directly from drugmakers to one of the Big Three to a pharmacy or hospital. Instead, the medicine was sold and resold in a swelling gray market of middlemen, passing through numerous hands, as each company took a wedge of the profit. These sales often went unrecorded or were accompanied by phony pedigree papers that obscured the origin of the medicine and left no way to ensure its safety.

This illicit diversion had become a multi-billion-dollar industry, Terrell L. Vermillion, director of the FDA's office of criminal investigations, estimated. From 1993 through September 2004, OCI had pursued 562 drug diversion cases and had gotten 642 convictions. Yet the activity of diversion—and even the word itself—no longer had a fixed meaning, because it so closely resembled the legal trading of pharmaceuticals and the laws governing it were murky at best. Even at the FDA, the two sides of the agency viewed it differently. The criminal investigators used the word to describe fraudulent sales that opened the doors to adulterated and even counterfeit medicine; the regulators often used it to describe a simple reapportionment of supplies within the distribution chain.

Michael Carlow and many others used this confusion to their advantage. They had state licenses, lawyers, accountants, and all the trappings of legitimacy. Their businesses embodied the spirit of "pure capitalism," as one of Carlow's lawyers described it. "Buy low, sell high, make money." But they had little incentive to maintain drugs in pristine condition.

In Florida it was laughably easy to become a pharmaceutical wholesaler. All you needed was a refrigerator, an air conditioner, an alarm to secure your products, $200 for a security bond, and $700 for a license. You needed no experience and no particular knowledge. You had to certify that you had no criminal record, but the state's pharmaceutical bureau did not actually check. Once established, there was little need to worry about federal inspections. State authorities alone regulated your business. And each inspector had some three hundred companies to look after. Almost any other licensed profession you could think of—doctors, mechanics, veterinarians, exterminators, plumbers, air-conditioning repairmen, citrus-tree growers, oyster shuckers—underwent greater scrutiny.

Florida's pharmaceutical wholesale companies proliferated like rabbits. They existed far beyond the need for them. By 2002, Florida had licensed 1,399 wholesale pharmaceutical companies—one for every three pharmacies in the state. Only 422 of the companies listed Florida as their principal address. The others were out-of-state companies licensed to sell medicine in Florida.

Some companies registered as out-of-state, such as El Paso Pharmaceuticals in Texas, were in fact Florida operations. The wholesalers set up "corporate headquarters" or "worldwide offices" by rerouting their calls and faxes to make it appear that they had offices everywhere.

One of Carlow's lawyers, Craig A. Brand, who mostly represented pharmaceutical wholesalers, opened a side business as a drug wholesaler himself. On his desk he displayed both sets of business cards, one for his law practice and the other for his company Global Pharmaceutical Services. Both shared the same Miami address. Brand's pharmaceutical company boasted of an international business providing foreign governments with cutting-edge HIV medicine.

The wholesalers licensed by Florida ranged from trained pharmacists, doctors, and lawyers to criminal kingpins and uneducated street thugs of all nationalities and ages. Some, like Fabian Díaz, who appeared to own El Paso Pharmaceuticals in Texas, were former drug dealers seeking greater profits and a safer line of work. Aided by lax regulations and Florida's large Medicaid and medicine-dependent elderly population, those trafficking in diverted medicine were making a fortune. In one eight-month period, for example, Michael Carlow cleared a profit of $2.5 million.

DAY AND NIGHT, TWO MEN MADE TRIPS TO CARLOW'S WESTON mansion toting duffel bags and old boxes that once contained computers, vegetable oil, and other goods. Now they contained a jumble of pill bottles, medicine vials, and bags of blood derivatives—all culled from different sources and some still bearing the labels of patients to whom they had been dispensed. The men, Fabian Díaz and Henry García, were known in certain South Florida circles as Carlow's "cooks." And their job was to acquire as much medicine as possible.

The medicine they collected was Carlow's lifeblood. He needed a constant flow of it, not just to grow his empire but to pay for the one he had already built. To make the kind of profit he wanted, the medicine needed to be cheap. Free was best of all. Ordering medicine and not paying for it, as MDI had alleged, was just one way to do this. Another was through the efforts of Díaz and García. The men were so productive that they had turned Carlow's laundry room and garage into a "pharmaceutical repacking operation," as a pharmacist who had dropped off medicine there observed. Wholesalers came to the mansion to pick up orders of medicine that had been boxed for them, or to go on "shopping sprees," the pharmacist told investigators.

In order to maintain the flow of discounted inventory, Díaz and García used an array of tactics. At the street level, they bought cancer and AIDS drugs from the patients treated at health clinics in Miami's slums. Those infected with HIV/AIDS, who also had crack habits or other drug addictions, sometimes preferred to get high than to get well. Díaz and García's foot soldiers waited for them outside the clinics and swayed them to sell their medicine for a few $20 bills.

Díaz and García also bought medicine from small-time businessmen, their business being that of professional patients who sold, rather than took, their medicine. One Medicaid recipient, Michael McKinnon, who had been diagnosed with HIV, played the part of a destitute and dying patient. In three years, Medicaid had paid more than $200,000 for his protease inhibitors, making him one of the state's top recipients of the aid. He sold the drugs cheaply to those like Díaz and García who were seeking cheap medicine. As he would later tell investigators, he kept only the medicine that he needed for himself. He had made $60,000 in a single year this way, lived in a comfortable house, and drove a sporty Acura.

Rounding up medicine patient by patient could be tedious. So Díaz and García had other methods, as the Horsemen learned. Sometimes they didn't need patients at all but could simply create them, by retrieving names and Medicaid numbers from pharmacies and treatment centers. If necessary, they also could steal drugs by breaking into warehouses. Through his shifting roster of companies, Carlow then resold the drugs to other wholesalers ranging from small local ones to huge national companies that distributed medicine to pharmacies and hospitals.

For years Carlow had sold his medicine through his licensed companies, including Quest Healthcare. After his arrest in June 2000, he continued to work as a wholesaler by creating companies in Florida and other states that appeared to be owned by others. He and his lieutenants staffed these far-flung "businesses" in part by hiring those who would sign leases, collect mail, require little payment, and ask few questions while appearing to be in charge.

To find these willing recruits, Díaz and García attended rehabilitation programs, where they befriended patients and offered them jobs. At Alcoholics Anonymous, Díaz approached John Bullock, a friendly, trusting sort who had recently been laid off, and offered him a trip and some cash in return for a favor. Bullock went to the Miami airport where another man met him, flew with him to Houston, Texas, put him up in a hotel room, fed him dinner, and gave him papers to sign that put him nominally in charge of a new company called JB Pharmaceuticals. Bullock earned $400 but was left bewildered by the trip. "It was very difficult for me to get information," he recalled.

Armed with these front companies, Carlow sold such a big volume of pharmaceuticals that an entire network of smaller wholesale companies relied on him almost entirely for their inventory. The Stone Group was one. Omnimed was another.

Even though Omnimed was a licensed company with an office in Boca Raton, the owners would often meet Carlow or his assistant at a Shell gas station off Interstate 595 to exchange medicine and checks or cash. At Omnimed's office, nail polish remover, lighter fluid, and the paint remover Goof Off cluttered the work tables and desks. The Horsemen believed that the employees used these products to remove patient dispensing labels and any other evidence of a product's origin.

But Carlow had not stopped at selling to small, obscure companies. He had developed what every small wholesaler dreamed of: a lucrative relationship with one of the industry giants, Cardinal Health. In 2000, Carlow's company Quest Healthcare sold almost $2 million in products to National Specialty Services (NSS), a Cardinal division that was the nation's largest supplier of blood products, cancer drugs, and other specialty pharmaceuticals to hospitals.

Carlow had accomplished this, in part, by cultivating a personal relationship with the vice president of NSS, Neil Spence, who made frequent trips to Florida to cement relationships with his suppliers, Carlow among them. By doing business with Carlow, Spence secured a supply of discounted medicine—the price tag sure to please his company. And for Carlow, Spence was a ready buyer with vast reserves of cash and a bottomless need for pharmaceuticals.

THE HORSEMEN PIECED TOGETHER THIS PICTURE OF CARLOW'S operation with help from some in his inner circle, including Mark Novosel. At fifty, Novosel had lost none of his persuasive charm. He favored casual sports clothes and kept his thinning hair closely cropped. He had narrow, inquisitive brown eyes, a deep scar at his right temple, and a lopsided grin.

Once he had been an alcoholic, a drug addict, and a con man, with a lengthy criminal record to show for it. In 1998 in Youngstown, Ohio, he had bribed a judge and other court officials to refer alcoholics to the rehabilitation and mental health centers he ran. He ultimately pleaded guilty, became a government informant, and served six months for racketeering and for mail and income-tax fraud.

Then he reinvented himself as a diverter of medicine and became Carlow's right-hand man. In Fort Lauderdale, Novosel ran one of Carlow's shell companies; his salary alone was over $3,000 a week. On the side he ran his own business, selling blood products and other medicine out of his $4,500-a-month waterfront rental. He came to know every open window and back door in the pharmaceutical supply chain, having climbed through most. The Horsemen arrived at his house with a search warrant on March 28, 2002, after tracking bad medicine to his address. Months later he agreed to cooperate, saving himself once again.

Novosel had wanted out anyhow, as he later explained, because Carlow and the entire industry were "out of control." Previously, prescription drug dealing had been a gentleman's crime and one that was relatively easy, he said. But diverting medicine had become more profitable, more competitive, more dangerous, and more potentially deadly to patients than it used to be. He had grandchildren now, was getting too old for prison, and wanted the kind of moral clarity that came with going to sleep at night knowing he hadn't hurt anybody. He said he felt sorry for what he'd done.

To succeed as a pharmaceutical wholesaler, he explained, all you have to do is "beat the number" of the drugmakers' price. Whatever they sell it for, you sell it for less. When asked how to do this, Novosel emitted a dark, echoing laugh. Not by following the rules, he explained.

You must overlook, ignore, not ask, and not tell the origin of your best bargains. Your operation should be long on deal-making and light on overhead. The most successful wholesalers don't have a sales force per se. They have a few guys in a room making phone calls all day long, offering buyers—and soliciting from sellers—pharmaceuticals that have legitimate-sounding pedigrees for prices that can't be beat.

The buyers for the biggest regional and national wholesalers get bonuses for acquiring cheap drugs, even from shady smaller wholesalers who may offer them personal incentives like cash in envelopes to sweeten the deal, Novosel said. The big companies make billions this way, shaving down their purchasing costs in an industry known for thin profit margins.

The toughest part can be forging relationships with these

buyers, their identities usually secret, their names the coin of the realm. One day, a buyer for the huge distributor AmerisourceBergen accidentally faxed Novosel a list of the other buyers for her company. She could have been fired for the breach, he said. And he easily could have resold the list itself for $10,000. But he concluded that as his secret, the list had far more value.

Another good way to buy low and sell high is the so-called "Puerto Rican" turn, otherwise known as U-boat or export diversion. Because manufacturers typically offer large discounts to overseas buyers, including those in Puerto Rico, many wholesalers, even the Big Three, establish companies in Puerto Rico that buy the drugs at a discount. Cargo planes bound for Puerto Rico turn around mid-flight and resell the drugs to the states for a far higher price.

In 2000, Florida's pharmaceutical bureau fined a company in Guaynabo, Puerto Rico, J.M. Blanco Inc., for shipping drugs into the state without being licensed. The company's executives explained in writing that J.M. Blanco was a subsidiary of the huge wholesaler Bergen Brunswig, which was licensed in Florida, and the shipment had been an "intracompany transfer." They had not realized a separate permit was needed. The pharmaceutical services bureau accepted this explanation and waived the fine of $250. (A year later Bergen Brunswig merged with Amerisource Health to form AmerisourceBergen, the nation's largest pharmaceutical wholesaler.)

Novosel said his know-how has made him eminently employable in the vast quasi-legal world of pharmaceutical diversion. The job offers come practically every day. Recently, two First Amendment lawyers who represented Internet pornography sites begged him to come to California as a consultant, because they too wanted to become pharmaceutical wholesalers.

The promise of remarkable cures, and their astronomical cost, had created a veritable gold rush. The demand for pharmaceuticals at the top levels of the distribution chain, from the Big Three and their ever-growing customer accounts at pharmacies and hospitals, had led to a stampede at the bottom to form secondary wholesale companies to meet those demands, which is why you now had 1,399 wholesale companies licensed in Florida. Or as Marty Bradley at BioMed Plus put it, "If a man who stole car stereos didn't have a place to sell car stereos, he wouldn't steal car stereos."

9. Stealing Time

Late May 2002
Miami, Florida

MARTY BRADLEY USED TO LOVE LIVING AND WORKING IN Florida. "It's a privilege to have a license in the state," he would say in the upbeat patter he used as the CEO of BioMed Plus. But recently, he used words like "dark" and "scary" to describe the experience of running a wholesale medicine business in Miami. The criminal elements had seemingly hijacked his once-fine industry, the "degree-holding people" chased out by street thugs, he said. For evidence, he looked to his own back door. In the last five months, the white van had returned two more times, its occupants bent on looting his warehouse of blood products and medicine.

Just one month after the first burglary, the thieves returned late on a Saturday afternoon in February. They cut through steel brackets, removed the deadbolt locks at the back of his warehouse, and rolled up a set of metal doors. The doors did not lead to the actual warehouse but dead-ended into empty storage space. The new alarm blared and the intruders fled empty handed.

This time, Cesar Arias didn't even believe him when Bradley called to report the incident. Arias knew better than anyone that the situation in South Florida was bad. But it was not like Al Capone's Chicago or Mel Gibson's *Road Warrior*, with profit-seeking gangs roaming the streets. The story of the same white van returning exactly one month after its first appearance seemed unlikely. Arias suspected that Bradley had staged the break-ins for insurance money.

But Arias was wrong. Bradley had not even submitted a claim for the first incident, not wanting his already high insurance rates to go up; nor did he do so after the second break-in. Instead, he resumed the painstaking work of turning his warehouse into a fortress. Wild with rage, he spent $100,000 more on security, adding floodlights triggered by motion, reinforcing steel doors, increasing the number of video cameras, and extending hours for the armed security guard. In part, he felt defiant. "Let 'em try," he thought to himself.

Three months later on a Wednesday night in May, the white van came back a third time. The new cameras captured three men in black ski masks as they leaped from the van, one wearing shorts and flip-flops. This time they attached a two-by-four to the back of the van and drove it backward into the locked metal gate, smashing the door inward and triggering the alarm. Then they severed the cord of the security camera outside the warehouse's rear entrance.

As they sprinted into the dark warehouse, their motion tripping the lights, the cameras inside captured their grainy images. They raced across the warehouse to the freezers, ripped open the doors, and swept the contents of the shelves into garbage bags, which they pulled behind them like bad Santa Clauses, one man's silver watch glinting under the lights. They got away with more than $250,000 of NovoSeven, Iveegam, and other products that had been recently delivered. The entire heist took just sixty-three seconds.

Bradley felt that he was living a nightmare, one that he no longer had to play over and over in his head. This time, he could play the actual break-in over and over on the surveillance tapes that captured the entire incident. In his new bad-luck world, even he appreciated the irony that all his fancy cameras had done was let him watch his products being stolen.

By now, he concluded that Arias and Gary Venema would never catch the perpetrators. As far as he could tell, they had accomplished nothing and seemed totally outgunned by the problem. The Stone Group was still in business, even after admitting five months ago to buying his stolen drugs out of a felon's laundry room. Weeks after the first break-in, Venema had showed up at his office with a bag of Michael Carlow's garbage, exhibiting various spaghetti-stained documents. It was an exuberant show-and-tell, but nothing seemed to have come of it.

The two investigators certainly meant well. But their efforts were like trying to put out a forest fire with a squirt gun. And that is when Bradley began to think about relocating his business. "I wish I weren't in the state of Florida," he would say.

IN 1960, GENE ODIN BEGAN RUNNING HIS OWN SMALL PHARMACY in Buffalo, New York, working fourteen-hour days to make the business succeed. He knew all his customers, doled out advice, and kept an eagle eye out for any adverse reactions. He personally typed all the instructions for the medicine he dispensed, feeding the labels into a typewriter one by one. In the back of his store, he also compounded medicine, making powders, ointments, and various syrups.

But it was grueling work and after three years, his wife, Shirley, urged him to return to school, where he got his Ph.D. in medicinal chemistry. He spent the next two decades working for big chemical companies in Buffalo, his days a blur of office politics and clashes with his engineer colleagues, whom he didn't like or understand. The day he was fired from the Buffalo Color Company, where he had worked for nine years as the technical director, he returned home all smiles and told Shirley they were moving to Florida.

They landed in the sultry heat of South Florida in 1986 to begin their semi-retirement. Odin planned to work part time as a pharmacist, a profession always in demand given the state's ever-expanding elderly population. Shirley looked forward to gardening and reading after her years as a public school teacher. By 1987, they were living in Coconut Creek, northwest of Fort Lauderdale, and Odin was filling in part time at various Rite Aid pharmacies, remembering all the things he didn't like about the work. Then one day he received a letter that the state health department had sent to all South Florida's registered pharmacists. The Bureau of Statewide Pharmaceutical Services had an opening for a drug inspector.

Odin felt immediate interest, though he didn't exactly know why. Already beyond middle age, he felt that his real abilities had never been tapped. He had always wanted to embark on some sort of mission, but the world had never sent one his way. Something about the job seemed to promise this, so he played his interviewer like a "finely tuned Stradivarius," as he would later joke, and the job was his.

Most of his life, he had been a contentious and insubordinate employee, the opposite of a company man. He had a sharp, sly sense of humor and a tendency to proclaim his dislikes without reserve. He had a leaden hand at office politics and a blunt, literal style that he took from his scientific and medical training. All of this had led to a turbulent professional life. But these same qualities made Odin a superb drug inspector. Sometimes sardonic and self-effacing, at other times morally indignant, he loved being an irritant and driving his listeners crazy. He haggled with those he regulated. No detail was too small to fuss over, no lecture too overused. In all instances, his manner was suffused with intimacy. He even called the women at the businesses he inspected "bubbe" and, despite the often contentious relationship, many of them greeted him with a kiss and looked to him for regulatory and even moral guidance.

At first, Odin's new job seemed mellow enough. He enjoyed the travel and the interaction with those he regulated. Sometimes, Shirley would tag along on inspections and wait in the car, reading or knitting.

Later Shirley noted that Odin's job and his attitude about it seemed to change.

Across the country, the pharmaceutical trade had become a Roman market of sorts, fraught with dishonesty and excess. Many institutions, such as nursing homes and the pharmacies that supplied them, had begun to take advantage of the discounts they received from drugmakers by ordering extra medicine and reselling some of it for a profit. Even as early as the mid-1980s, Congress had begun to hear reports of abuses: how one twelve-bed hospital had purchased a forty-two-year supply of an anti-epilepsy drug, while a fraud ring in Indiana had established a fake chapter of the American Association of Retired People in order to purchase pharmaceuticals more cheaply.

A major federal investigation called Operation Gold Pill revealed that wholesalers were collecting medicine samples that drug reps had dropped off at doctor's offices, repackaging them to appear new, and reselling them, often to corrupt pharmacists. The investigation turned up mountains of loose pills and led to the arrests of over two hundred pharmacists. In South Florida, Arias and Odin helped with the probe and posed for photographs next to garbage bags filled with brightly colored pills, as though from a raid at some jelly-bean factory.

The frenzied trading and minimal regulation of this drift-ing medicine greatly concerned Odin and Arias. Odin would say that the dregs of the earth were down here, that people had no idea what was happening to their medicine, and that the bureaucrats in Tallahassee were doing nothing about the problem.

In the sleepy state capital near the Georgia border, where Spanish moss hung from the elm trees and a favorite drink was sweet tea, Odin's bosses saw little cause for alarm. Jerry Hill, the chief of the pharmaceutical bureau, explained that it was unusual in the early 1990s to find a case that rose to a criminal level. Yes, they knew that pharmacies were buying baggies of pills off the streets and ethnic grocery stores kept selling a diet drug banned by the FDA. But as for real crimi-nality, "I heard stories going back in the '90s," he said, "but no evidence. These were stories."

Once criminals began applying to be wholesalers, the pharmaceutical bureau was hamstrung by the limits of its au-thority, compliance officer Sandra Stovall recalled. It could not run fingerprint checks or get information from FBI files. All it could do was ask applicants if they had criminal records. The drug supply in Florida was run largely on an honor system.

From where Odin and Arias sat, the officials in Tallahas-see lacked the spine to address and solve these problems. If they saw no crime, it was because they hadn't looked or be-cause they didn't want to see. If the existing laws did not allow the bureau to protect the public health, as was its mission, then the laws needed to be changed. As Odin's dislike of those in Tallahassee grew, he became committed to their overthrow.

Partly, Odin and Arias were thwarted by a problem of timing. It was the height of the drug war. Crack cocaine was tearing apart America's cities and the law-enforcement community was bristling with newly formed task forces, S.W.A.T. teams, and commissions to battle it. There seemed to be no penalty great enough for the major traffickers and kingpins. Against this backdrop, a little hustling in prescription medicine hardly seemed like a problem. It wasn't wrecking neighborhoods. It wasn't killing people.

But to Odin's mind, no one in law enforcement actually knew what a drug was or what the consequences of mishandling one could be. In one instance, the Collier County sheriff's office called him after learning that a piñata shop in Immokalee was dealing in the antibiotic Ampicillin. After seizing the pills and testing them at a police laboratory, the cops told Odin, "This is not a drug," meaning cocaine, heroin, or marijuana. Even when the inspectors and the cops agreed that a crime had taken place, there was almost no one to prosecute it.

By the early 1990s, things seemed destined to improve. In 1988, the United States Congress had passed the Prescription Drug Marketing Act, which aimed to end the recycling of drugs through the supply chain. The act prohibited the sale of drug samples and restricted the resale of pharmaceuticals by hospitals, pharmacies, and other end users. It prohibited the re-importation of American-made pharmaceuticals that had been sold more cheaply overseas and required wholesalers to disclose the source of the drugs they purchased. It required states to license pharmaceutical wholesalers and established the first criminal penalties for drug diversion.

Key portions of the PDMA would become stalled for years by industry lobbyists. But if nothing else, the law made clear that crimes could be committed with medicine. Some sheriff's offices formed pharmaceutical diversion units, though mostly focusing on the narcotic painkillers that were abused by addicts and sold on the streets and that ultimately fell under the jurisdiction of the federal Drug Enforcement Administration.

In 1992, partly in response to the law, the U.S. Food and Drug Administration created a new investigative division to catch those who tampered with the nation's medicine and food. This office of criminal investigations, or OCI as it became known, opened six branch offices around the country, including one down the road from where Odin and Arias worked.

Excited, the two inspectors went over to introduce themselves, looking forward to a collaborative relationship. Within months, Odin decided that the agents were not a solution to their problem but an extension of it. Many had retired from the Secret Service or had formerly been with the FBI and CIA. While they knew little about pharmaceuticals, they were schooled in control and secrecy and how to "manage" their state counterparts, as one former OCI agent put it. Odin quickly felt that he was being mined for information while the agents shared almost nothing in return.

One day, after the agents refused to update Arias on the progress of a case he had given them, Odin and Arias walked into the Miami field office of the special agent in charge, Kent Walker. Odin demanded, "This is the kind of shit we get from you guys? You tell us your guys can't tell us anything?" For years after the confrontation, a story circulated among Walker's agents that someone had angrily overturned a bag of medicine vials onto the floor. Regardless, the encounter set the tone for Odin's relationship with the agency. Walker recalled, "He had a New York way about him."

While those at OCI who worked with Odin knew he had an excellent antenna for bad medicine, they found his moral indignation exasperating and his grasp of the practical realities of crime fighting sorely lacking. In one case, Odin was outraged when he got a letter from the United States Attorney's office in Miami, swearing him to secrecy in a case that he had given to OCI. As it turned out, the letter was a standard directive for anyone who had access to sensitive information being heard by a grand jury. It was not an effort to push him out of the case, as he believed.

Rightly or wrongly, Odin felt continually slighted, with one OCI agent after another shutting him out of investigations that he had developed in the first place. To him, the very acronym OCI was like a red cape waved in front of a bull. "Whenever he had to interact with them, he just went nuts," Shirley recounted.

His reaction was not just that of an aging inspector hankering to be included. It was more part of the mentality that he'd brought home from the Marines, which he'd dropped out of high school to join in 1946. Though he had never seen actual combat, he joined in battles, standoffs, and flare-ups wherever he might find them. He even carried a 9-millimeter pistol under his car seat, ready to defend himself if necessary. "He just doesn't give a damn," Shirley explained with a sigh, "so I let him do what he wants."

As Odin's struggles with OCI continued, so did the "trading" within the pharmaceutical industry. Medicine passed through numerous hands in deals that increasingly resembled those in the narcotics trade, with drop-offs and sales in back alleys and parking lots and legitimate businesses fronting for illegal activity, with huge profits being made. This was no accident. As the government's war on drugs grew hot, narcotics traffickers looked for a safer line of work—one with lower penalties and similar, if not greater, profits. And the nation's pharmaceutical supply, a haphazardly regulated hinterland, proved to be very inviting.

By 2001, Odin and Arias began to feel that their work was urgent, that a public health crisis was unfolding before them. But everything from the state's weak laws to their bureau's inertia seemed to work against them.

In June of that year, the police responded to a report of two men brawling in a private house. Inside, they found a refrigerator full of an injectable growth hormone, Nutropin A.Q., used to treat dwarfism and kidney failure. Having no idea what it was, they called the police narcotics unit, which called Arias. The medicine turned out to be part of a counterfeiting scheme, in which vials of far less expensive human insulin, potentially deadly to those not in need of it, were being relabeled as the Nutropin A.Q.

The police seized the medicine, but some had already reached an Orlando pharmacy that had shipped it to patients across the country, including an orthopedic surgeon in Michigan who unwittingly administered it to his son, who suffered a burning sensation but survived.

That same year, another South Florida man who they suspected of trafficking in both counterfeit Retrovir, an AIDS drug, and Viagra from an unlicensed channel entered into business negotiations with an undercover officer and was actually arrested, a rare triumph. But later the officer was charged with corruption and all his cases were thrown out. The Tallahassee bureaucrats then issued a wholesale license to the suspected trafficker's wife against the inspectors' objections.

As Odin turned seventy-one in 2001 and yet another date for his retirement approached, he felt farther from retirement than ever. Everyone else he knew was out playing tennis and golf. But Odin felt that he was standing between the public and some incipient disaster. A family tragedy, more than anything else, may have impelled him to stay on the job.

In February 2001, Shirley's brother Seymour, a pharmacist in Buffalo, died of colon cancer. He had been taking injections of Procrit and had been operated on in November. But in February he developed a sepsis infection and in a matter of days had lost his battle. Odin felt instinctively that something was wrong, beyond just his disease. He asked Shirley, "How can we know that the things that Seymour was taking weren't illegal, misbranded, counterfeit?"

Odin's sister-in-law, already coping with the loss, did not seem particularly interested in revisiting the question of what her husband's medicine had contained. But for Odin, the open question brought home the reality of what he was seeing every day: that no one had any way of knowing for sure whether their medicine was good, bad, or indifferent.

MAXINE BLOUNT HAD ALWAYS HAD A KNACK FOR SURVIVAL. SHE had been born without a rectum, and doctors rebuilt her insides through successive surgeries that her father paid for by selling the truck he relied on for work, delivering cattle and hogs to stockyards in East St. Louis. Afterward, he changed jobs and ran his father's general store, and Maxine loved to spend hours there, pumping gas outside from the big gravity feed pumps. At age eighteen, in an era before seat belts, she crashed her Chevy drag racing. The car rolled over several times, yet she walked away, unhurt.

Even after her diagnosis of breast cancer in 1998, Maxine still had a will to survive. When the doctor who informed her of the cancer said without conviction, "We will do what we can," she changed doctors immediately, to someone who seemed to believe she'd get better. She also knew that her best hope of living an extra day, week, month, or year lay in her medicine, and she said as much to her brother, Max Butler. For any patient with cancer, he observed, "medicine is the primary hope you have, even if it doesn't do you any good at all."

Once Maxine learned that her medicine had been counterfeit, she lost this hope. "The setback really deflated her," Max recalled. "I could see an immediate change in her attitude."

Even though the decline came quickly, her sense of fun returned during her final weeks at the hospital. The friends who had left the flamingos on her lawn finally revealed their identities. They had debated whether to do so, since the mystery was intriguing and kept Maxine guessing, which maybe kept her alive. For a brief moment, the revelation lit her up so much that even her best friend, Patti Silvey, felt a tinge of jealousy that she had not been part of the group that had thought up something so delightful to Maxine.

A young woman who worked at the casino Maxine frequented and had become a friend came to visit. When one of Maxine's IV drips emptied and a machine began to chime, Maxine jumped, obviously startled, but then laughed and laughed after the woman asked her, "What's wrong, Maxine, did you think you'd won a jackpot?"

While in the hospital, Maxine described in great detail for Max the order of her funeral service and the things she wanted said: personal recollections for each of her children, her husband Ed, her closest friends, and her brother too, who faithfully took her dictation, including the sentence, "He is supposed to write this the way I asked him to, I hope I can trust him. Ha." She also asked him to read "The Dash," a poem about the dash between the dates of our birth and death, and how a life should be measured by its quality and not its length.

Maxine died on a clear, sunny day in October at age sixty-one. People from all over came to the funeral. Overwhelmed by emotion, Max was unable to read the poem after all. Maxine was laid to rest on a verdant hill overlooking St. Charles. Her friend from the casino placed $100, $50, and $25 tokens inside the coffin.

For months after she died, Ed's world seemed barely screwed on straight. He cried often. His delicate and watchful manner intensified. At home, he sat in the dark, pondering the effect the weaker medicine may have had on the progress of her disease. Patti believed that with the right medicine, her cancer might have gone into remission. The counterfeiters, whoever they were, had stolen life and time away from her. Maxine's son Bill, who for months could not think of his mother without breaking down in tears, believed the counterfeiters had diminished the quality of her life. In that way, they had stolen time from the entire family. "I lost time," he would say, weeping at the memory of her lying in bed, unable to get up.

Ed did not know what to think. In the great, big kitchen in Harvester, he kept the evidence of what had happened in the refrigerator with a rubber band around it. On one occasion, a hush fell as he placed it on the table and everyone leaned forward while one of the family's lawyers, Gretchen Garrison, pointed out the nearly invisible strike-throughs in the tiny zeros of the lot number, P002384. The vial was so small that it disappeared from view beneath Ed's thumb. The tiny numbers were printed in white lettering on a band of blue along the side of the label.

Flustered and disturbed, Garrison felt that she was in some labyrinth, trying to put together a case in which every alley was a blind one. While hardly naïve enough to expect the drugmakers and wholesalers to rush to their aid, she was dismayed by their response to the lawsuit. They claimed that they had no obligation to prevent, or even to warn patients about, the crimes of unknown third parties.

Their denials made Garrison angry. "They all take the position, 'It's not my product.' Son of a bitch," she declared one day as she barreled down the highway from the Blounts' home. "That's part of the problem."

In Florida, the Horsemen learned incidentally about Maxine's ordeal. But their investigation gradually brought to light the path that her and other patients' medicine had taken, one painstaking fact at a time.

Part Two

Part Two

10. "My Son Is Not a No One"

January 2002
New York City

"LISTEN, WE THINK WE HAVE A LIVER FOR YOU NOW."

Kevin Fagan heard the words and thought that he was dreaming. From a cot on the floor of his son's hospital room, he struggled awake to see a doctor framed in the doorway, the hall lights behind him spilling in to the murky room. Again, the doctor repeated the words, "We think we have a liver."

Kevin's sixteen-year-old son, Tim, woke up too. He wanted more than anyone to return to normal life. The blue-eyed and once-athletic teenager was now gravely ill, his complexion chalky. He wanted to play lacrosse again, to pet the dog that his parents had gotten him just before he went into the hospital, to worry about his tests and whether girls liked him. Above all, he wanted to sleep in his own bed at home in Deer Park, Long Island, a verdant neighborhood of comfortable homes with inviting front porches.

Tim had only one way back to that life: His diseased liver had to be replaced with a healthy donated one. So far he had spent three-and-a-half weeks at the hospital, sharing a room with a boy who had brain cancer, waiting for his name to top the donor list and for this ghastly vigil to end.

Now, at 1 A.M., as Tim and Kevin emerged from the fog of sleep, the time had come. Kevin, an imposing man with a ruddy face and a goatee, called his wife, Jeanne, at home. She awoke on the first ring. They all knew this could be their godsend.

By 4 A.M., their bleary-eyed hopes had turned to sharp disappointment. At the last minute, their surgeon told them gently that the liver had gone to someone who needed it even more urgently.

At home, Jeanne lay awake as the family's two young daughters still slept down the hall. The forty-two-year-old mother stared with bright blue eyes into the dawn, imagining the days, weeks, and months ahead. Before this challenge, her Christian faith had always helped her find reason and strength. But this morning she felt totally lost. It was far easier to be faithful, she realized, when disaster hadn't struck.

TIMOTHY FAGAN HAD BEEN ADMITTED TO NEW YORK UNIVERSITY Hospital in Manhattan after his liver, covered by a mysterious web of scar tissue, had all but stopped working. Instead of filtering proteins from his blood stream, his liver had let them back up into his spleen and esophagus. He urinated blood and then began to vomit a bloody mixture that resembled coffee grinds. The doctors didn't know why his liver was failing. They called his disease cryptogenic cirrhosis—a scarring of the liver for no known reason.

Before Tim got sick, the Fagan household had been joyful and chaotic, with children and animals underfoot and a constant traffic of school projects, Girl Scout meetings, and piano lessons, as well as work emergencies. It was a "nut house" even on slow days, as Kevin described it.

Both parents had hectic lives. Jeanne studied nights and weekends to earn her Ph.D. in education, while still teaching an eighth-grade class full time and volunteering at church and as a Scout leader. Kevin, a supervisor at Con Edison, oversaw a team of eighty-four people at the utility company that responded to New York's gas-leak emergencies. Working no less than sixty-five-hour weeks, he would beam out orders on his Blackberry, handling everything from a gas leak at Gracie Mansion, New York City's mayor's residence, to a carbon monoxide leak in a Manhattan nightclub that killed five. After the September 11 attacks, he spent five straight weeks at ground zero as gas fires raged.

Jeanne and Kevin thought they knew how to handle stress. But nothing in their work or home lives had prepared them for Tim's illness.

In 2000, Tim first noticed that he felt lethargic while playing lacrosse. He had gained weight, too. A routine physical and blood work revealed elevated liver enzymes. From there, the family learned that Tim's liver was enlarged. In December, he underwent a needle biopsy so that doctors could determine the exact cause of his disease.

The results were hardly conclusive. Some of the doctors who analyzed his case or saw slides of tissue suspected Neimann-Pick disease, a neurological condition that causes liver failure and eventual death. But others weren't sure.

Then one day in January 2001, Tim emerged from the shower to find his legs had turned perfectly blue, almost the color of an Oxford men's shirt. Doctors at the local emergency room were stumped. A renowned specialist at Bellevue Medical Center in Manhattan said she had never seen anything like it. His doctors at New York University sought advice from experts across the country and around the world, unable to pinpoint a diagnosis.

Jeanne, desperate for answers, turned to medical journals and the Internet, only to become frightened by what she read about Neimann-Pick and other fatal diseases. While Kevin told her, "Let's be happy for today. We have Tim today," Jeanne's mind raced anxiously toward a future without her son.

On Christmas Eve of 2001, Tim urinated blood. "Don't say anything to Mom," he begged his father during mass. "I don't want to ruin Christmas." But at home, Kevin felt obligated to tell his wife. Jeanne sat on the kitchen floor and sobbed.

With the encouragement of their New York doctors, the family went to the Mayo Clinic in Rochester, Minnesota, in January 2002 and spent a week there while a team of doctors evaluated Tim. Uncertain as to the cause of his disease, the doctors concluded that his own liver likely would last another ten years, and that he could get by with the regular medical monitoring of his NYU doctors.

But within weeks, Tim began vomiting blood and was admitted to NYU, his name on a national waiting list for a liver. The Fagan family's odyssey—from Tim's first symptoms to an initial misdiagnosis to his recent rapid decline—had been devastating. Now the brief elation at the prospect of a donor liver had given way to the agony of more waiting.

The Fagans returned to the excruciating routine that had replaced their normal family life. Kevin and Jeanne traded off hospital vigils, one of them by Tim's side at all moments, the other either working or at home, taking care of their two daughters, ages seven and fourteen.

"It just ripped you apart," Kevin would later say, "but you could not just show that side to anyone. You had to keep it as light as you could and sometimes it was very, very, very difficult to keep your spirits up and keep them positive."

About ten days after their first hopes had ended in disappointment, their surgeon, Dr. Lewis Teperman, alerted Jeanne to a second possible liver. This time, the prospect seemed so good that Tim was actually prepped for surgery. Kevin arrived from work to see doctors wheeling his son into the surgical bay.

Shortly afterward, though, Teperman emerged looking crestfallen. The liver was not a perfect match for Tim and it needed to be. Once again, they had to wait. Over the weeks, a tender relationship had developed between the surgeon and his patient, one that was almost like father and son. That afternoon, Teperman broke down crying in front of the family. Tim cried, too, and they hugged one another.

About two days later, on February 15, Jeanne was again at the hospital when the doctors notified them of a third possible donor liver. Again, it looked very good. By the time Kevin got to the hospital, Tim was already in surgery. Family members gathered quickly, along with some of Kevin's co-workers.

As they waited for word from Dr. Teperman, both Jeanne and Kevin were reminded of when Tim, at age five, had undergone heart surgery, the result of a birth defect that had created a hole in his heart the size of a quarter.

Then, the surgeon had been cold and remote. As they waited at the end of a long corridor, they had watched the doctor walk toward them slowly, his head down and his manner deliberate. Only at the end did the doctor pick his head up and offer a tepid smile to indicate that their son had pulled through.

But Teperman was different. Even after nine hours of surgery, he strode down the hallway beaming, his broad smile and waving hand offering them instant relief. The surgery had been so successful that the new liver immediately began its work within thirty minutes, and Tim's swollen spleen began to shrink noticeably.

Overwhelmed with emotion, Kevin—who, unlike Jeanne, was not particularly devout—wandered alone into a hospital courtyard and offered a prayer to the still-unknown donor who had given his son a chance at life. Through an eventual exchange of letters, the Fagans learned that the donor had been a pilot and a family man in his early forties who had died of brain cancer and donated all his organs. Remarkably, he had lived four blocks from the Fagans and the two families' daughters had been friends at school. He and Tim had even frequented the same local delicatessen, where they both enjoyed the chicken salad.

These coincidences could not help but brighten the Fagans' view that despite all the hardship, anxiety, and suffering, something akin to a miracle had occurred. The donor's widow, also deeply religious, had returned such a gracious and profound letter that Jeanne kept it in her pocketbook and read it every day. The woman wrote that she felt overjoyed some good could come out of tragedy and was comforted to know that part of her husband still "blessed this world."

Tim had to take six different drugs so that his body would not reject the new organ, and the healing was long and difficult from the 108 stitches in his chest, groin, and lungs. Even as Tim struggled to recover, the Fagans felt that he had come through the worst.

ON MARCH 20, AROUND MIDNIGHT, JEANNE WOKE TO HORRIBLE screams and sat up in bed. She shook Kevin and the two of them raced to Tim's room, Kevin thinking with dread that someone must have broken into the house. They found their son convulsed on his bed, his body racked by radiating pain, a sensation that he could only describe as a charley horse all over his body. The spasm subsided several minutes later.

The next morning Tim's doctors had no idea what was wrong. They considered the possibility of some adverse reaction to his drugs. But *which one?*

The newest drug prescribed for him was Epogen, to help remedy Tim's anemia, a typical transplant reaction. Tim's red cell count had plummeted and he had become weak and pale, his eyes raccoonlike. He slept almost all the time.

Jeanne learned how to administer the injections that would help strengthen him. A single dose came in a tiny vial, about the size of a man's thumbnail. The cost for just one injection of the strongest dose, 40,000 U/mL, was $470. He needed one a week.

As the shots continued, so did Tim's terrible reactions. One night, as Kevin dashed again to his son's room, he noticed his seven-year-old awake, her hands over her ears to block out the screaming. He wondered about the cumulative toll on his family's mental health.

Though Jeanne continued the shots for eight weeks, Tim's red blood cell levels did not rebound. Cramps wracked his body after each injection. He remained exhausted, his legs barely able to carry him through a school day. Each afternoon, he would come home and plunge into an hours-long sleep that never restored his energy. Doctors on his transplant team, which had its own full-time pharmacist, were baffled. They had never before seen the medicine fail or provoke this kind of reaction.

Each week Tim hoped that his mother would forget the dreaded shot that left him convulsed in agony. But Jeanne, dedicated above all else to helping her son recover, never did.

On May 14, the day after Jeanne phoned their drugstore, CVS Procare, to order a refill of the Epogen, a staff person there called back to warn about counterfeits. Six days earlier, the drugmaker, Amgen, had announced that it had discovered vials of counterfeit Epogen in lot number P002970. The vials contained medicine one-twentieth the strength of what appeared on the label. The label looked almost identical to the real medicine except in one respect: It lacked the two tiny degree symbols next to the recommended storage temperatures of 2° to 8° Celsius.

The pharmacist assured her that the prescription just mailed to them would be legitimate. The pharmacist was wrong.

Their refill *was* from the affected lot, the label missing the telltale degree symbols. Amgen's warning had come just in time. But now Jeanne wondered about the previous batches she had injected, the vials all used and discarded by now. Had they been adulterated too? Had they been from the tainted lot? And then she remembered that she still had an empty vial that she had put in a bag in the freezer along with some other items from a family trip.

The empty vial was from a different lot, P001091. No one had identified this batch as a problem. But as she scrutinized the tiny vial and the complex patterning of numbers and letters on the label, her heart stopped. There, in the upper right-hand corner by the storage information, were the words "Store at 2 to 8 C" with no degree symbols either. She immediately called back CVS.

Jeanne Fagan had just uncovered another counterfeit batch of Epogen that no one, including the drugmaker, the government, and the wholesalers distributing it, had announced. And worse, their son had been injected with it for at least four straight weeks.

Kevin, who had been at work when Jeanne called him, could barely understand what she was saying. How could the drugs have been counterfeit? What was the liquid that they had injected into their son? Trying to help him, had they harmed him instead?

Frantic for answers, Kevin got on the phone. A CVS official faxed him a copy of Amgen's warning letter and also suggested that he call AmerisourceBergen, the wholesale distributor in

Chesterbrook, Pennsylvania, that had sold the product to the pharmacy chain. Kevin had never heard of the company.

An official at Amerisource told him to call the California drugmaker, Amgen Inc., which would be able to determine through testing what the counterfeit contained.

At Amgen, an official told him not to worry. Tests showed that the drug had been authentic Epogen, but a less potent version. As for how this had happened or where the counterfeit vials had come from, only Amerisource could answer that question, the official said.

Kevin was furious when he got off the phone. They had gone from not knowing Tim's diagnosis to dozens of different doctors and tests to waiting for a transplant to trying to keep Tim calm and reassured. And though Amgen officials were now effectively saying, "I guarantee you it's nothing," how did they know? The finest doctors at the Mayo Clinic had said his liver would last another ten years and two weeks later it had failed. Kevin was not in any frame of mind to accept such a guarantee.

He then called the FDA. An official said the agency couldn't discuss the matter because it was under investigation. Kevin explained that he wanted an outside laboratory to analyze the remaining dregs in the single vial and to make sure that other patients with that same lot number would be notified. Instead, the official told him that the agency would be back in touch if it deemed testing necessary.

As Kevin got off the phone, he wasn't sure what to do next—but knew he had to do something. All he had gotten so far were evasions, non-answers, and classic buck-passing. He thought to himself, *I'm not an employee, I'm not a number, and my son is not a no one.*

He called back Amerisource and snapped at the woman who had taken his call, "If I don't get some answers, I'm going to come down to your corporate headquarters, bring my sick son, call the media, and sit on your front lawn until you tell me where this medicine came from."

"Uh, please don't do that, sir," she said, taking down his phone number. "Someone will call you back."

That someone was the company's senior director of security and regulatory affairs, Chris Zimmerman, who said that he couldn't divulge much because the matter was being investigated.

"How could something like this happen?" Kevin demanded.

Zimmerman began to explain, delicately, that while the company purchases 98 percent of its medicine directly from the manufacturer, the remaining 2 percent comes from other distributors who may be overstocked with product.

"What controls do you have in place for the 2 percent?" asked Kevin.

Zimmerman, now speaking haltingly, said that there were controls, but evidently they hadn't been sufficient.

Zimmerman did not mention that just a week earlier, Florida's Bureau of Statewide Pharmaceutical Services— where Arias and Odin worked—had fined his company more than $63,000 after one of its distribution centers in Orlando purchased medicine from two out-of-state wholesalers not licensed to do business in Florida. Inspectors found that the company had failed to request pedigrees for the medicine and had stored the drugs in unsafe conditions. One of the drugs, an AIDS medicine called Retrovir, had turned out to be counterfeit.

As Kevin got off the phone, the enormity and complexity of the problem began to unfold in his mind. He and Jeanne, as parents, had unwittingly injected an adulterated liquid into their son that had made him very sick. They didn't know what was in it. Though supposedly a weaker version of the drug, this did not explain the agonizing and persistent symptoms they had witnessed at close range.

The people who had allowed this to happen would not tell him what had gone wrong, either because they knew and refused to say, which was reprehensible, or because they didn't know, which seemed even worse. If the latter were true, why was information so scarce? And why should he trust Amgen or Amerisource or anyone to tell him that his son would be okay, when no one had tested—or would be able to test—the medicine that Tim had already taken?

11. Two Streams Become One

April 2002
South Florida

WHAT HAD BEGUN LITTLE MORE THAN THREE MONTHS AGO AS the investigation of a warehouse break-in had morphed into an all-out chase down blind alleys for bad medicine that might harm patients. Under Florida law, the investigators could detain drugs with fake pedigrees. But first they had to find them. The medicine moved like lightning in the overheated gray market, touching down only briefly at wholesale companies. The drugs were sold and shipped quickly, often hours before the Horsemen arrived.

In late March, they learned that a Utah company had sold a huge volume of questionable medicine to several Florida wholesalers. But there was no cross-country shipping involved. It turned out that a Fort Lauderdale courier service had picked up the medicine from an elegant rental home on one of the city's historic canals. Michael Carlow's deputy, Mark Novosel, lived there and operated the supposed Utah-based company, Optia Medical, from a room off his kitchen.

By the time the Horsemen arrived with the search warrant that led to Novosel's cooperation, the medicine was gone. Novosel had distributed 1,004 boxes of Procrit and 664 bottles of the blood product Panglobulin from his house. All of it was technically adulterated since the pedigree papers accompanying it were false.

Recovering the medicine before it reached patients proved difficult. The investigators found that the medicine already had been hotly traded and scattered across an array of middlemen, who then sold it to national companies that sold it to end users. The Horsemen stormed over to a Miami company that had bought the Procrit. One of the owners later recalled that Venema burst into the warehouse with his badge out and a gun on his hip, declaring that he was going to take away Michael Carlow's business and, by extension, the man's own. "It was a colon-loosening experience," the owner said. "They scared the shit out of everybody in the company."

Venema had not worn his gun that day. He always kept it locked in his trunk. But apparently his zeal, demeanor, and tough declarations had telegraphed *gun* with equal power.

But the Procrit was already gone. The Miami company had sold it to a company in Texas called Bindley Trading, which was a division of the huge drug wholesaler Cardinal Health. The medicine had crossed state lines—beyond the Horsemen's reach—and entered the nation's supply. All the Florida investigators could do was call the Texas health department and try to impart a sense of urgency, born of the fact that the medicine could be made of *anything*.

In 2001, counterfeit versions of three injectable drugs—Neupogen for cancer, Serostim for AIDS and Nutropin A.Q. for dwarfism—made their way into the nation's pharmacies. With public attention focused on the more sensational crimes of the rogue Missouri pharmacist Robert Courtney, FDA officials played down the counterfeits. "This little spate of cases is highly unusual," William Hubbard, the senior associate commissioner at the FDA told the *New York Times* in June 2001, in an effort to reassure patients that the nation's drug supply was safe and secure.

But the Horsemen began to see a very different picture, one in which a current of diverted, degraded, and expired medicine, traded by felons and accompanied by false paperwork, lay right below the surface of the so-called legitimate supply chain. It was not simply that the two streams merged on occasion, by accident, but that the legitimate supply was routinely polluted by inventory from dangerous sources.

Everybody bought from everybody. The biggest, most established wholesalers, including Cardinal and Amerisource, vetted their vendors in advance. After that, price was the guiding criterion—the lower, the better.

Once suspect medicine entered the Big Three's warehouses, it became intermingled with—and inseparable from—medicine purchased directly from manufacturers. No markers or labels distinguished it. Any of the wholesalers' customers, including pharmacy and hospital chains, could receive it.

Publicly the Big Three stated that they bought only 1 to 3 percent of their inventory from the secondary market. They contended that separating this medicine and documenting its origin would be too costly and inefficient—"an unbearable burden" to undertake, as their trade group wrote to Florida's pharmaceutical bureau in January 2002.

Consumers had no way to distinguish a clean drug from one that had been recycled or worse. In the end they both looked identical, moved through the same warehouses, and wound up in the same pharmacies. The Horsemen began to see their mission as an almost impossible race to keep these two streams of medicine apart.

THE FIVE MEN WORKED FROM A CONFERENCE ROOM IN VENEMA'S office, an enormous grid of wholesale companies taped to the wall and festooned with Post-It notes. They made intricate graphs in an effort to follow the medicine that on paper had originated across the country but in fact did loops around South Florida before being shipped to other states. Everywhere they looked turned up more suspect transactions, more dubious medicine, more "bad boys," as Venema called them, his appetite for hunting them down seemingly insatiable.

The Horsemen codified their identity in late March when Randy Jones distributed five black polo shirts that a friend designed. The image of a grim reaper holding a scythe amid a cluster of horses adorned the front of the shirts, along with "The Three Horsemen of the Apocalypse" in white letters.

Never mind that they had been three and now were five— or that there had been four horsemen in the New Testament's Book of Revelation. The investigators, not a particularly literal group, loved their shirts and began to wear them while serving search warrants or attending meetings.

The Horsemen talked to each other on walkie-talkies day and night. They were never without their phones, even taking them into the bathroom at home. "Hey man," Odin would answer instantly.

Venema's phone rang continuously as he coordinated their efforts, checking everything with Feldman to make sure they weren't driving off some cliff. Arias wouldn't speak at all, just chirp back with his thumb on the buzzer to acknowledge he was on the line.

They developed code names. Venema's was "Ice Station Zebra," and he would sign out with "10-4 rubber duck." Odin named himself "Midnight Rider," though Venema called him "Gene, Gene the Sex Machine," presumably because of his eye for the ladies. They liked to rendezvous at Hooters, where they plotted strategy.

They razzed each other too. When Venema learned that Petri had gone to Starbucks for a coffee, he howled over the walkie-talkie, "This latte shit is getting a bit too sophisticated. Pretty soon it's going to be wine and cheese and art showings."

Venema liked hamming up the contrast between their modest means and the lifestyles of their suspects. While serving search warrants, Venema took great joy in seizing fancy computers with flat screens, plunking his water bottle down on a gleaming antique table as a maid watched in horror, or tying his shoe with one foot up on the leather interior of a sports car, without so much as an "oops." He could not help but point out that one suspect paid $17,000 in tax alone for his new Bentley—$10,000 more than the cost of Venema's new used car.

They kept profile sheets on their growing list of suspects in ever-expanding binders. Arias's profile sheets lay scattered about his Buick, while Petri's files were impeccably organized, not a stray paper inside the truck in which he practically lived.

As their workload grew, so did their esprit de corps. Odin postponed his retirement, sacrificing the state-funded payout of $36,000, so he could keep working the case. Jones also put off the early retirement he had planned. Instead, he and Petri sat in their trucks, sometimes through the night and into the morning, as they watched the homes, offices, and secret warehouses of suspects. Slowly, links between seemingly unconnected suspects and drug enterprises, the intertwined root system normally hidden, came into view.

AS GARY VENEMA MOWED HIS OVERGROWN FRONT LAWN ONE Saturday afternoon, his cell phone rang. A deep Southern voice asked, "Mr. Venema? Edward Hardin. I'm a lawyer in Alabama." The formality, the new state, all got his attention. Venema, though most comfortable doing several things at once, actually stepped away from his mower to listen.

Hardin explained that he represented an Illinois company, Caremark Prescription Service, which managed large mail-order prescriptions, including those for Florida's civil servants. He wanted to follow up on a conversation that Venema had with a Caremark employee a few days earlier regarding a Davie company called First Choice Pharmaceutical Wholesalers.

Caremark, a $7 billion publicly traded company with a reputation to preserve, had bought $2.4 million worth of medicine from First Choice in the last year. The purchases included $1 million of an AIDS drug, Serostim. And now Hardin wanted to hear why Venema suspected that all this Serostim had been either stolen or diverted from the streets.

Venema recounted the tangled story for the lawyer: Several weeks ago, the Horsemen had visited a wholesale company called Atlantic Diabetic Supply. Its name had appeared on the pedigree papers of medicine moving through Mark Novosel's house in Fort Lauderdale. When the Horsemen arrived at the company, they found an employee preparing to leave for the weekend with $14,000 of cancer medicine in his car trunk. The drugs had no pedigree papers.

The employee explained that he had just returned from a buying trip with a friend from the company First Choice. Together they had purchased the medicine from a man named José L. Benitez. The diverter, a primary target of Operation Stone Cold, was the one who had sold medicine to Michael Carlow but had created invoices for construction equipment.

The Atlantic employee said that twice a week, his friend at First Choice called to say that his "guy," Benitez, had "product." The Atlantic employee then drove to Davie, picked up the First Choice employee, and headed south with him to Hialeah, where they met Benitez in the parking lot of a pharmacy. There, the Atlantic employee said he paid Benitez by check for a box or a cooler packed with high-cost drugs, most of which previously had been dispensed to patients and still bore their prescription labels.

Back at Atlantic Diabetic, the employee used lighter fluid and nail polish remover to "clean" the medicine of any labels, glue, or other markings. He then invented pedigree papers and invoices for each prescription and drove back to First Choice, where he "sold" the same medicine back to his friend in imitation of a legitimate transaction. His friend, the First Choice employee, paid him up to $100 per box of medicine for his role, which was essentially to generate "paper" to conceal the product's origin.

In four months, Atlantic Diabetic used this method to "sell" almost $1 million worth of cancer and HIV medicine to First Choice. In turn, First Choice sold this medicine to Caremark, the giant company in Illinois that shipped it across the country.

First Choice had not been accused of wrongdoing and had fired the employee involved.

Venema now explained to Caremark's lawyer that First Choice had two sources for much of its cancer and HIV medicine: Atlantic Diabetic Supply, which was supplied by José L. Benitez, and Brazil-US Trading LLC, which was the company of Benitez's half brother, a convicted marijuana smuggler. In short, First Choice's medicine had come from the Benitez family. And therefore it was bad.

Attorney Ed Hardin fell silent. Then he explained that Caremark had sent out a letter to all Florida wholesalers, advising them of the problem with First Choice. The company also planned a nationwide recall of the Serostim, some of which had already been dispensed to patients.

"How can we buy from any wholesaler and be sure they haven't bought adulterated or stolen medicine from First Choice?" Hardin then asked.

"You can't," Venema responded, thinking as he got off the phone that First Choice was only a small piece of the problem. If anything, Hardin's fixation on one supplier seemed quaint, as though First Choice alone had somehow degraded the nation's otherwise safe supply. But the lawyer had inadvertently lit on the big question that troubled Venema and his team. How could anyone, anywhere, know where their medicine had been or whether it was safe?

WAITING IN THE BAKING SUN ON APRIL 4, VENEMA PEERED DOWN the empty side street in North Miami. Everything seemed to reinforce the seedy and illicit nature of the deal he was about to make. "The delivery should be here, I just called them," the short, bearded man next to him announced as they both looked out for any sign of an approaching car.

Posing as a buyer for an export company that wanted to ship medicine to the Middle East, Venema was waiting for a delivery of Epogen with a fishy pedigree and a surprisingly low price. He assumed the medicine would be dirty, recycled from the streets of Miami through a long compromised chain.

The bearded man, Sheldon Schwartz, worked for a local wholesaler, AD Pharmaceuticals, and had brokered the deal for one hundred boxes of Epogen today plus the seventeen boxes of the AIDS medicine that Schwartz had delivered the day before. Venema had agreed to cut him a check for $509,000.

As they waited, Schwartz, shifting his weight nervously, reviewed the ground rules. No one was going to exchange the cash right here. When the medicine arrived, Venema would inspect it in the warehouse right behind them. If he liked what he saw, he could leave an associate to guard the drugs, and then head to a nearby bank where he would get Schwartz a cashier's check. In turn, Schwartz would give a check to his supplier, "Brian," who would be arriving with the drugs. Everyone would get his cut on the spot. The medicine was far below the drugmaker's lowest price even after traveling through five different wholesale companies, each one making a profit of some $20,000 by raising the price that much for the next buyer.

"I don't want to move anything until we go down and you have your check and you're a happy camper," Venema bantered casually. "I'll just look over to see if the dates are cool and everything," he said of the drugs. "I don't think you're going to ah . . . vanish while we go to the john."

Schwartz nodded.

Although the sting had no direct connection to the pursuit of Michael Carlow, the newly minted Horsemen were elated at the opportunity it presented. For one, they were fishing in a tainted lake, and were sure to draw out at least more information if not diverted drugs. And Venema got to play the role of sleazy wholesaler, such a good use of his fulminating energies that the exercise nearly lowered his blood pressure.

Arias, Odin, Petri, and Jones watched and waited silently in cars and trucks positioned around the parking lot. This sting represented a team effort: pooling of their knowledge and expertise, combining their slim resources, applying crime-fighting tools to combat dubious medicine. It seemed to represent everything that had been missing from their work lives—until now. Finally, as the temperature climbed well into the 80s, a silver Mercedes crept down the alley.

CESAR ARIAS HAD LEARNED ABOUT THE DRUGS FROM A PHARMA-cist in downtown Miami, Amjad Aryan. Amjad, along with his brother Aiman, ran a phenomenally successful pharmacy chain, along with a small wholesale company. Several days earlier, AD Pharmaceuticals had called the Aryans offering the Epogen at a price too good to be true—each box $100 less than the manufacturer's lowest wholesale price. In an indus-try that runs on fractional margins, with profits made only on immense sales, the price raised an immediate red flag. They called Arias, who instructed them to agree to the purchase.

Young, handsome, and hard-working, the Aryan brothers embodied a true American success story. They had resisted the temptation to become dispensing mills for frequently abused drugs or repositories for gray-market and substan-dard medicine.

"If I wanted to be in business for a year or two and then run, I could do it," Aiman had told Arias on one occasion. "But we have a lot of expectation from a lot of people includ-ing my father." Their father, a pharmacist, had been the health minister of Jordan.

The Aryans had been a constant, quiet wellspring of in-formation for Arias. They called with tips about dubious dis-counts, suspected prescription drug abuse, possible Medicaid fraud, and other concerns.

Their cooperation was smart business. The family's phar-macy chain, Robert's Drug Stores, would have been an obvi-ous target for investigators who did not know them. Situated in the poorest neighborhoods of Miami, their pharmacies were the state's largest billers of Medicaid and the biggest dispensers of a much-abused drug, Oxycontin. In order to avoid drawing suspicion or slipping into a gray area, they not only documented everything and called Arias regularly, but they also called the FDA and the Drug Enforcement Administration.

Arias often met the brothers in the Wynwood barrio near downtown Miami, in the back of one of their drugstores, where the ants ran a steady patter along the wall. On one occasion, they showed him a pile of photocopied drivers' licenses of every single patient to whom they had dispensed Oxycontin. Though not required, they did this to deter obvious abusers. Aiman Aryan told Arias that filling so many prescriptions for painkillers made him feel dirty. "This is not why I went to school," he complained.

The Aryans had allowed Venema to pose as a buyer for their wholesale company, RDS Export. Even though the products had been offered to a licensed company buying legal products, its records open to review by state regulators, the transaction resembled an illegal drug deal. Venema knew almost nothing about the medicine's origin or quality or who would deliver it. The supposed deliveryman, "Brian," was probably as much a fiction as Venema's own alias.

FROM THEIR STRATEGICALLY PARKED CARS, THE INVESTIGATORS followed the progress of the Mercedes, their walkie-talkies crackling to life. As the car glided into view, Arias felt chills. He knew the man behind the wheel. He was Brian Alan Hill of Jemco Medical International. Two years ago, he and Odin had spent a week in Hill's warehouse with a leased Xerox machine, photocopying all his records and making his defense attorney rich. They had investigated the man for years, but could never find an explanation in his records for his huge success. Now here he was, playing a crucial role in this transaction.

Hill, with his bloodshot hazel eyes, blond-streaked hair, and full-moon face, appeared a contrast to his dark, wiry, and intense passenger, his shipping manager, Claudio Boriminoff. As the two men climbed out of the car, all jovial banter came to a stop. With no smiles or handshakes and all but ignoring Venema, Hill popped open the car trunk and there, baking in a cardboard box without benefit of a cooler or other protection, was the Epogen, almost certainly degraded by the extreme heat and the turbulence of the ride.

The men went inside the warehouse and Venema inspected the medicine as warehouse workers looked on. He had expected to find worn-looking boxes that had been recycled through the system, their lot numbers unmatched and cobbled together through a patchwork of street sources.

But as he pulled out the boxes of high-dose Epogen and inspected them one by one, they all shared the same lot number: P002970. It was a number that Kevin and Jeanne Fagan would come to recognize. Not one had the sticky residue of medicine that had already been dispensed. As far as he could tell, the drugs appeared pure. As prearranged, his walkie-talkie buzzed and he uttered the words that signaled the backup units to move in.

The men poured from their cars so quickly—several other Miami cops who had come along to help even drew their guns—that Odin found himself alone in John Petri's truck. Once again, he had missed the bugle call.

Inside the warehouse the police separated everyone, placing them against the wall and searching for weapons. Sheldon Schwartz yelled, demanding an explanation. Staying in character, Venema feigned outrage and surprise.

Suddenly they all heard a loud knock on the front door and a voice called out, "Gene Odin. Department of Health." One of the men opened the door to find the silver-haired regulator standing in the entrance. "I'm here to do an inspection," Odin announced loudly, as if all the different parties had descended on the warehouse at once by coincidence.

Schwartz—oblivious to the multilayered cover stories—appeared relieved to see Odin, who had regulated him for years. "Gene, I'm just doing a deal here," he said. "What's going on?"

"I'll try to find out," Odin said, in such a straight manner that Schwartz bought it. Venema almost burst out laughing. His colleagues quickly hustled him from the warehouse before he blew everything. Hill, now silent, looked shaky and stunned.

Together, Arias and Odin studied the medicine. They expected it would be bad.

For starters, AD Pharmaceuticals was run by Chantal Banatty, who pleaded guilty in 2000 to selling stolen medicine through her husband's wholesale company. She was still on probation when Arias's bureau ignored his objections and granted her the license for AD Pharmaceuticals. In 2001, her company had purchased a cancer medicine, Neupogen, which contained nothing but saline. The state finally revoked her license in January. Now, three months later, her supposedly defunct company was shepherding drugs into Venema's hands.

The paperwork was also an easily discernable lie. It showed that the drug's manufacturer, Amgen, had sold the Epogen to a Houston company, which in turn sold the medicine to a firm in Dallas. But a single call to Amgen's security department revealed that the drugmaker had never sold Epogen directly to either of the Texas wholesalers. Neither company was licensed to do business in Florida.

In a plaintive tone, Hill admitted that he had purchased

the drug from a company in Doral, west of Miami, L&L Distributors, which had purchased it from one of the Texas companies. "There's nothing wrong with the sale," he said. "L&L is licensed. I've got all the documentation at my office."

Arias and Odin recalled that the owner of L&L, Ricardo Lamas, had been convicted of health-care fraud the previous year and was on probation. But as Arias and Odin examined the medicine, it looked perfect even to their practiced eyes.

WITHIN AN HOUR, THE INVESTIGATORS ARRIVED AT THE L&L warehouse. Though large enough to hold six tractor-trailers, the warehouse was nearly empty that afternoon. The refrigerator inside it was bare. The president, Ricardo Lamas, seemed surprised and unhappy to see them.

As Venema, Arias, supervising FDLE Agent Michael Mann, and the others listened, Lamas explained that he traveled abroad continuously and knew nothing about the Epogen. His mother worked as the bookkeeper and she was home sick, leaving him with no access to the records. Perhaps his business partner might be able to shed some light on the deal, but he was out, Lamas said.

With Venema glaring at him and the others beginning to poke through the warehouse, Lamas called his partner, Javier Rodriguez, on his cell phone and the two men had a brief discussion. "He could get you those records tomorrow," Lamas offered.

"You get his ass over here now, and by that I mean right now," Venema said with disgust. "Not tomorrow or next Thursday. I'm losing my patience. And I'm not too patient to begin with."

While waiting for his partner, the investigators interviewed the few warehouse employees they could find, none of whom remembered a shipment from Texas. They searched but found no shipping records.

Rodriguez finally arrived in his burgundy Jaguar, proffering a stack of sales records that showed the transactions between the companies. He said he could not find the interstate shipping document that proved the drugs had actually been in Texas and had been mailed here.

As Venema half-listened, he noticed a surprising plaque on the wall behind him: "Javier Rodriguez, best wishes from all your friends at the FBI Miami."

It turned out that Rodriguez had formerly worked as a clerk in the FBI's organized crime squad. He now lived in a four-bedroom house by the water and made a great deal more money in pharmaceuticals.

As Michael Mann watched his agents at work, his cell phone rang and a voice fumed, "You need to cease and desist." It was L&L's attorney, Craig Brand, who was returning from vacation when Lamas reached him. "I am coming over there right now. You are circumventing the search warrant procedure and I don't want you looking anywhere."

"I don't agree with you," Mann said calmly, "but you're welcome to your opinion. . . ."

"I've handled people like you before," Brand said, cutting him off. "I am the premier attorney handling these cases and I'm just trying to save you your job."

Mann said evenly, "If you come here and interfere with this inspection, I'll have you arrested for obstruction of justice."

As the inspectors continued their work, a man in a T-shirt and shorts with close-cropped hair, trimmed eyebrows, and stylish wire-rimmed glasses strolled into the warehouse with a take-charge attitude.

Craig Brand introduced himself by dropping his business card into Venema's shirt pocket. As Venema looked down to inspect the attorney's card as though checking for germs, Brand declared, "You need a search warrant. You're not allowed to get those records."

He waved his hand dismissively at the records that Rodriguez had already given Venema. "The pedigree papers are just a matter of semantics."

The thirty-four-year-old lawyer liked to promote his specialty as defending health-care clients from overzealous government regulators. As his firm literature spelled out in bold print, WE STAND READY TO CHAMPION YOUR CAUSE AND FOIL THE GOVERNMENTS [sic] PLOT. He represented dozens of secondary wholesalers, whom he described as running mom-and-pop businesses. His clients, he told a visitor on one occasion, were "do-gooders who want to be in business forever and pass the businesses along to their children." His client list included Michael Carlow and José L. Benitez.

Mann, who had been quiet until now, stepped forward. "We are well within our right to do a regulatory inspection. I'll allow you to sit there so long as you don't interfere." He pointed to a small air-conditioned office.

"I have a right to talk to my client," Brand glowered, "and I'm going to file a lawsuit. You're exceeding your authority."

"I'm going to count to ten," said Mann, "and if you're still here I'm going to have you arrested for obstruction. Gary, give me your handcuffs. I have to lock up an attorney for obstruction of justice." Venema passed them over.

"One, two, three, four, five . . ."

As Brand retreated and took up his post outside, Venema declared with satisfaction, "I love it when they think they're slicker than whale shit."

The rules of the game had changed. In the past Brand

could have enlisted a lobbyist or low-level politician to make knees shudder and feet turn to clay. Mann's agency was not going to play along.

As they left, Mann told Brand, "I've been threatened by better people than you so I'm not impressed when you tell me you're going to have my badge. You act like a gentleman in the future, I'll treat you like a gentleman. You act like an asshole, I'm going to arrest you."

Mann's bosses at the Florida Department of Law Enforcement ignored the letter Brand sent them shortly thereafter, which proclaimed that Mann had abused his authority. In his practice, Brand relied often on these thinly veiled threats to sue. Just a month earlier, in an e-mail to the health department's general counsel, he wrote on behalf of another client, "I believe the department overutilizes the term 'adulterated.' I also believe that DOH misapplies or strictly construes the pedigree laws. I am willing to bet that a jury agrees with me."

The following October, Brand filed a class-action lawsuit against the health department for using "technical errors" in the pedigree papers to seize medicine it unfairly deemed adulterated.

As the Horsemen would later discover, Craig Brand did not just serve as Ricardo Lamas's lawyer. The men also were colleagues. Brand owned the wholesale company, Global Pharmaceutical Services, and Lamas served as his chief operating officer.

The interconnections, the slippery nature of the industry, and the lies convinced the men that nothing was as it seemed. "What it is *is*," Odin would say as the preamble to almost every discussion, as though even reality needed translation. Whatever was going on seemed so festering and involuted that even old-fashioned corruption—something akin to influence peddling in Tallahassee, for example—might have come as a welcome relief, a sign that even illegality had an order to it.

Meanwhile, the cache of Epogen seized in the sting remained a puzzle.

Arias wrote to one of Amgen's security officials, Jon Martino, asking whether the company had sold one hundred boxes of high-dose Epogen with the lot number P002970 to a single buyer in the last year. Martino wrote back that one hundred boxes of high-dose Epogen would have been too big an order for anyone in the country. Amgen had sold the largest order of that strength and lot number—eighty-eight boxes—to AmerisourceBergen, the nation's largest wholesaler.

This led Arias to wonder: If no one had bought one hundred boxes of the high-dose medicine from a single lot at one time, how did they end up grouped together? And who could afford to buy that much? The medicine had a market value of almost $500,000.

12. The License Shrine

Spring 2002
Pembroke Pines, Florida

BY ALL APPEARANCES, JEMCO MEDICAL INTERNATIONAL WAS A growing company with big designs. Located just outside the semi-industrial town of Pembroke Pines, in a small corporate park bordered by sawgrass, Jemco fit in well with its surroundings. Many companies had relocated to this area west of Fort Lauderdale after 1992, when Hurricane Andrew devastated areas nearer the coast. Such firms as American Express and BellSouth also had brought their workers here.

Jemco had run into trouble with regulators in 1998, when the company still operated in a low-rent neighborhood of Hialeah Gardens. FDA investigators uncovered twenty-four bottles of counterfeit Retrovir, an AIDS medicine, in the company's warehouse. The labels and seals on the bottles were poor imitations, lacking the design of the real medicine.

Other run-ins with authorities followed as the company expanded and moved north. By 1999, Jemco was thriving and held its Christmas party on a chartered yacht with a lavish catered dinner and dancing. The next year, the office staff enjoyed an all-expenses-paid vacation to Acapulco. The company invited its medicine suppliers to come along too. Jemco's prosperity could be seen in the company parking lot: The sports cars parked there included two identical white Mercedes. The matching pair was owned by chief operating officer Brian Hill and company president José Castillo, who shared a home in an exclusive gated community surrounded by a moat.

In his visits to Jemco's tidy strip of offices, Gene Odin repeatedly found deficiencies such as expired and mislabeled drugs and medicine purchased from companies without Florida licenses. The resulting fines amounted to a mere $2,000.

On the morning of April 12, Odin and a colleague arrived in Pembroke Pines for a routine inspection. Though usually cool and well prepared, José Castillo appeared uneasy this time. Since his partner Brian Hill had gotten caught up in the undercover sting eight days earlier, the ground seemed to be shifting beneath their feet.

Castillo swept into the reception area, polite as ever. Urbane and well-groomed, with a subtle goatee and white peppering his dark hair, he extended a manicured hand to Odin and his colleague, Mary Ghabrial. Odin, in his black polo shirt and khakis, must have appeared to him even older and more befuddled than on his last visit.

He trailed behind the inspectors through the orderly, air-conditioned space. One wall of the reception area was covered with plaques and certificates, including a copy of the company's state license. Next to it a gleaming plaque boasted the company's membership in the Healthcare Distribution Management Association, or HDMA, the nation's largest trade group for drug distributors. Membership conferred some legitimacy: The Big Three, Cardinal, McKesson, and Amerisource, were also members. Another plaque from the United States Chamber of Commerce recognized Jemco as well.

Odin strolled back to the cool, well-organized warehouse, a hive of activity with workers using forklifts to hoist pallets of medicine.

"Is there any other warehouse that you use?" Odin asked as he poked around, examining medicine bottles that appeared new and clean.

Castillo shook his head. "This is it."

"Are you sure you don't have another warehouse?" Odin asked as inspector Mary Ghabrial took notes.

"This is the only one," Castillo insisted, talking pointedly as though Odin were deaf.

This adjacent warehouse was the only one registered with the state and therefore the only warehouse in which the company could legally store pharmaceuticals.

"I just want to make sure," said Odin, a picture of confusion and forgetfulness, "that this is the only warehouse I need to inspect. My hearing aids aren't working too well."

Fighting against his impatience, Castillo again confirmed that Jemco had just this warehouse.

"Could you come outside to my car?" Odin asked after a few more minutes. "There's some papers I need to give you."

As Castillo walked grudgingly toward the car that Odin had deliberately parked some distance away, Michael Mann, three other Horsemen, and an investigator from the Attorney General's Medicaid Fraud Control Unit drove up in a swirl of cars. Arias, Jones, and Venema also wore their black polo shirts bearing horses and hooded, scythe-wielding figures.

As Castillo confronted the men, Odin said happily, "We need the key to your other warehouse. Can you get it?"

Castillo's face betrayed his apprehension and disgust. "What other warehouse?"

"You'd better get it right now," Mann told him.

Odin and Arias had long suspected that Jemco maintained a second warehouse and kept two sets of books. They dis-

missed the company's great show of certificates—a license
shrine, as the inspectors called it—as a decoy. But they had
no proof until John Petri and Randy Jones staked out the
Pembroke Pines office park for several days. The detectives
watched Castillo go back and forth between the registered
warehouse and another one directly across the parking lot. The
fact that this other warehouse had been leased in Castillo's
name had given the investigators enough evidence for a search
warrant—the true purpose of Odin's visit today.

"I don't know where the key is," Castillo offered sourly as
he headed back to the office, the short distance appearing
interminable.

Back inside the legitimate warehouse, with Arias and
Mann right beside him, Castillo moved from desk to desk asking his baffled employees, "Do you have the key?"

They looked at him in confusion, not sure why or what he
was asking.

"You know, *the key*," Castillo intoned until someone finally handed it over.

Key in hand, Castillo trudged across the blazing parking
lot as though walking an asphalt gangplank. He slowly opened
the metal door and the men walked into a stuffy reception
area strewn with trash. The space had a weak air conditioner
and was filled with costly medicine of every imaginable kind:
Zyprexa, Viramune, Vioxx, Videx, Norvir, Trizivir, Crixivan,
Viracept, Combivir, Epivir, Ziagen, Kaletra, and Zocor. Bottles
of lighter fluid and old rags sat amid the medicine.

One old box contained ninety bottles of Zyprexa, an anti-
psychotic medicine, and a handwritten list of medicine with
prices on it. Another paper showed that Jemco had done
$650,000 in business with the Stone Group in Boca Raton.

As Castillo hovered in the doorway, the inspectors filed

into the room behind the reception area. It was a raw, airless, filthy warehouse. It looked big enough to house a medium-sized airplane. Enormous pallets of medicine stood along the walls on shelves that reached nearly from floor to ceiling. More medicine sat on the floor amid drifts of garbage and discarded packing materials.

The inventory rivaled a government stockpile. As the investigators moved through the warehouse they spied pallets of Albuterol products for asthmatics and antibiotics, including penicillin. They found rows of boxes bound together by industrial-strength plastic and filled with isopropyl alcohol, vaginal specula for performing gynecological exams, and Lactated Ringers, an intravenous solution used to rehydrate patients.

Many of the boxes bore directions clearly stamped on the outside: store the medicine between 35 to 77 degrees Fahrenheit and protect it from light. But the warehouse was hot—at least 90 degrees, some of the Horsemen estimated.

Even as they stood there, a truck from Ricardo Lamas's company, L&L Distributors, backed up to the registered warehouse across the way and Jones swung his video camera around to capture the enormous truck.

"Where did all this product come from, pal, and where are all the documents for it?" Venema asked Castillo.

Castillo, whose olive complexion looked unnaturally pale, explained that he didn't know how the products had ended up in this warehouse. He also said he had no documents to prove that the medicine had been legally purchased.

The Horsemen had to secure this mountain of medicine, all of it technically adulterated. They counted and sorted for hours in the humid, filthy space. Michael Mann actually climbed into a forklift, having learned to operate one years ago when he had driven a bakery truck in New Jersey. Odin stood on a pallet of boxes to direct the work. Arias reviewed the sorting, his bifocals propped onto his forehead. And Venema wandered through the maze of medicine, offering wild and profane exclamations.

AT HOME IN DAVIE, SANDY VENEMA WAS PUTTING THE FINISHING touches on a spontaneous romantic dinner. In a moment of rare extravagance, she had bought fresh shrimp, a bottle of white wine now chilling in the fridge, and some elegant tapered candles for the dining room table.

Venema had called about an hour earlier to explain that they had seized a load of bad medicine, but that he would be home as soon as possible. As Sandy waited she felt herself growing angry.

She loved Gary and could appreciate, perhaps better than anyone, the positive effect this case had on his spirits and outlook. But a manic energy had replaced his depression. He worked fourteen-to-sixteen-hour days, then joked that Sandy had failed to put enough Zyprexa, a medication for schizophrenics, in his breakfast cereal. At home he was either on his constantly buzzing walkie-talkie or busy documenting everything the team did, as Stephanie Feldman had ordered.

At night, Sandy struggled to follow the ever-changing roster of suspects, the connections between them so convoluted as they spilled from her husband's lips that sometimes she only pretended to listen. Even in the middle of most nights, when at least they could have slept side-by-side, he often disappeared for hours to sift through the trash cans of gated communities. She almost never saw him anymore.

She planned the dinner to change that—if only for a

night. But more hours passed. The shrimp grew cold. Sandy blew out the candles. She had never bothered to open the wine. That would have been her husband's job.

BY NIGHTFALL, THE HORSEMEN HAD FILLED TWO U-HAUL trucks with medicine and were headed south to Miami, where Arias had arranged to rent a warehouse.

Most times, the men thought nothing of driving for hours. They did it all day, every day, as they tracked suspects all over South Florida. But after unloading all of the evidence into the new warehouse and driving another hour back up to Pembroke Pines to retrieve their cars, Venema felt so exhausted he found himself nodding off behind the wheel, the truck bouncing toward the side of the road until he jolted awake.

He sensed from his brief telephone exchanges with Sandy that he had screwed up something badly. Instinctively seeking protection, he said to Arias and Odin, "Why don't you guys come back with me to the house?"

This struck them as extremely odd. Why show up at the house of their new friend when it was almost 10 P.M. and they were exhausted, dirty, and smelled like goats?

Yet Venema insisted with such vigor that they finally acquiesced, following him in through his garage as he tried to sneak into his own home. They found Sandy seated at the dining room table amid the wreckage of the uneaten meal. She looked up at the men without any trace of a smile. Venema instantly saw the melted candles and evidence of a dinner that had required extra effort. Arias and Odin—who between them had known seventy-six years of married life—recognized the mood and started spewing excuses, racing to see who could make it to the door first. "They must have thought I was a witch," Sandy said later.

Venema knew he was in trouble and did his best to apologize. But nothing could diminish his sense of triumph as he lay down in bed that night. They had just seized almost $2 million worth of pharmaceuticals.

Arias, too, felt pleased as he climbed into bed. Yet his thoughts drifted to all the bad medicine that had gotten away—not just from Jemco but from every corrupt company he had investigated over the years before meeting Gary Venema. The following month, both men would celebrate when the state pharmaceutical bureau revoked Jemco's license and the wholesaler filed for bankruptcy.

IN MID-APRIL, ARIAS CONFIRMED THAT THE PEDIGREE RECORDS for the one hundred boxes of Epogen they had seized in the undercover sting were phony. According to the Texas health department, the Texas companies on the records actually belonged to men in South Florida. Their names were Eddie Mor and Carlos Lorenzo Luis. They would soon become critical parts of the shifting investigation.

The two men had been under scrutiny before. Luis's name had surfaced in a federal investigation into the recent counterfeiting of an AIDS drug, Serostim. Mor co-owned a Florida wholesale company that had distributed counterfeit Serostim and was not being allowed to renew its license. The news from the Texas health department meant that the Epogen had originated in Florida and not Texas, as the phony pedigrees stated, and was therefore technically adulterated.

Arias had seen out-and-out counterfeits, medicine that had been purposefully tampered with for profit, on only a few occasions. Each time, the counterfeits had been relatively easy to detect. In 1994, he and Odin had found capsules of counterfeit Feldene, an arthritis medicine, while inspecting a

wholesaler. Not only did the labels look worn and "nasty"—
Arias's word for medicine that looked impure—but the seals
over the tops of the bottles had been visibly reglued.

Last May at AD Pharmaceuticals, the four boxes of counterfeit Neupogen, a cancer drug, also had been easy to spot.
The outer boxes had been dented, their surfaces smudged, the
ink runny. It had been clear that something was grossly wrong.

It was a rare day when things were actually worse than
Arias imagined. But that day came in late April when Amgen's
security specialist, Jon Martino, called from California regarding the one hundred boxes of high-dose Epogen seized in
the sting operation.

Wasting no words, Martino, a former Los Angeles cop,
told him, "It's bad."

The drug really was Epogen and came from Amgen,
Martino said. But it was not the high-dose Epogen—the
Rolls-Royce of anti-anemia dosages—as labeled. It was the
plain-Jane and comparatively cheap, low-dose medicine one
twentieth the strength, worth $258 for a box of ten vials.
Someone had glued on counterfeit labels, making each box
worth $4,700.

In the parlance of the drugmakers, the medicine had been
"uplabeled." The scheme troubled Martino particularly because of how expertly it had all been done. The counterfeit labels were indistinguishable from the real ones except for two
tiny degree symbols missing in the words "Store between 2 to
8 degrees Celsius." Even some of the manufacturer's packaging specialists had not been able to distinguish the difference.
Amgen was sending out a warning letter to physicians and
pharmacists nationwide, identifying the counterfeit lot, showing pictures of the fake vials and the real ones, and urging patients who suspected counterfeit medicine to promptly call
the FDA.

The news floored Arias. He had inspected the medicine himself and it had appeared in perfect condition. All they knew for sure was that the affected vials shared the same lot number: P002970. The Florida inspectors and Amgen officials were confronted with some tough questions. Who had done this? How much medicine had the counterfeiter uplabeled? If more existed, where was it? And how could they find it before it reached patients?

Now Arias could make sense of the confusing information that Amgen had never sold one hundred boxes of the high-dose Epogen to anyone. Whoever did this apparently had bought one hundred boxes of the *low-dose* Epogen and re-labeled them. Arias dashed off letters to the three biggest wholesalers, McKesson, Cardinal, and Amerisource: Had they sold one hundred boxes of 2,000 U/mL Epogen to anybody in South Florida in the last year?

13. A Do-or-Die Cause

Summer 2002
Deer Park, Long Island

KEVIN FAGAN GREW ANGRIER AND MORE SUSPICIOUS AS DAYS, weeks, and then months passed with no answers to his questions. The true nature of his son's adulterated injections had never been explained, leaving Kevin with a bad taste of corporate indifference from his brief interactions with Amerisource-Bergen and Amgen.

He had practically begged officials with the FDA to come test the five vials of counterfeit Epogen that remained in the family refrigerator. But three months had passed since he first contacted the agency and no one there had even bothered to call back. Though he didn't know it, his calls had entered a regulatory maze and had not yet been referred to the agency's office of criminal investigations, which had already collected samples of the counterfeit Epogen from other parts of the country.

So Kevin began to write letters. Almost every night after work, he sat at his home computer, the screen saver a brightly colored American flag, composing e-mails to those he viewed as responsible. He wrote repeatedly to the companies that had handled Tim's medicine, including Amgen and Amerisource-Bergen, but also to the FDA and other regulators.

On August 25, he wrote to an official at the Department

of Health and Human Services, the federal agency that oversees the FDA, explaining that no one from the FDA had come
to test their vials of counterfeit Epogen. "Despite our pleas,"
he wrote, "they still refuse to talk to us or take the vials for
toxicology studies."

He noted that his family had identified the extra counterfeit lot in the first place. "I find it ironic when we ask the
F.D.A. for help, all they can offer is for us to check the Amgen
Inc., website for info on what the vials may have contained,"
the letter continued, "similar to 'asking the fox what happened to the chicken!!!'"

By then, even newspaper and television reporters had
begun to contact the Fagans about the counterfeits Tim had
received. But he heard nothing from those who he assumed
knew the truth. In September, he wrote to the legal department of Amerisource: "What is the status of your internal investigation or that of the authorities investigating, as to how
this counterfeit drug entered the supply chain? We need all the
information available to ascertain what our son was injected
with and how this has effected [sic] his health as a result."

Still he got no response. More than anything, Kevin despised what he felt to be the indifference of these companies.
It would have been one thing if some mom-and-pop drugstore
going out of business had bought cut-rate medicine. But it
was another thing to be dealing with AmerisourceBergen, one
of the nation's largest pharmaceutical distributors, and Amgen,
the world's largest biotechnology company.

A week later he wrote again, this time to Amerisource's
chairman, Robert Martini, detailing how the company had ignored his last e-mail and how four months had passed since
he had heard from anyone there, despite promises to keep him
in the loop. "I would think that a company that had $39 bil-
lion in annual operating revenues has the resources to answer

an e-mail in a timely fashion to a parent concerned for their child's health," he wrote, adding, "Any response . . . if even to say we received and read the memos and are in the process of replying, would be refreshing."

The next day, he heard back. Chris Zimmerman, the company's senior director of corporate security and regulatory affairs, called him at work to tell him not to contact the company again. If he wanted information, Zimmerman said, he should call the FDA agent in Florida who was in charge of the investigation. He gave him the cell phone number of Agent Luis Perez.

To Kevin this seemed strange. Amgen was in California. Amerisource was in Pennsylvania. What did Florida have to do with it?

THE FAGANS DECIDED THAT THE COMPANIES WOULD DIVULGE the truth only if someone forced them. And so Kevin went on the Internet and began their search for a lawyer the best way he knew how—by finding every personal injury lawyer in Manhattan with a Park Avenue address. And that is how he found Eric Turkewitz, Esq., who had an office at 99 Park Avenue.

It had been the end of May when Kevin walked into the modest suite of offices where Turkewitz, a solo practitioner, shared space with six other attorneys. Too distracted to even notice that he was not in a big-name firm, Kevin instead spotted a framed sketch of a judge to the side of Turkewitz's desk.

"That's John Sirica," he blurted out, referring to the federal judge whose persistent quest for the truth during Watergate contributed to President Nixon's resignation. Kevin, an amateur government buff who had majored in political science at Queens College, felt hopeful then that Turkewitz could help his family.

Kevin's rough edges and plain speech notwithstanding, his aptitude surprised Turkewitz, who had never seen anyone identify Sirica from the sketch. As Kevin laid out his son's story, Turkewitz immediately saw the case's merit, despite the obvious difficulties of taking on pharmaceutical companies that could afford battalions of top lawyers.

The forty-two-year-old lawyer was a risk taker. After several years as a litigator at a medical malpractice firm, he had quit and spent a year traveling around the world. On his return, he went into business for himself. He set up a home office and made his first stationery by taping his business card to a piece of white paper and photocopying it. He ran ads in the *New York Law Journal* and accepted even the most menial per-diem work, freelancing for the big firms by taking depositions from car-crash victims. Slowly he worked his way up and into the modest offices on Park Avenue.

As his business grew, so did his family. His wife was five months pregnant when Kevin arrived at his office, and Turkewitz felt particularly sensitive to the notion that someone had harmed Kevin's child. It seemed clear to him that the companies that had handled the counterfeits needed to be held accountable. He also figured that there must be other victims, most of whom probably were unaware they had been victimized. If the evidence got destroyed by injection and the patients failed to get better, everyone would chalk it up to their underlying disease. Turkewitz took the case. In the months and years to come, he liked to remind skeptics that the landmark abortion case of *Roe v. Wade* had been won by a solo practitioner.

Kevin, meanwhile, continued his letter writing. First and foremost, he wanted answers: What had they injected into

Tim's body? How exactly had counterfeits reached their pharmacy? Who was going to fix this broken system?

He sent these questions to senators, congressmen, and governors, to New York Attorney General Eliot Spitzer, and to various regulators. He sought out politicians willing to sponsor legislation and persuaded a Long Island congressman to get involved. Representative Steve Israel held a press conference one Sunday on the Fagans' porch to announce legislation he was sponsoring to require pedigree papers for all drugs.

That day, Israel spoke of his confidence in passing the act. "I have Timothy Fagan with me," he said. "I have a kid who woke up in agony in the middle of the night, taking medicine that shouldn't have been in his body, that shouldn't have even been in trash cans. We have facts and the Fagans on our side."

Kevin even tried to enlist First Lady Laura Bush in this fight. As he wrote to her:

> *Today, society is suffering from a moral breakdown where huge companies look only to the bottom line and not to what is the right thing to do. The Enron and WorldCom stories, as well as this, depict this only too clearly. I ask that you, a parent as I, do whatever you can to bring forward legislation to require drug companies to document the shipment locations of prescription drugs from the point of origin to the end user. . . . the general public.*

He added:

> *We hope this will prevent other families from going through what mine endured!*

As a general matter he viewed silence as an invitation, often writing two or three follow-up letters to get a response.

Finally, in mid-September, a staff person from Senator Hillary Clinton's office interceded at the FDA. Two days later, an FDA agent showed up at the family's home—more than four months after Kevin Fagan's initial phone call—to collect the five counterfeit vials. Now doubting everyone's motives, the Fagans kept one as proof of what had happened.

Jeanne believed in and supported her husband's efforts. In part she viewed them as a productive outlet for his anger and fear. She also knew they couldn't stay silent, because what had happened struck her as so immoral. "It's just so wrong it hits you in the gut," she said, "and if anything positive can be done out of this disaster then let it be."

The discovery of the counterfeits and the fear of more had turned their son Tim into a "freak," by his own description. He dreaded taking his medicine, which still included regular injections of Epogen and Neupogen. Gone was any sense of a healing ritual. He studied the labels on his medicine bottles until he could barely see straight, not knowing what to look for except something he wouldn't expect to see.

He had become gloomy and fatalistic and prone to bouts of anger. Alternately blaming himself and his family's genes, he viewed another health crisis as inevitable. Sometimes he would say, "I don't know if I'll live to see forty." Other times when taking his medicine, he would say, "I'm going to die so what's the difference."

Tim's ordeal also changed the family dynamic forever. The Fagans' eldest daughter had stepped in to take care of her little sister and to make dinner all those nights when her parents had been at the hospital. In Jeanne's view, she had been forced to grow up too fast.

Kevin felt that some of the responses he got minimized the horror of what had happened. "If people say, 'What's the big deal?' the big deal is, this is just one more thing over the stream of years that rips this kid and this family apart," he explained one evening.

Despite all the letters he wrote and the questions he asked, his biggest question remained unanswered: How come nobody, from the manufacturer to the wholesaler to the pharmacy to the government agency supposedly regulating all this, could tell him where his son's medicine had been or how this had happened?

A PARTIAL ANSWER LAY IN HISTORY THAT KEVIN FAGAN KNEW nothing about, a vortex of law-making and lobbying hundreds of miles away in Washington, D.C.

Fifteen years earlier, Congressman John D. Dingell introduced the Prescription Drug Marketing Act, a bill that aimed to create the kind of transparency in the drug distribution system that Kevin had described in his letter to Mrs. Bush. As Dingell declared at the time, "The only people who will oppose this bill are the fast-buck artists and shady dealers who now profit enormously from these various nefarious practices."

In fact, almost everyone in the sprawling pharmaceutical industry—from pharmacy chains to wholesalers to drugmakers—opposed it for one reason or another. Almost immediately, the docket at the FDA grew thick with industry protests: The new regulations would strangle innovation, swamp small businesses with unnecessary paperwork, and restrict trade, the middlemen insisted. As the bill became law in 1988 and the battle over it raged on, few patients even knew about it or the middlemen who handled their medicine. The skyrocketing cost of drugs had not yet become a front-burner issue. Epogen and other costly drugs like it had yet to be invented.

The law required each state to license and inspect drug wholesalers by 1990, and to enforce minimum standards for storage and record keeping. The wholesalers, who had faced little scrutiny until then, objected most strongly to one provision of the law, which specified that wholesalers must maintain a "pedigree" that would identify the drug's previous buyers and sellers. This provision aimed to create a paper trail that investigators could follow if drugs of questionable quality turned up on the market.

This pedigree was no more than what was required of dog breeders or car dealers. Nonetheless, the industry launched a withering salvo to defeat the requirement, organizing trade groups, writing letters, and wooing politicians to help. The wholesalers contended that such a rule would force them to reveal their sellers to their buyers, which, in turn, would reduce the competition that kept drug prices low for the consumer.

They expressed their concerns in a forest of form letters that arrived, daily, at the FDA. "If the proposed regulations require us to reveal the sources of our products to our customers it would permit them . . . to buy directly from our vendors, effectively putting us and other wholesale distributors out of business," one of the letters stated.

Manufacturers, too, protested the requirement. Smith-Kline & French Laboratories argued that the list of distributors authorized to sell its drugs was confidential business information and any rule requiring disclosure constituted "taking of property without just compensation, a violation of the Constitution."

Only independent pharmacists, who at the time dispensed more than 70 percent of the nation's prescriptions, liked the rule. They viewed the audit trail as an essential protection because it was the only way to trace the origins of substandard medicine.

Straining under pressure from powerful interests, the FDA sought cover in a compromise. It decided to review the pedigree regulations, thus putting them on hold. In their place, the agency issued a guidance letter that temporarily limited the scope of the pedigree. It stated that pedigrees did not have to trace a drug's path all the way back to the manufacturer. Instead, the paperwork only had to trace the drug to the last "authorized distributor"—defined as any company that bought drugs directly from a manufacturer twice in the previous two years.

This meant that the largest wholesalers, such as Cardinal, McKesson, and Amerisource, were not required to maintain pedigrees. On one level this made sense: *Since these authorized distributors almost always bought from manufacturers, a pedigree would be redundant.* But in reality, this allowed the nation's biggest wholesalers to buy from the smallest wholesalers (including the likes of Michael Carlow) with no record of having done so. Even if a drug moved through six different middlemen dealing out of their car trunks before an authorized distributor purchased it, the record of the drug's origin was wiped clean on the presumption that it had been purchased directly from the manufacturer. It was a huge loophole that allowed any big distributor to launder the origin of substandard medicine.

For the next ten years, this guidance letter became the de facto law. With no effective federal pedigree requirement, the frenzy of buying and selling escalated. The trade became even more profitable as drug prices rose throughout the 1990s and phenomenally expensive miracle drugs came on the market to combat new strains of disease. A steady stream of pharmaceuticals traveled back and forth from the gray market of secondary wholesalers to the large reputable distributors, many of the transactions undocumented. The two streams of medicine merged indistinguishably in the ever more efficient warehouses of the nation's biggest wholesalers. And once merged, these companies argued, the streams would be too difficult to separate.

In 1999, eleven years after Congress had passed the Prescription Drug Marketing Act, the FDA finally moved to lift its temporary guidelines and enforce the pedigree regulations. The agency faced a lobbying battle that made the earlier one seem quaint. Many health-care companies had become huge conglomerates with deep pockets and lawyers and lobbyists. Large retail chains such as CVS and Wal-Mart—which opposed the regulations as unreasonable and impractical—were fast displacing the mom-and-pop pharmacies that had supported the pedigree requirements.

The wholesalers also put forward a new argument. If the status quo had worked well for the last eleven years, why change it? In a formal petition, the wholesalers argued, "We are unaware of any significant health or safety concerns" that would justify imposing the pedigree requirements. They warned that the change would "drive thousands of small wholesalers out of business, disrupting the supply of prescription drugs and increasing prices."

They hired lobbyists to arrange high-level meetings. This time, the Clinton administration's Small Business Administration appeared sympathetic to these complaints, writing in a report in 2000 that "authorized wholesalers (even large ones) are not now able to and could not, at any reasonable cost, provide pedigrees to those to whom they distribute drugs."

The FDA again issued a one-year stay of the rule. Over the next four years, the FDA delayed imposing the final rule five separate times as Congress traded charges with the agency over its failure to enact the law's provisions. The wholesalers had succeeded in their do-or-die cause: to derail the pedigree rules.

14. A Bad Lot

May 2002
Grapevine, Texas

TWENTY MILES OUTSIDE DALLAS, ALBERT J. HOKINS, JR., DROVE down an industrial road and stopped before a warehouse complex belonging to a company called Bindley Trading. Despite the sultry heat, Hokins, a drug inspector for the Texas health department, had brought his down jacket so he could stay a while in the refrigerated storage area. At sixty-two, he no longer worried about appearing tough. He'd done that for thirty-two years in the air force, from which he had retired as a full colonel.

A tip from Florida's pharmacy bureau had brought him here to inspect 1,004 boxes of Procrit with an irregular pedigree: The Utah company from which it originated actually operated out of someone's house in Florida. Since the medicine had originated in another state, Hokins had brought an FDA agent with him. The Procrit all bore the same lot number, P002641, according to the Florida inspectors. It was unusual to see such a large amount from the same batch sold to a single buyer.

These were the same drugs that Mark Novosel had sold from his home to a Miami wholesaler, who had sold it to Bindley.

Hokins had not yet been here for a routine inspection because the warehouse had opened less than a month earlier. Cardinal Health Inc. had acquired Bindley, a large medicine wholesaler, with great fanfare a year earlier. Bindley's mission for its new parent was to do what it did best: look for discounted medicine within the secondary market.

Inside, Hokins was surprised and pleased to find that his job had been made easier. The Bindley managers had already sequestered the medicine in question. They led him into a conference room and brought in the drugs—886 boxes—which Hokins photographed with his digital camera. The 118 other boxes had already been sold.

He then had the idea to compare the suspect lot against a good one. "Could you get me a couple of good boxes so I can photograph them both?" he asked the managers, who returned with several boxes from a lot that was not in question: P002384. Side-by-side, the boxes appeared indistinguishable. He photographed them and e-mailed the photos to his boss in Austin, John Gower, with a note explaining that there was no obvious discrepancy in packaging between the good and the suspect lots.

The Bindley managers continued to be helpful, voluntarily packing up some samples of the suspect lot and express-mailing them to investigators in Florida and the FDA to be tested. Gower asked Hokins to retain a sample for the department as a control against what they sent to the FDA.

Gower knew little about Procrit. He was nervous about keeping the sample, especially after Hokins called and said, "You all be careful with that. It costs $500 a bottle." Gower wondered, what if it turned out to be good? How would his budget-strapped department compensate Bindley for the vial?

AFTER AN UNEVENTFUL MEMORIAL DAY WEEKEND, HOKINS WAS at his home office the following week when Gower called from Austin. He had just received the results of Amgen's test of the

Procrit: The suspect lot, P002641, was counterfeit. Hokins was surprised. The vials had looked perfect. With the naked eye, they could not be distinguished from the other lot he had photographed. As with the Epogen in Florida, it appeared that someone had taken low-dose vials and had glued counterfeit high-dose labels onto them.

Hokins had assumed when he saw Bindley's purchasing records that the price the company had paid for the Procrit—$1,718 a box—was the going rate. Now he was beginning to understand that this price was about $50 below the wholesale cost. The company had overlooked a troubling paper trail and jumped at the discount.

A senior manager from Cardinal's headquarters came down to the Texas warehouse to oversee the disaster. He stared at the boxes as they were about to be turned over to the FDA. The small roped-off mountain with taped signs looked every bit like the crime scene it had become. The manager asked, "Who pays for this?"

"You bought it, didn't you?" Hokins told the man. "We do not reimburse for bad business decisions." Disgusted, he could not help but lecture the group about the consequences of their cost savings. "You and your purchasing people need to have a little more integrity," he snapped. The men fell silent.

Hokins and his wife had a close friend with advanced ovarian cancer. She had taken Procrit and it hadn't done a thing for her. Now he was almost certain of the reason why.

She had survived her disease through the skill of her doctors and the power of prayer, he believed. Though a born-again Christian, Hokins began having dark, un-Christian thoughts about what he'd like to do to whoever had counterfeited the Procrit.

———

THE FDA AGENTS WHO SAW HOKINS'S PHOTOGRAPHS OF THE TWO lots side-by-side were struck by the subtle ways in which the good lot, P002384, resembled the bad one. The positioning of the text on both boxes seemed a little off. A small area at the bottom of the boxes appeared unvarnished instead of shiny. They resolved that the good lot needed to be tested too.

Hokins would later say that it was either dumb luck or pre-determination that had led him to request those other Procrit boxes for the purpose of comparison. The sixty-two boxes from lot P002384 turned out to be counterfeit too. In Florida, Arias's supervisor, Gregg Jones, circulated Hokins's photograph of the two boxes in an e-mail titled "Incredible Counterfeit."

It was the perfect crime. The packaging looked exact because it was genuine, from the plastic lids and rubber stoppers to the foil seals over each vial. The medicine also would have passed a first-blush laboratory analysis, since it contained the right active ingredient (just in far lower quantities). If the patient didn't improve or got worse, their disease, not the medicine, would be blamed.

Hokins had flown his B-52 on fifty-three combat missions over Vietnam. He had dropped over two million pounds of bombs. He had done intelligence work in thirty-three countries, including Lebanon, Saudi Arabia, and Turkey. He had seen terrible things that for security reasons he wasn't even allowed to describe. But in his view, he had never seen anything as terrible as this.

BY THE END OF MAY, VIALS OF COUNTERFEIT EPOGEN MARKED AS lot number P002970—the same as the one hundred boxes that Venema had bought undercover and that CVS had mailed to Tim Fagan on Long Island—had moved through Georgia, Kentucky, Michigan, Illinois, Wisconsin, North Carolina,

Ohio, and California. The medicine slipped in and out of warehouses, always seemingly a step ahead of the FDA agents, state inspectors, and others trying to seize it.

As Amgen's warnings circulated, calls came in from pharmacies, hospitals, and patients who had gotten the counterfeits. A hospital in Alton, Illinois, received the counterfeits from a division of AmerisourceBergen in St. Louis. In Kentucky at an Amerisource warehouse, FDA agents seized 1,617 boxes of the Epogen—$6.5 million worth. It was counterfeit too. New names of wholesale companies appeared on these pedigrees: Consumer Services Group in Knoxville, Tennessee, and Premier Medical Distributors in Gainesville, Georgia.

In the former mining town of Iron Mountain, Michigan, Gary Lindeman, the pharmacy manager at Dickinson Memorial Hospital, discovered that his shelves were stocked with Epogen from lot P002970—and that some of it already had been dispensed to patients. Less than three weeks later, another warning from Amgen came across Lindeman's fax, this one about two more lots of Epogen, P001486 and P001091, the counterfeits spreading as inexorably as the diseases of the patients who needed it. Lindeman was amazed. The FDA often warned Americans about the dangers of foreign drugs, yet these counterfeits had come from inside the country.

Several times a day, Lindeman received flyers from companies he had never heard of, some of them in Florida, offering Epogen and other medicine at cut-rate prices. He threw them out almost as soon as they arrived. The hospital only bought from large wholesalers and well-established national buying groups. Now he wondered if even that was safe: He had no way of knowing where those companies got their drugs.

———

FOR YEARS, THE FORMER COPS AND G-MEN WHO WORKED SECU-
rity for the nation's drugmakers had been hidden in back
rooms. Given the manufacturers' reliance on an image of purity, few wanted to acknowledge that they even needed a security department. Amgen, which manufactured Epogen and
Procrit, was typical in this regard. It employed thousands of
people in marketing and sales, but less than two dozen people,
including security specialist Jon Martino, worked in its security department.

Martino, a Los Angeles cop for eight years, had subsequently joined the security detail of an international oil company and for eighteen years traveled the world, arranging
armed security and armored vehicles for executives and as-
sessing threats of kidnapping and assassination.

In 2002, his job at Amgen seemed to involve one cataclysm
after the next. The previous year, someone had counterfeited
the company's Neupogen—a drug that helps cancer patients
fight infection—by substituting the medicine with tap water.
The scheme had been a clumsy one that had been detected
quickly, but not before it had reached patients.

The more insidious Epogen counterfeits compounded
the company's problems, raising the specter that either the
Neupogen counterfeiter was at work again or someone else
had had the same idea but with the means to execute it on a
greater scale. He needed to use all the resources he could
muster to help the government find the counterfeiter and
to intercept the bad medicine before more of it reached
patients.

IT WAS NOT ENOUGH TO LOCK UP BAD GUYS IN FLORIDA, GARY
Venema believed. The Horsemen needed to make "true

believers" of investigators in other states. The need for this would become obvious when the Horsemen tracked some of the counterfeit Epogen to a small wholesaler in Yadkinville, North Carolina, Medi-Plus International, which sat next to a tire shop on Main Street. This turned out to be the new base of the now-defunct drug wholesaler Jemco Medical International, whose secret warehouse they had raided.

In December 2002, Venema, Odin, and Arias went to Yadkinville and found Brian Hill, Jemco's former president, in the one-room office of Medi-Plus International, where he claimed to be a "consultant," not an owner.

Hill's arrival in Yadkinville convinced the investigators that they were displacing crime, but not stopping it. Two other wholesalers being investigated in Florida had also set up storefronts in North Carolina, where only two drug inspectors with little investigative authority covered the entire state. On seeing the Florida investigators, Hill looked alarmed and furious and then ordered them to leave, saying that they had no authority in North Carolina, which was true.

Hill, who had been waiting for a ride, finally left on foot in an effort to shake off his Florida visitors. Hoping to follow him, Venema enlisted the help of two Yadkinville cops lounging in a cruiser outside a car dealership. They weren't too interested in the idea that Hill might be dealing in stolen medicine. They revved their engines to follow him only after Venema explained—in an effort to make them care—that Hill was a homosexual. Nothing came of this odd pursuit.

Given the need for help in other states, Venema concluded that he had to educate others as Arias and Odin had educated him. Believing that secrecy would only slow them down, Venema readily handed off his reports on Michael Carlow, José L. Benitez, Eddie Mor, Carlos Luis, and others to investigators from Delaware to Texas to New Hampshire. His relentlessness paid off. Investigators in other states called regularly with information and questions as they helped track the far-flung companies and medicine shipments.

One New Hampshire investigator checked out an address linked to three pharmaceutical wholesale companies, one of them registered to Michael Carlow's wife, Candace. On his first visit, he found the Nashua, New Hampshire, branch of Accucare to be a dilapidated home with cats roaming everywhere, doors without knobs, and a front yard piled with trash. The woman who lived there claimed to be nothing more than a friend of Candace Carlow's.

On a second visit, the investigator found the house transformed, with new stainless steel appliances, a granite countertop in the kitchen, marble in the bathroom, and a new Ford Expedition parked by a tidy front lawn. As it turned out, the woman there got a percentage of the deals in which Carlow used her New Hampshire address on the pedigree papers.

Every day, Venema took calls with this type of information. He would rattle off his evidence and promise to send files. "That's enema with a V in front of it," he told one investigator. "Don't laugh. If you go to Holland, we're everywhere—I hear." (He had never been anywhere but Canada, once.) At the end of each call he declared that arrests were imminent. "We got something lined up for them—called prison."

To some he became an inspiration. To others he appeared crazy. He had e-mailed Chris Zimmerman at Amerisource-Bergen, urging that his company discontinue business with a Kansas wholesaler that had refused to cooperate with their investigation. He told Zimmerman that he had threatened the company's lawyer "that I would tell everybody on planet earth that he is an asshole and is tainting the entire nation's supply of HIV/cancer drugs—now I am following up on that promise."

"Thought you might be able to pass on to the honest guys, the few that there are anyhow," that the Kansas company is "with the Taliban," he wrote. "I am worried about my own prescriptions now!!!!!!"

He invited Zimmerman to reach him on his cell phone, since he would be out of the office for a week.

Zimmerman wrote back a pale and perfunctory e-mail saying, "Hopefully you will be on vacation." In fact, Venema was headed to the Panhandle for his wedding anniversary, and not a moment too soon. The months of nonstop investigation had taken their toll. His blood pressure was climbing. He was still shaky from a recent incident in which a raccoon the size of his dog had come crashing through a dumpster as he sifted through it in the pre-dawn silence. He also was embroiled in a pissing match with the feds, who suddenly wanted his help.

"GARY, THIS IS SPECIAL AGENT LUIS PEREZ WITH OCI. I HOPE THAT we can work together and help each other out. Can you please get in touch with me?"

Perez, an experienced investigator who formerly worked for the United States Army in its criminal investigation division, had been at the FDA's Miami office for less than a year and was a lead agent in the federal investigations of Epogen and Procrit counterfeiting. He left his phone, cell phone, and e-mail contacts on Venema's answering machine. Perez felt besieged by the pressures of the case and the frenzied second-guessing at Amgen and other drugmakers, who wanted arrests made.

From what the FDA agent could tell, most of the counterfeit medicine now moving in countless tributaries across the country had originated in South Florida. He had no idea who made it or where all of it was now. All he and his colleagues knew for sure was that a Florida cop, Gary Venema, had been telling state investigators around the country that he was about to make arrests. But Venema refused to tell the FDA who he planned to arrest and flatly turned down the agency's proposal to work together. He told several FDA agents in succession that he had no interest in cooperating, no intention of ceding control, would not share any information, and had no need of their help. And he didn't tell them nicely. "A couple were insulted pretty good by him," Perez's boss, Doug Fabel, said. "He used a couple of racial slurs."

In Perez's office, the agents began calling Venema Gary "Enema," because he was always running off at the mouth about something. Fabel finally called his own supervisor in Washington to report that there was a crazy Florida investigator who had become turf-obsessed and he had never seen anything like it.

Venema stared at the wall of his office as he thought about Perez's phone message. Just a year ago, it had been Perez not returning his phone calls on the over-the-counter pharmacy theft case. And that was after Venema gave him everything: photos, charts, reports, undercover deals, informant statements. Perez hadn't even dignified him with a response. Venema resented and distrusted the FDA as a result.

The same qualities that made Venema a relentless investigator also made him tactless at times and an impulsive judge of people. Among his peers, Perez was highly regarded as a diplomat and a gentleman, which Venema would only later appreciate. For now, he leaned forward in his chair and typed furiously, his e-mail message to the Horsemen recapping his assessment of Luis Perez's latest message:

> *Alas, he now wants to work this thing together, and we can help each other out in our respective cases, we each have a lot of info that could help the other bring the cases*

> *forward—onward Christian Soldiers!!!! I am getting an*
> *inner warmth rising from my loins—I feel a real*
> *brotherhood developing—I am floating in a state of*
> *ecstasy—my Federal brothers and I will strike the dark*
> *forces and we together will slay them!*
>   *. . . . Should I call Luis Perez back???????*
>   *I DON'T FUCKING THINK SO!!!!!!!!*
>   *Que Come Mierda!!!!!!!!(Randy—that's Cubano*
> *for consume feces)*

He pressed the send button.

INSIDE THE FDA OCI HEADQUARTERS IN ROCKVILLE, MARYLAND, and its field office in Miami, agents fumed about "the Venema problem." He was no longer just idiosyncratic or stubborn; his refusal to cooperate had become downright dangerous, if not illegal, in their view. Once tainted drugs moved across state lines, the FDA had jurisdiction. Not only had Venema refused to share his findings, but he appeared to be actively circumventing the agency. Instead of reporting possible federal crimes to them, he had been approaching other states' investigators piecemeal.

Those same investigators began to call the FDA, asking about Venema, "Who is this guy?" They explained that Venema had asked for their help, but also warned them not to tell the feds since they might steal or impede the investigation in some way. Amgen officials also called the FDA, reporting strange conversations with Venema that inevitably began with "Make sure you don't tell them." The FDA officials pondered their options, which ranged from asking the Florida governor's office to run interference to threatening Venema with arrest for obstruction.

Jon Martino at Amgen flew to South Florida, as did James Dahl, assistant director of the FDA's office of criminal investigations. As the counterfeiting investigation grew, OCI's Venema problem became more untenable, involving even the agency's director, Terry Vermillion.

Vermillion called down to OCI's special agent in charge in Florida, asking, "What the hell is going on? What are the bad relations?"

Vermillion almost never involved himself in such prosaic details. But in this case, he even offered to fly down to Florida and deal with the matter himself. All the supervising agent could tell him was that they had no idea what was going on.

"I don't even care anymore," Vermillion told him, "so long as it's repaired."

But the FDA continued to get the same answer from Venema: "We don't need you."

"It was just real weird," Vermillion recalled. "I was worried we had done some horrendous thing. I was thinking, there's got to be something, because no one would act this crazy."

As the feds prepared an undercover agent to make a buy of adulterated medicine in Miami, the agents worried that he might end up buying from a state undercover agent. This prospect seemed dangerous, raising the possibility that two opposing undercover agents could wind up in a shootout.

When the FDA agents called Venema to check if he had people undercover they might stumble on, he still refused to answer. This time, the feds determined that if the guy they bought from was with FDLE, they were going to arrest Venema for obstructing a criminal investigation.

In a quiet moment, Martino told Venema that he needed to put aside his personal agenda: "You can't let this happen. Someone is going to get hurt."

Finally, Venema agreed to a meeting, which Michael Mann

set up as more of a duel. He told Perez's boss, Doug Fabel, over the phone, "We'd be happy to look over your files. Why don't you bring what you've got and we'll talk about it." Fabel and his agents, Perez and Chuck Kimmel, showed up with a few old, unrevealing documents, a perfect snapshot of the kind of "partnership" that Mann had been expecting from them. Despite their talk of cooperation, all they had wanted was to "suction" the state investigators' information. And to protect their own information, the Horsemen were not budging, either.

Nonetheless, Fabel made a pitch to combine their teams, even proposing to pay Venema the overtime that FDLE couldn't afford. In turn, Venema explained his reluctance in a typically blunt way: Two years ago, he felt that the FDA had ignored his investigation into over-the-counter pharmacy thefts and he still held a grudge. It was that simple.

Once again, Mann came away with a bad feeling. The Horsemen were out practically living on the streets, pulling trash at 4 A.M. In seven months, they had served twenty-five subpoenas and executed nine search warrants. They didn't need the feds.

Nonetheless, Mann told Venema, "Give them something." So Venema called Perez and offered to share the first letter of the first name of the suspected counterfeiter they were pursuing. Perez also forked over a letter. It was a different letter. Venema tried to stay in touch, but within the month, Perez had stopped returning his calls.

Venema burst into Mann's office, looking crazed. "Those fucking feds! The fucker isn't calling me back. Do we need to work with them?"

"No, fuck them," Mann said with a laugh. "They did exactly what we thought they would do. Nothing."

"I thought we had to work with them," Venema bellowed.

Mann suggested, "You need to take a Xanax."

15. Rats in the State

Late spring 2002
Fort Lauderdale, Florida

FOR A TWENTY-EIGHT-YEAR-OLD WITH MINIMAL EXPERIENCE prosecuting major cases, Stephanie Feldman had remarkable confidence and self-possession. In matters of law enforcement, she had already proved to be far tougher than the Horsemen could have imagined, facing down opponents who made more seasoned public servants balk. If she succeeded, it was not necessarily through political or professional calculation, but because she burned with an inner sense of outrage about the "silent murder," as she put it, that they had begun to expose.

Their first public face-off with the industry they had been investigating came on April 30. Feldman took a 6 A.M. flight to the small airport in Tallahassee, looking all business in a checked suit with a bright silk shirt and pearls. The Horsemen had donned conservative ties and jackets, except for Venema, who had paired a neon purple tie with a blue shirt and wrinkled gray trousers.

Together, they headed to the state health department on Bald Cypress Way for the first public meeting of the Ad Hoc Committee on Pedigree Papers. Though obscurely named, the state committee had been convened to tackle a central problem that the federal government had left unsolved for more than fourteen years of legislative wrangling and FDA inaction. Could the state require complete documentation of a medicine's origin without imposing too onerous a burden on the wholesale industry? Feldman had been invited to speak but had not bothered to bring notes. When she walked into room 301 that morning, she knew exactly what she was going to say.

The conflict over pedigree papers had ignited the previous November after Arias's bureau sent out a two-page memo to every wholesaler in the state, simply reminding them of Florida's regulation that wholesalers pass on a complete pedigree with the medicine they sold. The document, no more than a restatement of the existing law, sent shock waves through the wholesale industry and came to be known as the "November memo."

Previously, the industry had viewed Florida as a permissive place in which regulations were suggestions, not steadfast rules. It was assumed that legislators were open to persuasion. Many of the attorneys in Tallahassee doubled as lobbyists, and the reference guide listing their names was entitled "The Third House of Florida." As State Senator Walter "Skip" Campbell, a Democrat and a self-made millionaire known for his independence, put it, "This is a government for the healthy, wealthy, and good-looking, and if you don't fit in those qualifications, you can chuck it."

Those in government who kept their jobs took a relaxed view of the state's authority. This was certainly true at the Bureau of Statewide Pharmaceutical Services, where Cesar Arias's boss, Jerry Hill, explained, "I don't think government can come in and make changes that are going to have a fiscal impact" on an industry without the industry's cooperation.

Though Hill was no visionary, his defenders viewed him as a realist. He presided over a bureau that had about as much glamour as the engine room of a tugboat and considerably less power. Situated in a small, run-down strip mall in the state capital, the bureau had two floors: The lower level had worn hallways and mostly empty vending machines and a warehouse that stocked medicine for the counties' health programs; the upper floor housed the modest offices of senior management. Those requesting public records from the bureau were shown politely to a stack of sealed boxes and given an X-Acto knife.

Ideas for change did not move far in this environment. In May 2001, Gene Odin had suggested to his Tallahassee bosses that the state restrict the amount of medicine wholesalers could sell sideways to one another, in effect requiring them to straighten the channel through which drugs passed. When the bureau's compliance director sent the idea to Jerry Hill, he wrote back, "Can we do this legally? Restrain[t] of trade??" The idea went no farther.

In Nevada, however, officials facing similar problems did just that: They imposed a regulation that wholesalers must sell 90 percent of their medicine to end users, sparking the very battle over restraint of trade that Hill foresaw. As Nevada officials unveiled their plan, Hill's deputy, Gregg Jones, grew inspired. He e-mailed the senior leadership of his bureau, declaring: "Nevada has tamed the Wild West of Drug Wholesaling. The time for Florida to stand up and be counted among public health protection minded governments is now."

Hill wrote back to Jones that he liked the idea and hoped it "picks up steam." It didn't.

In November 2001, one month after Nevada's bold move, Jerry Hill did take a stand. He sent out the November memo.

The wholesalers responded with an all-out lobbying effort to avert the enforcement of a regulation that had existed for eight years. A series of lawyer's letters and meetings at the pharmaceutical bureau followed. As the industry's resistance escalated, so did the political hazards. In late January, the main trade group for wholesalers, the Healthcare Distribution Management Association, wrote Hill accusing the department of exceeding the scope of the law in demanding that its members supply pedigree papers for any medicine not purchased directly from manufacturers.

The group complained that to separate out and document the origin of any medicine not purchased directly from a manufacturer would require new storage and warehouse systems. "Consequently, each distributor would have to weigh the cost and inefficiencies of complying with such a requirement against the benefits of participating in the secondary distribution market," the group wrote. The two streams of pharmaceuticals, some of which came from manufacturers and some of which came from small wholesalers, were merged and could not be separated without a "devastating effect in terms of cost and efficiency."

The bureau's compliance director, Sandra Stovall, faxed the letter to the health department's lawyer, marking "Urgent!" on the cover sheet.

Within weeks, in an effort to build consensus, the state's health secretary, Dr. John O. Agwunobi, formed the nine-person Ad Hoc Committee made up of wholesalers and regulators. He asked the members to consider all possible ways to prevent diverted and counterfeit medicine from entering the supply chain and to weigh not just how to interpret the current law but to recommend changing it if need be. The members included Sandra Stovall and Gregg Jones of the pharmaceutical bureau, as well as AmerisourceBergen's Chris Zimmerman, the senior director of corporate security and regulatory affairs. It was Zimmerman who, in less than a month, would return the frantic phone call of Long Island father Kevin Fagan demanding to know how his son had received counterfeit Epogen.

———

BY 9:30 A.M. THE CONFERENCE ROOM WAS PACKED, EVERY SEAT taken, for the first public meeting of the Ad Hoc Committee on Pedigree Papers. Feldman and the investigators took their seats. Each sector of the wholesale industry had sent lawyers and lobbyists who now crowded the room.

Salvatore Thomas Ricciardi, the nation's de facto leader of the secondary wholesale industry, had arrived quietly with his trusted and ever-present deputy, Bruce Krichmar. The two men had come from South Florida the night before and had their Tallahassee attorney and lobbyist, Ross McSwain, at their side.

Ricciardi was on the Ad Hoc Committee. He worked in Boca Raton as the chief operating officer of Purity Wholesale Grocers Inc., a conglomerate of twelve companies throughout the country and Puerto Rico that resold prescription medicine, health-care, and beauty aids. While it was technically one of the "small" businesses threatened with extinction by the regulations, *Forbes* magazine had ranked Purity as the 160th-largest private company in the United States, with annual sales of $1.6 billion. A local magazine, *Florida Trend*, listed the Boca Raton company as the fifth-largest in the state (NASCAR being the third).

He was also the director of a trade group he formed in 1999, the Pharmaceutical Distributors Association. Though few had ever seen a list of its members, Ricciardi claimed to informally represent four thousand small prescription drug wholesalers nationwide. The sole purpose of his group had been to quash or delay the federal pedigree rule, even though it required less strict documentation than Florida law. Ricciardi had argued, successfully, in letters to the FDA that the burden of generating comprehensive pedigrees for all medicine would destroy "thousands of small, family-run businesses" and displace "countless employees."

Ricciardi now took his seat at the horseshoe-shaped table.

He had brought a detailed PowerPoint presentation showing charts of the nation's drug distribution system, to explain why any further regulation of his industry was unnecessary and could have a devastating effect not only on the jobs of his employees but also on the patients who needed his medicine. Venema could not help but notice his suit, which looked very expensive, and an elegant watch that flashed from beneath a sleeve.

THE PROCEEDINGS BEGAN SLOWLY WITH A DEFINITION OF PEDIgree papers and a restatement of the conflict over them. And then it was Feldman's turn. She stood.

"I am here to inform this industry that you have a problem," she began. "You have a rat in the state. You have many rats."

Instantly, those in the room tempted to doze became alert. There was an audible stirring. The rats, she explained, were "dirty wholesalers" who diverted and adulterated medicine for profit, harmed patients, and drained taxpayer dollars. They also manufactured pedigree papers, which remained crucial whether phony or not, because they left a critical footprint for investigators to follow.

The paper trail could not be eliminated, Feldman insisted, and if anything needed to be stricter. At the very least, stricter regulations would force dirty wholesalers to "be creative and invent a company name or some explanation" for the circuitous routes of the drugs they sold.

The movement and noise in the room sounded like a storm building. "Mr. Zimmerman is shaking his head," she said, looking directly at the Amerisource executive, "and that's because he knows in the past two weeks, his company has experienced a $3.3 million recall because of a dirty secondary wholesaler." She was referring to some of the counterfeit Epogen that the company had recently recalled. Murmurs shot through the room. Ross McSwain thought to himself, *This is a trial lawyer.*

"Does anyone here take a prescription?" Feldman continued, her tone fiery and preacherlike, tears sparkling in her eyes. She held up a vial of the counterfeit Epogen. "A person suffering from cancer is being treated with hemophiliac-strength drugs. That person will die if being treated solely with this drug and the doctor administering this drug will never know why his patient is so sick."

She closed by saying, "Keep the pedigree papers and strengthen them in any way that you can."

The room erupted, one man visibly red-faced and furious. The members of the pharmacy board looked on the brink of applauding. Arias marveled that with no notes, Feldman had said it all.

Though Ricciardi would follow the prosecutor with the smooth PowerPoint presentation that helped him back the United States Congress into a corner, Feldman had already trumped him with her direct emotional plea.

After meeting over seven months, a divided committee would recommend requiring full pedigrees for thirty of the drugs most vulnerable to counterfeiting, tougher standards for licensing wholesalers, and increased criminal penalties for violating the state's health laws.

SALVATORE RICCIARDI WAS NOT USED TO LOSING. TALL, DARK, AND stocky with a glowering air, he had a thick neck and bristly crew cut that brought to mind a bulldog. In lobbying against the less strict federal pedigree requirements, he had drawn fifteen senators and congressmen to his side and launched an effective whispering campaign in Congress to set aside the law.

Over the years, Ricciardi had not just argued against regulations. He had built a legal, economic, and intellectual framework in support of what he called "parallel" markets. In a footnoted ten-page letter to the Federal Trade Commission, he explained that secondary distributors do not knowingly traffic in counterfeit or stolen goods and that so-called diversion was nothing more than taking advantage of fluctuating prices and supply-and-demand and, in turn, giving consumers the opportunity to purchase "first-quality branded products at discount prices."

In letters and testimony before congressional committees, he also decried what he claimed was "inappropriate use of the intellectual property laws to restrain or otherwise inhibit legitimate competition in vibrant downstream markets."

Those who fought the surprising clout of Ricciardi's group became skeptical of its claims and viewed the rhetoric of the secondary wholesalers as a smokescreen to conceal illegal trading. "I'll show you some vibrant downstream markets here on the streets of D.C. about 10 P.M.," said Donald deKieffer, a Washington lawyer whose firm, deKieffer & Horgan, specialized in international trade law and represented pharmaceutical companies.

Ricciardi had honed his arguments since the mid-1990s, when he lobbied against federal legislation that would have restricted middlemen from removing manufacturers' product codes and other identifying marks from consumer goods. The manufacturers sought to protect their products with inviolable tags. From behind the shield of a group called the American Free Trade Association, which also did not disclose its members, Ricciardi and others succeeded in defeating this anti-tampering legislation. Ricciardi had argued that a vote for the legislation was a vote against the consumer.

John Bliss, a Washington lawyer who had represented the

manufacturers during that fight, had come to view the secondary market as dominated by diverters who trafficked in stolen and illegally acquired goods. "Basically we were dealing with crooks," he said. "It was an odd and unsavory experience to have all these people in my law firm's conference room."

As the battle over the legislation heated up, he recalled, "I got some gruff-sounding calls at my home saying, 'You have no idea what you're messing with.' Some folks involved in this have alarms on their houses and unlisted phone numbers," he said of his colleagues who got similar phone calls.

The presence of middlemen may indeed have created more competition and brought down prices, allowing consumers to buy cheaper razor blades and blue jeans. Yet prescription medicine was another matter altogether—one in which the potential harm to patients should have trumped all other concerns. However, here Ricciardi had succeeded in arguing that the industry already was heavily regulated and faced all kinds of record-keeping requirements beyond that of the pedigree. In August 2001, Senator Charles Schumer from New York was among a number of legislators who wrote to the FDA on the wholesalers' behalf, arguing, "The regulations in question took eleven years to finalize and the effective date has already been extended twice. It is highly unlikely that a further extension would pose a health or safety risk to the public."

The FDA bowed to the cumulative pressure, granting a year-long stay of the rule and calling for further review. The agency would do so on five separate occasions.

AT THIRTY-SEVEN, DR. JOHN O. AGWUNOBI, FLORIDA'S SECRETARY of health, was a rising star in the backwater and sometimes boorish culture of the state capital. Composed and elegant, with dual Scottish and Nigerian heritage, he came from a long

line of distinguished British-trained physicians. In addition to
his M.D., he held an M.B.A. from Georgetown University in
Washington D.C., and prided himself on team building.

Appointed in 2001 by Florida governor Jeb Bush, Agwunobi had assumed his post precisely at a moment when ter-
rorist threats and potential breaches of homeland security
loomed largest. Less than a week after Agwunobi became
health secretary, an employee at a tabloid newspaper in Boca
Raton died from inhaling anthrax, the first victim of the nation's mysterious bio-terror attack. Agwunobi weathered that
crisis, only to learn that the state's poorly secured drug supply could become a vector for even greater catastrophe.

Directly after the pedigree meeting, Feldman, the Horsemen, Michael Mann, Jerry Hill, and several others trooped
into Agwunobi's small ground floor-office for a meeting that
Feldman had requested. She wanted to personally advocate
for enforcing the pedigree requirements and changing weak
state laws. As Agwunobi listened attentively, she and Venema
described in detail the scope of the problem they had unearthed, showing him a vial of real Epogen next to one of the
counterfeits they had seized, the two impossible to distinguish
from one another. Venema also could not resist explaining that
the diverters drove new Mercedes-Benzes while his own used
car had hundreds of thousands of miles on it. Mann smiled:
Each time Venema told this story, he cranked up the mileage
on his car by a few thousand as a way of making his point.

"We have to do what is right for the public," said Agwunobi intently.

Certain that Arias and Odin's bosses had never relayed the
inspectors' concerns up the chain of command, Feldman persisted, "These problems did not occur overnight."

"What I really want to know is how things got to this
point," said Mann.

Agwunobi looked coolly at Jerry Hill without bothering to address him by name. "For that, we'll have to ask my colleague across the table."

Hill looked flustered, as those in the meeting recalled.

IN THE DAYS FOLLOWING THE TALLAHASSEE MEETING, FELDMAN's phone rang continuously. News of her speech had traveled quickly, and managers from the nation's large and regional wholesalers called her for information: Which were the bad companies and which the safe ones to buy from? Though she could not answer their questions—that would have meant revealing who was being investigated—she urged them to scrutinize the pedigree papers that came with the medicine they were buying: Was the information in the documents true? Were the companies they bought from actually licensed? Were the prices suspiciously low?

The state's top defense lawyers were calling too. Some already had been hired by the wholesalers targeted in the investigation. Others were looking for work and sensed there might be a lot forthcoming. Feldman knew that something big was happening: Too many people were interested, and too much money was being spent to battle her.

One call came from Feldman's boss in Tallahassee, statewide prosecutor Melanie Ann Hines, who had already received angry phone calls, one from a lawyer in Washington, D.C. "Just want to know what you're doing down there, kid," she said. "I have to answer to people."

FIVE TIMES A DAY FOR FOURTEEN YEARS, FELDMAN HAD INJECTED herself with lifesaving insulin, rotating from arms to legs to stomach and back again. Together, she and her family had shared the complex daily work of keeping her alive. In early June, her father, a nursing-home owner and practiced patient himself, noticed a spot on her leg that appeared to have changed in color and size. He immediately sent her to a dermatologist, who diagnosed the spot as stage-one skin cancer.

Suddenly, Feldman went from fighting to protect cancer patients from bad medicine to being a cancer patient. In the oncology waiting room, she studied the drawn, anxious faces of the patients, some of whom would survive and some of whom wouldn't.

Sunny posters on the walls—all starting with the words "I am grateful for . . ."—reminded them of why they should want to live.

Within forty-eight hours of her diagnosis, her doctor scheduled surgery, every aspect of her care complicated by her diabetes.

In her weekly meeting with the Horsemen, she tried to be low-key. "I'm going to be out of the office for a week because I need to have surgery," she told them quietly. "I've been diagnosed with skin cancer."

Venema and the others were stunned. Odin could not help but be impressed by her courage. Arias sensed fear beneath her calm demeanor. Venema, never one to spend time parsing the emotions of a given moment, felt emotional himself. The next day, Odin drove from Boynton Beach back to Fort Lauderdale to present the young prosecutor with an important gift—a Kabbala bracelet made of multicolored strings and glass beads for protection against evil and bad luck. "I know the work you're doing is going to keep the evil eye off you," said Odin shyly, "but just in case."

Feldman immediately put on the bracelet, the gift giving her some sense of relief. In part, the Horsemen had become *her* talisman and magical protection, and now she wore them on her wrist.

When Feldman awoke from surgery shortly before Father's Day in late June, her leg bound in gauze and her father and stepmother standing above her, the first phone call she got was from Venema. "When you coming back to work?" he asked. On hearing that the operation had gone well, his relief was palpable. No longer afraid of the answer, his second question was, "So, how ya' doing, kid?"

SOMETIMES IN HER RELATIONSHIP WITH THE INVESTIGATORS, Feldman felt like Dorothy in *The Wizard of Oz.* The men had become her second family and she didn't want to leave them or their mission. But for months she had been commuting to see her fiancé, who had taken a job in Atlanta as a hotel manager. Although they had known each other since she was twelve, they had become romantically involved only last fall and wanted to start their lives together. A job offer at Emory University Law School as a legal writing instructor had come through, and Feldman had a growing desire to end the couple's week-long separations.

She broke the news to Venema first over his walkie-talkie as he was barreling down a highway and ranting about daily developments in the case. He ignored her.

"You know I'm leaving the investigation," she repeated as Venema rattled on.

"Did you hear me?"

"I heard you," said Venema coolly. "You are not leaving. That is bullshit."

The men struggled to understand her decision. Odin conjectured that the melanoma "made her realize that she was mortal, which made her grasp at that chance to get married."

Petri concluded, "Love is blind."

There was no doubt in the men's minds that Feldman had started a revolution. She had fired a rocket-propelled grenade into an industry used to flouting the rules. At a good-bye luncheon, Petri stood and made a speech on the men's behalf, saying that in his twenty-seven years in law enforcement, he had never worked with a better prosecutor. Though Venema had been unable to come, he had written a letter to Feldman that Petri read aloud.

Jones, who shunned official events, showed up wearing a suit. He had spent hours preparing joke props for Feldman's new life as a professor, and these included a laminated ID card that said X PROSECUTOR. He sealed each in an evidence bag. They had also made her a beautiful cherry wood plaque with a gold gavel bearing their names and the words TO STEPH, THE BEST PROSECUTOR, LUCK ALWAYS.

Feldman wept. Odin and Arias became teary. Jones, perhaps the toughest among them, looked away so that he wouldn't cry.

Afterward, they missed her sorely. She had moved the unwieldy machinery of government that they each, separately, had been straining against for years without success. In the months to come, nothing pleased them more than remembering her public takedown in Tallahassee of men who felt a bit burdened by the law. "Five-foot nothing and she leveled the room!" Venema would hoot.

Part
Three

Part
Three

16. Crazy Money

2002
Goulds, Florida

IN 1992, HURRICANE ANDREW DEVASTATED MIAMI'S SOUTHERN-
most tip, leaving behind a featureless landscape without trees,
houses, or even street signs. Those lucky enough to track their
way back to the remnants of their homes set up guard with
shotguns and handwritten signs saying, "You loot, I shoot."

What remained became a wild frontier. Ten years later,
after the crooked plumbers, contractors, and electricians had
moved on, the area of Goulds still looked deserted. Bulldoz-
ers toiled alongside a desolate highway dotted with auto repair
shops, warehouses, and nude bars. The entire area had a root-
less, even menacing feel, and Playpen South fit in perfectly.

The strip club's pink canopy, with sketches of naked girls,
boasted of "Nude Revue" and cheap beer and tequila shots.
No bouncer stood outside and there was no cover charge, the
freestanding building on the side of the highway as naked as
the employees within.

Inside, a mirrored hall led to a central room about the size
of a small family restaurant. A disc jockey blared rock music
from behind a wall of opaque glass blocks. Since the ceiling was
too low to accommodate a stage, the center of the room was a
well with a single brass pole running from floor to ceiling.

The girls who danced here would likely not have been hired at any upscale strip club. Some of the dancers were fat. One, so thin and pale that she appeared to be a drug addict, was missing front teeth. A couple of others were beautiful, though they looked underage. On a typical night, the men who watched them included migrant farm workers, out-of-place college students, and a few dangerous-looking business types.

In between performances, the dancers marched among the audience. They performed lap dances and collected dollar bills in their garters. Some of the patrons looked like they had little money to spare.

The cash register at the bar opened and closed sporadically. A greasy grill kitchen along one wall served an occasional meal. The business didn't seem to take in much money. The strip club's most lucrative business took place from behind a series of gray doors marked in red: AUTHORIZED PERSONNEL ONLY.

NICHOLAS JUST AND DR. PAUL PERITO OPENED PLAYPEN SOUTH in September 2001, paying $600,000 in cash to buy the business. The men seemed unlikely partners. Perito, a prominent urologist, lived behind an ornate gate on a leafy, expensive street in Coconut Grove. At his thriving practice nearby, he treated well-to-do patients. He had attended medical school in Maryland and completed his residency at Jackson Memorial. A small-time philanthropist, he raised thousands for AIDS orphans in Africa by organizing cocktail receptions at chic Miami restaurants. His friends, a group that included models and jet-setters, brought their checkbooks and contributed to his cause. A strikingly handsome divorcé, the forty-two-year-old doctor appeared to live a bachelor's dream life.

Before he and Perito went into business together, Nicholas Just, at six-foot-three and 280 pounds, worked as a bouncer at another strip club, where he earned $12,500 a year and had a reputation for violence. The purchase of Playpen South fulfilled his longstanding dream to own a strip club, a lawyer for Just later said. The police came to suspect him of burning the car he bought for his girlfriend, a dancer at the club, after she threatened to leave him.

The doctor and the bouncer shared certain preoccupations. Both men abused prescription medicine. Just took Lortabs, a narcotic painkiller that contained hydrocodone, and was sometimes seen popping dozens at a time (two tablets being the recommended dose). Perito snorted Ketamine, or Special K, an animal tranquilizer that can cause hallucinations. On weekends, Perito could be found at Club Space in Miami with catatonic female companions to whom he'd given the drug.

That Perito and Just were out of control seemed evident to those who worked at the Playpen. A bartender there, scheduled to have Perito perform a medical procedure on him the morning after one of the doctor's binges, claimed that he chose to undergo it without anesthesia because he wanted to be alert throughout. The club manager, Benjamin R. Ojeda, a convicted cocaine distributor who had spent four years in prison, began taking notes on the men, in case one day he'd be forced to answer someone's questions about their activities.

Ojeda claimed that he did not want to be accused of partaking in a conspiracy and that Playpen South was used by Perito and Just to conceal their "crazy money"—the profits from their illicit activities. They sold cancer medicine from an oversized cooler in one of the club's back rooms, he said.

———

CARLOS LORENZO LUIS WAS A REGULAR PATRON. EVEN AT lunchtime, the fifty-four-year-old businessman regularly drove down to the small, dingy strip club from his medical supply company in South Miami. Well dressed with a fringe of silver hair around his tanned bald head, Luis had a compulsive interest in naked women. But he mostly went to Playpen South to buy medicine from Just and Perito.

The Horsemen had begun to watch Luis because a pharmaceutical wholesale company he owned, Medix International, was listed on the pedigree papers of the one hundred boxes of counterfeit Epogen that Venema had bought undercover. On paper, Luis's company operated in Houston, Texas. In reality, Luis ran the company from his medical supply business in Miami and his home.

Gary Venema suspected Luis of being the counterfeiter, because the money trail on the bogus Epogen had begun with Medix International. The medical supply salesman had no criminal record, though he had been arrested in the early 1980s in a cocaine case that got dropped.

Three months earlier, Cesar Arias had gone to Luis's Miami business, Medical Support Systems, to do a regulatory inspection. There, Luis told him he dealt only in latex gloves and other medical supplies. Nonetheless, in a back room Arias found $100,000 in cancer, HIV, and anti-psychotic medicines, as well as blood products, all prescription medicine that Luis had no license to sell. Arias seized all of it. The Horsemen returned with a search warrant a few weeks later and seized more medicine, as well as lighter fluid and Goo Gone paint remover—the diverter's stock-in-trade. They also found evidence that Luis owned more wholesale companies and was applying for licenses in North Carolina and Tennessee.

That same day, investigators searched his home and found hundreds of drug samples stored in garbage bags piled in the hall closet, medicine boxes on the family table, and heaps of drugs in the garage and on the bedroom dresser. This did not include his wife's cancer medicine. She opened the door to the Horsemen, clearly ill, her head almost bald from ongoing chemotherapy. She kept her Neupogen in the refrigerator, where it belonged.

Later, Luis's civil defense lawyer, Dave Ryan, said FDLE had taken "obnoxious liberties" with the facts and the truth, with Venema even calling Luis a murderer to his face. "They are like wanna-be federal agents," he complained.

Shortly after that search, Arias and a Medicaid fraud investigator questioned Perito at his medical office. The doctor explained that Luis was a loyal customer at the Playpen. He acknowledged that Luis's company had paid him more than $600,000, but said it was an investment in a health club he was starting. The Horsemen suspected the payment was for medicine.

As the Horsemen learned, huge amounts of cash got counted and recounted in the Playpen's back room. On one occasion, Just asked several of the staff, including a female bartender he liked, to count out $82,000 to make sure it was all there. He paid the woman $500 for her help. Another time, a boat builder associated with Luis dropped off $100,000 and left with a cooler of medicine. A Texas man and a doctor in Virginia also bought medicine from Perito and Just after their source—a man they knew as "Tony"—dropped it off at the club.

ON APRIL 2 AT AROUND 7 P.M., LUIS AND JUST MET IN THE Playpen's back room and confronted their first real sign of trouble. A customer of Luis's was displeased. The man had returned an order of Procrit, claiming that the product was bad. And so Luis had brought it back to the strip club where he'd bought it.

News of counterfeit Procrit was spreading, and no one wanted to be caught with it. So now as the bar manager Ojeda watched, Luis and Just turned off the lights in the room, huddled together, and gazed at the boxes of medicine beneath an ultraviolet light. Behind them, $219,000 in cash was stacked on a table, the deal of the evening put on hold as they studied the details of the packaging. Some of the club's employees, who were also in the room, could not be sure whether the men were examining the boxes to detect if they were counterfeit or to see whether they were good-enough counterfeits to resell.

Just reassured Luis that he would make sure the boxes were better in the future. That night, Just exulted, "I love making $100,000 per week." Perito, too, was netting this much. Ojeda—who had initially thought the men were selling steroids to muscle-heads—also was struck by a phrase the men used to describe the medicine: "colored water."

MICHAEL CARLOW—WHOSE CHARMED LIFE CONTINUED SEEMingly without interruption—had a tangential relationship with the men at Playpen South. He didn't know them and didn't need to. Some of his suppliers did, and they bought from them. Consequently, investigators found, some of the drugs that poured from Playpen South moved through Carlow's shell companies. Carlow and others like them made their money by trafficking in diverted, adulterated, and counterfeit medicine of the kind that Perito and Just sold. If it was cheap, they bought it. The medicine sold by Perito and Just was dirt cheap, less than a quarter of the retail price, and appeared authentic enough to resell.

Carlow and his associates had become intoxicated by their income. He and his wife were living in "La La land," as Mark Novosel said, planning for ever-greater expenditures, even at times when they couldn't meet payroll because of their astronomical overhead. They continued to throw lavish parties, including inviting two hundred guests to Carlow's Hawaiian-themed birthday bash with organized hula dancing. They ordered expensive luxury goods from mail-order catalogs and paid an architect to draw up blueprints for an even larger mansion they imagined building one day soon. Carlow hired a private guitar teacher to come to the house. "They were a legend in their own minds," said Novosel, "like the Beverly Hillbillies."

BY MAY 2002, DESPITE THE HORSEMEN'S BEST EFFORTS, CARlow's shell companies operated in six states. With some of his closest associates, Carlow likened himself to the "Teflon Don," certain that nothing and nobody could stop him from doing what he did best.

While Carlow carelessly tossed sensitive documents into the garbage, others around him exercised greater caution. From her impeccable home in Lake Worth, where she served as the bookkeeper for her son-in-law's empire, Marilyn Atkins shredded the most sensitive documents, reducing them to white confetti. Before doing this, however, she faxed each document to Carlow for his review. He blithely threw the paperwork in the trash, unwittingly allowing the Horsemen to retrieve his business records.

As the investigators unmasked his shell companies—including BTC, Accucare, and El Paso Pharmaceuticals—Carlow opened new ones with barely a break in his step. He established two new businesses in Maryland—G&K Pharma

and Complete Wholesale—their offices next door to one another in Odenton and empty much of the time.

The more successful Carlow became, the farther his tainted
medicine reached. In early May, a pharmacist in Phoenix, Arizona, was unpacking newly delivered medicine when she discovered a patient dispensing label on a box of AIDS medicine.
The box, which showed obvious signs of wear, had been dispensed three months earlier to a patient at a Fort Lauderdale
pharmacy. After Arizona authorities called Florida, the Horsemen tracked the box to one of Carlow's shell companies.

AS CARLOW'S WEALTH GREW, HIS CONTACTS EXTENDED EVEN
into the same law enforcement agencies that were supposed to
be watching him. Robert Chaille, a lieutenant in the Miami-
Dade Police Department, was like-minded in his pursuit of
medicine. Chaille had a wild streak and a taste for the good
life. The stocky, brown-eyed fitness buff went regularly to the
islands with friends, bringing back photos of booze-soaked
junkets on yachts and entertainment by strippers. He brazenly
showed these to his colleagues.

"That was his lifestyle," recalled his supervisor, Captain
Alan Mandelbloom, who declined invitations from Chaille to
accompany him. "Policing was just a hobby. I used to tell him,
'This should be your primary focus.'" But it wasn't and never
would be. Instead he focused on parties and friends, pulling
some in the department into his orbit, and bragging that he
could get them any kind of prescription medicine they wanted.

For years his supervisors had lauded Chaille in evaluations as self-motivated, well-groomed, and punctual. They
described him as a problem solver, as well as funny, likable,
and persuasive. In an essay Chaille wrote in his 1984 applica-

tion to the force, he described policing as a "personal challenge," adding, "The job offers many attractive benefits such as good pay, retirement and security that are important to a young man starting a lifetime career."

By 1999, he had lost his interest in the job, and his supervisors noticed the change. His evaluations began to describe a loss of initiative, longer lunch breaks, late arrivals, and unexplained absences. Captain Mandelbloom wrote in one evaluation of "outside interests that sometimes detract from his duties." Chaille wrote back in his rebuttal: "I feel that my personal activities have no bearing on my dedication to my profession and the department."

Chaille's increasingly bad behavior culminated in the early evening of January 31, 2000, when he flew a private helicopter onto the roof of the police station in Kendall. He landed it without notice or approval or being qualified by the department as a pilot. He had two friends flying with him. The department disciplined him. While he apologized, he did not repent.

In October 2002, Chaille did something even wilder, unfathomable to almost any police officer. After eighteen years on the job, just seven years away from his scheduled retirement and a lifetime pension, he submitted his resignation. Word whipped around the department that he had come into a vast inheritance, a rumor that Chaille fed.

In fact, he was leaving to become a drug wholesaler. The day after his resignation became official, Chaille opened a pharmaceutical wholesale company in Wilmington, Delaware, called New Horizons Network.

As Gary Venema became an increasing irritant, Carlow explained to associates that the cop had a "hard-on" for him, and they should sell their medicine in states other than Florida if at all possible. According to these associates, he asked Chaille if he could place Venema under surveillance, thereby turning the tables. But Chaille, a veteran of surveillance techniques, explained that a rolling stakeout of the FDLE agent could require upward of a dozen people in order to avoid detection. It was an investment that Carlow declined to make.

CARLOW AND VENEMA HAD SOMETHING IN COMMON: EACH wanted to watch over his adversaries. The Florida Department of Law Enforcement had a tight budget and was unwilling to commit more money to help Venema track a growing number of suspects. So he cobbled together an army of cop buddies who volunteered to drive by the homes of suspected bad guys, call in tips, and scan the streets for new associates and license plate numbers.

While this system expanded Venema's reach, it also greatly increased the hours he worked, as he inevitably pursued the promising tips that came in from the volunteers, regardless of the hour. One night, a cop friend raised Venema on his walkie-talkie to report that a moving truck had pulled into the driveway of a suspected Carlow confederate who lived in the same gated community as the cop.

"Thanks, buddy," Venema barked. While racing down I-75, weaving across lanes, Venema ticked off the suspect's fancy assets. The $60,000 Hummer. The Porsche convertible. A new waterfront home south of Miami. Meanwhile, Venema had bought his used Ford Focus for $7,000. "I'm not complaining," he muttered. "But it frosts my ass."

At the Isles of Miramar, the suspect's gated community, Venema flipped open his badge for the security officer, then drove past the waterfall sculpture at the entrance and into a nearby driveway so he could examine the moving truck. Taking in the entire scene, he concluded that the suspect was not fleeing, but was simply relocating to his even fancier new home.

Since it was Thursday at 11 P.M.—garbage night in Hialeah—Venema gunned his truck toward the home of another suspect who lived there. He had headed out without his rag-picking outfit of tattered shorts, a T-shirt, and latex gloves. Venema drove through a web of side streets until he spotted a modest house with a truck from a tire company parked outside. That was his man. Swiftly, Venema pulled into a neighbor's driveway, left the truck running, loped across the street, grabbed the garbage bag between two fingers, ran back, dropped it quietly in his truck bed, and was off, before anyone appeared to notice.

Sometimes he returned the bag but tonight decided against it. Instead, he circled the back roads before pulling up next to a Dumpster outside an all-night pharmacy, where he tore a hole in the side of the bag and began to sift amid the old food.

Out came a $3,000 cell-phone bill that could prove valuable. He could subpoena the records for that number. The find also added to the database of phone numbers he was building to analyze who was calling whom. Little else in the bag interested him. Fishing out an invitation to a car showing from a Lexus dealership, he declared, "They didn't send me one."

17. A Special Price

June 2002
Miami, Florida

WHEN ROBERT PENEZIC WAS PUT IN CHARGE OF OPERATION Stone Cold, he had worked as a prosecutor for only three weeks. Before that, the thirty-seven-year-old lawyer spent six years as a public defender and as a defense attorney in his own practice. With his soft eyes, off-kilter smile, and haircut that looked like he did it himself, Penezic seemed every bit like the idealist he was. He had such a self-effacing manner and high regard for the truth that some thought him a very awkward lawyer indeed. But his longing for justice, his new fatherhood, and his willingness to work long hours made him a strong replacement for Stephanie Feldman.

Feldman did her best to prepare Penezic for his first meeting with the Horsemen.

"They're not going to like you so much," she warned, "but I told them you'll protect them."

Venema in particular glared at Penezic with crossed arms, prepared to destroy anyone who got in his way. The words *former defense attorney* did not help either. In Venema's Manichean worldview, once an enemy, always an enemy.

As Penezic listened to them describe the case, he found Arias to be cordial. In his typically honest way, Penezic told them, "You just described a problem. I'll prosecute it, but I have to tell you that I have no idea what you're talking about."

Warming to his favorite project, Arias said, "We'll make you understand it."

Odin and Arias's new student made rapid progress. "Do you mean that any time I go into a pharmacy I can't know where my medicine has been or whether it's safe?" Penezic asked them one day, the light bulb going on. That question told them that he understood, shared their outrage, and was ready to be of service as the Horsemen went after a pharmacy they suspected of being the main conduit for the Epogen used in the counterfeiting.

CESAR ARIAS FIRST NOTICED THE PHARMACY AS HE REVIEWED data in his corner office, bare except for a yellowing map of Florida on the wall. AmerisourceBergen had sent a list of all its customers in South Florida who had bought low-dose Epogen, the raw material used in the counterfeiting scheme. The printed orders showed that a pharmacy in Miami, J&M Pharmacare, bought far more than anyone else. It had ordered two hundred boxes of low-dose Epogen at least once a week for about a year. Every so often the pharmacy also purchased a single box of the costly high-dose Epogen.

J&M was a small walk-in pharmacy run by Jesús Benitez and Maria Castro, their first initials forming the business's name. Located in a building of doctors' offices off a busy thoroughfare, the pharmacy, from its small reception area to its shelf of sundries, was entirely unremarkable—except for the size of its Epogen orders, enough to stock a chain of oncology hospitals.

Why did a hole-in-the-wall pharmacy with no particular

cancer specialty order 3,363 boxes of cancer medicine? Arias believed there could be only one reason: J&M was illegally "back-door-ing" it.

WITH JUST A FEW CUSTOMERS WAITING, THE SMALL PHARMACY at Ninety-seventh Avenue already seemed crowded as Arias approached the counter with Jack Calvar, an investigator for the State Attorney General's Medicaid Fraud Control Unit. Calvar, a towering, bald man who smoked menthol Kools and resembled a Latino Telly Savalas, presented his badge to an elderly man behind the counter. He told the inspectors that owners Jesús Benitez and Maria Castro were vacationing in North Carolina.

The pharmacist on duty explained that the owners handled all the Epogen dispensing. Arias asked for the dispensing logs and received records for three dates from January and February 2002. The logs listed five patients and indicated that all of them had paid for the medicine in cash, shelling out thousands.

As Calvar would discover later that day, three of the patients were enrolled in Medicaid, the government health insurance program for the poor. Calvar had spent twenty-three years as a cop, eleven in homicide, and just one as a Medicaid fraud investigator. But all he needed was logic to know that such poor patients were unlikely to pay cash for their medicine, particularly when they had government insurance covering it.

Typically, a corrupt pharmacy acting as a Medicaid mill might try to lure as many poor patients as possible—in order to bill Medicaid for their medicine, whether the patients needed it or not. What Calvar could not figure out was why any pharmacy would dispense medicine to Medicaid patients and then circumvent the government program altogether.

FRAIL AND SICK AND WEIGHING LITTLE MORE THAN ONE HUN-dred pounds, a young woman opened her front door and looked quizzically at her two visitors, her husband standing behind her. Arias and Calvar stood outside the humble but tidy Liberty City apartment in a dingy complex and explained that they had come to check whether the woman had bought a two-month supply of Epogen, as J&M's dispensing logs had stated.

"Epo-what?" she asked. She had never heard of the medicine and had no idea what it was for. She explained that she took no injectable medicine, only pills for the HIV infection that had weakened her, and that she had never heard of J&M Pharmacare. Usually she just went to the Walgreens around the corner to fill her prescriptions.

"The pharmacy's records show that you paid $2,078 in cash for the medicine," Calvar said, watching her closely.

She looked aghast. "I don't have nowhere near that kind of cash," she exclaimed. One look at her modest apartment demonstrated the truth of that.

"How did those people get my name?" she asked, bewildered and concerned. Arias and Calvar had no answer for her.

Not a single person listed by J&M as an Epogen patient actually took that medicine. Some had gone to the J&M pharmacy in the past. Others had not. In each case, the cash payment attributed to them was far beyond their means. One elderly patient lived in a single-room efficiency in a dilapidated building in Homestead, a television set his only valuable. The government programs of Medicaid and Medicare paid for his prescriptions. "I don't have $2,000 to pay for medicine or anything else," he told the investigators, explaining that even a TV dinner was beyond his reach.

Less than a week after their first visit to J&M, the inspectors returned to the pharmacy for a discussion with Maria Castro, who was back from North Carolina. "Why has your pharmacy bought so much Epogen 2,000 U/mL in the past year?" Arias asked.

"We have a lot of patients from Venezuela who bring their prescriptions to the pharmacy," Castro said, appearing unconcerned.

"Why would a patient [on Medicaid] pay cash for her prescription?" asked Calvar.

The color drained from Castro's cheeks. She appeared ashen and began to stammer. As they showed her documents indicating that other patients also were enrolled in Medicaid and would not have paid cash, her hands trembled.

"My partner, Jesús Benitez, handles that part of the business," she said. As her words hung between them and the inspectors remained silent, Castro added softly, "I have to contact my attorney and ask him for legal advice."

"This stuff is getting counterfeited," Calvar said. "If anyone dies because they took it, I'm coming back here and arresting you for murder."

Outside, Petri and Jones watched the pharmacy. Within an hour, the two detectives saw Benitez's black Audi shoot out of the parking lot with Castro in the passenger seat. The two headed to the office of their attorney. Calvar later reflected that he had interviewed murderers who did a better job of staying cool.

In the days that followed, the picture of J&M's activities grew. The investigators learned that the pharmacy had also bought 8,931 boxes of low-dose Epogen from Cardinal Health.

The total amount that J&M bought constituted more than half of the low-dose Epogen that Amerisource and Cardinal had sold in the last year in Florida and Georgia combined. The pharmacy had purchased more than 125,000 vials of the low-dose medicine and had not actually dispensed it to a single patient. If J&M had indeed been the conduit for a massive counterfeiting effort, medicine worth $2.8 million had been transformed into $50 million worth—enough medicine to mistreat 31,000 patients for an entire month.

The seizure of 13,000 vials of the counterfeits just weeks earlier in Texas, Michigan, Illinois, and Florida—which investigators had viewed as a triumph—had removed just a fraction of the counterfeits out there. Another 112,000 vials, a month's supply for 28,000 cancer patients, had vanished into the distribution system.

In the months that followed, this revelation ended up drawing the Horsemen and the feds together. Venema realized that there was so much bad medicine out there that no one team could track it down alone. He had come to see the OCI agent Luis Perez as exhibiting grace under pressure. And the feds came to grudgingly acknowledge that Venema and his team, though strange, were making swift progress in uncovering the truth.

NEITHER THE OWNERS OF J&M NOR THEIR LAWYER FELT MUCH like talking. But in late July a veteran Miami defense lawyer, Norman Moscowitz, called Penezic to explain that he represented someone with extremely valuable information who was looking to make a deal.

Moscowitz knew the rules of the game. In cases with many potential defendants, where blame and criminal charges could spread quickly up the food chain, the best deals went to whoever contacted prosecutors first. He also sensed that the case was big. And so his call to Penezic came before the Horsemen even knew that his client existed.

Shortly after his call, Venema and Arias listened in Moscowitz's conference room as the attorney explained that the man they *really* wanted had approached his client about a year ago, looking for a source of low-dose Epogen. The man told his client, who had worked on and off in the health-care field, he planned to sell the medicine overseas. That man was a "rough character" who owned firearms and had stashed them at his client's house, either as a matter of convenience or to inspire fear. And his client, whom Moscowitz identified as Armando Rodriguez, would be willing to cooperate if he received full immunity.

The lawyer explained that Rodriguez had approached Jesús Benitez at J&M Pharmacare to see if he could procure the medicine. At first, he asked Benitez to order just a few thousand dollars of the medicine each week. The orders increased until they reached more than $70,000 a week.

Rodriguez had learned the true nature of the deal only after Arias and Calvar began to ask questions, Moscowitz said. Rodriguez had been at J&M when Arias called requesting a meeting. That is when Rodriguez had realized that cancer patients across the country were getting weaker medicine than had been prescribed, Moscowitz said.

Arias and Venema found it hard to conceal their desperation for a break in the case. The two men, ragged and exhausted, were ready to reach across the table and grab the lawyer by his silk tie, but Moscowitz continued his waltz. His client wanted to help and he wanted immunity, Moscowitz said, but he would not do the first until he was guaranteed the second.

———

AS STATE ATTORNEY PENEZIC WORKED ON AN IMMUNITY DEAL, Petri and Jones began their surveillance of Armando Rodriguez, who spent time at auto repair stores around Miami, often beneath the steering wheels of various cars. The Horsemen quickly figured out that they were seeing him at his day job—manipulating odometers and decreasing the mileage on cars that people were planning to resell. The purchase of medicine was apparently a sideline—but one that would lead the Horsemen to the mysterious "Tony," the source of the counterfeit medicine being sold at Playpen South.

AMERISOURCEBERGEN'S THIRTY-FIVE DISTRIBUTION CENTERS, some of which exceed 300,000 square feet, fill the nation's orders for prescription drugs every day at all hours. Conveyor belts that span the length of several football fields wind through the vast mechanized hubs, as medicine drops automatically into crates bearing electronic bar codes. The largest warehouses process more than eighty thousand orders a day. The company's customers—which range from the pharmacy benefit management company Medco Health Solutions, to the hospital system Kaiser Permanente, to the pharmacy chain Duane Reade—buy medicine by the ton and spend billions.

In effect, Amerisource's buying and selling habits had played a role in the dissemination of the counterfeit Epogen. As a seller, it sought to move product quickly, selling enormous quantities of the low-dose Epogen to a small pharmacy that could not possibly have used that much medicine. As a buyer, it sought the deepest discounts and had jumped at the chance to purchase the high-dose Epogen cheaply, despite its dubious pedigree. In both roles, the company had asked too few questions. Consequently, it bought back the exact same low-dose Epogen it had originally sold, except the drug had been counterfeited in the interim to resemble high-dose Epogen.

The dangerous trading that led to this was neither rare nor secret nor an exception to the drug trade. It *was* the drug trade. Chris Zimmerman, AmerisourceBergen's security director, had made this clear in his presentation at the public pedigree meeting in Tallahassee on April 30. He explained that in order to stay profitable, companies like his own bought from secondary wholesalers who offered far greater discounts than the manufacturers did. How those who acquired the medicine could sell it more cheaply than those who manufactured it was just one of the secrets of the pharmaceutical trade.

The wholesalers operated on a thin margin—the fractional difference between what they could buy a drug for and the slight "upcharge" for which they could sell it. These slight upcharges might account for three-fourths of a company's gross margin. Steeply discounted drugs, often of uncertain origin, became an essential way to increase profits. Each major wholesaler ran trading divisions that sought out deals in the secondary market. In August 2001, Amerisource Health merged with Bergen Brunswig, a wholesaler known for its pharmaceutical trading, to form AmerisourceBergen. From then on, Bergen subsidiaries served as Amerisource's scouting parties in the secondary market.

AmerisourceBergen knew from experience that discounted medicine could be dangerous. In February 2001, Bergen Brunswig had conducted a widespread recall of counterfeit Serostim, all of which the company had purchased from secondary vendors.

In June 2001, Florida's pharmaceutical bureau fined Bergen's Orlando operation $2,000 for buying medicine from twenty-seven companies not licensed in Florida and for failing to maintain pedigree records. Bergen's general counsel, Nicholas W. Ghnouly, sent the bureau a four-page, single-spaced letter challenging the fine and arguing that the law requiring the company to verify which companies were licensed was "unfair, impractical, administratively burdensome, and costly."

Ghnouly wrote, "Of course, there will always be instances where suppliers make false representations to Bergen, however it is commercially unreasonable and impossible for Bergen to document, verify, substantiate, and investigate the truth and veracity of all supplier representations."

The attorney wrote that checking up on whether potential vendors were licensed at the health department's Web site also was too burdensome: "Bergen would have to hire numerous personnel whose sole task would be to do nothing other than sit in front of a computer all day and look up whether particular suppliers are licensed or not with the state of Florida," he wrote, adding that the state was better equipped to undertake "such policing and monitoring activities." He even noted that the state's requirement was *unsafe* because it could cut off the supply of much-needed medicine to patients, simply because a wholesaler was unlicensed.

Bureau Compliance Director Sandra Stovall wrote back to Ghnouly that Bergen's "business perspective troubles me immensely" given the "increasing incidences of counterfeit products and diverted drugs entering the market place." She also wrote that the "articulated business practice of Bergen" could facilitate the entry of counterfeits into the drug supply. More than twenty instances of buying from unlicensed suppliers indicated a "pervasive disregard of the requirements and indicates a significant potential health threat as expressed in this letter." She concluded, "Rhetorically, is any system too costly to prevent adverse and possibly fatal consequences from patients taking counterfeit or diverted (adulterated) drugs?"

In a small concession, she reduced the fine to $1,500. Ghnouly sent the check in October 2001 on behalf of the newly merged AmerisourceBergen.

Before the merger, Amerisource also had run afoul of regulators for buying adulterated medicine from secondary wholesalers. Pharmaceutical bureau inspectors tracked counterfeit Retrovir to the company's Orlando warehouse in June 2001. Amerisource had saved $8 a vial by purchasing the medicine from a secondary wholesaler instead of from the manufacturer, Glaxo Wellcome. The bureau fined Amerisource $63,500 for buying repeatedly from companies not licensed to do business in Florida and for failing to get pedigree papers for the medicine, as required by Florida law. In May 2002, the merged company paid a reduced fine of $50,000 to settle Amerisource's earlier regulatory violations. Company officials wrote to the bureau that their new policies would reduce the chance of receiving or distributing counterfeit medicine.

When the counterfeit Epogen surfaced, the companies involved dealt with the matter as a trade dispute. In August 2002, Amerisource sued one of the secondary wholesalers, Dialysist West in Arizona, from whom it bought $8.6 million of the Epogen. Dialysist West sued its suppliers, who in turn sued theirs. One supplier that Dialysist sued was a Fort Lauderdale wholesaler, AmeRx Pharmaceutical. In 2001, AmeRx had been a direct supplier for Amerisource until it sold the giant wholesaler counterfeit Neupogen and Serostim, which reached nine of Amerisource's distribution centers. In response, Amerisource terminated AmeRx as a supplier.

Less than a year later, the newly merged Amerisource-Bergen bought counterfeit Epogen that came with pedigree papers showing that the medicine had moved through AmeRx.

The pedigree papers were a red flag alerting Amerisource-Bergen, as a potential buyer, to danger. The warning was overlooked, said an AmerisourceBergen spokesperson, because a Bergen division had placed the order and the two merged companies had yet to combine their lists of disqualified vendors.

SUSAN CAVALIERI RAN AMERX PHARMACEUTICAL FROM BEHIND A bolted door in Oakland Park. After hours the blonde pharmacist sat in her office reception area, legs in torn pantyhose swung over the side of a beat-up leather armchair and two golden retrievers at her side. Only her watch, a platinum and diamond-encrusted treasure, sparkled, a remnant of her pharmaceutical wholesale business. In 2001, her company had done almost $12 million in sales. By 2003, she agreed to let her license expire, relinquish $900,000 in profits, and become a co-operating witness for the Horsemen.

With the blinds drawn as they always were these days, in the office where she still ran her closed-door pharmacy for home-bound patients, she spoke for hours to a visitor about how she was duped and misled; how the market favored the big guys and shut out the little ones; how she, as a licensed pharmacist and proud member of a trusted profession, was still worthy of that trust. She graduated from pharmacy school at the University of Florida in 1977 in the same class as Cesar Arias, but took a different path.

She was a naïf when she opened her wholesale company, she explained, and soon found a business mentor who always knew about the best deals: Michael Carlow.

"You had to like him," she said. "He was quite a ladies' man."

Cavalieri craved financial success, but also comfort, praise, and attention. Carlow sent her flowers when her father died.

Gene Odin had admonished her "like a dad, 'You're not trading with Michael Carlow?'" But she was.

In Florida her reputation as someone who would buy anything grew quickly. William Walker, a convicted narcotics trafficker who used an array of aliases, sold her three hundred vials
of counterfeit Serostim that the FDA seized. Walker promised
to make up the loss by supplying her with more medicine.

Eddie Mor, who owned the Texas wholesale company that
the counterfeit Epogen traveled through, showed up. She said
that he presented himself as a drug industry rep who could
bring her medicine at garage-sale prices being sold off by the
large wholesalers Bindley Western and McKesson. With one
eye closed, his story made sense to her.

Because drugmakers offer secret and widely varying discounts, everyone suspects that someone, somewhere, is getting
a better deal. So when cheap medicine came along, Cavalieri
sought to rationalize its origin. The price could mean that
someone broke a contract or some company made a sacrificial
sale to close out merchandise. Someone else's breach of contract was not her problem, Cavalieri figured. And so great
deals could be legitimate, sort of, Cavalieri's logic went.

From Mor, she bought Nutropin A.Q., a growth hormone
that turned out to have nothing but insulin in the vials. The
Viagra he sold her came from a hijacked truck that had vanished with $5 million in pharmaceuticals inside. He also sold
her Gammimune, a medicine for immune deficiency, that was
diluted with saline and one-tenth the labeled dose. Every
single type of counterfeit pharmaceutical that moved through
South Florida in 2001 and 2002 passed, at some point, through
Cavalieri's hands. Some of the counterfeit Epogen from lot
P001091, which Tim Fagan received on Long Island, passed
through her company.

Cavalieri moved this medicine into the nation's distribution chain, in part through her relationship with Amerisource. Once the company approved her as a direct vendor in early 2001, she began to ship medicine to ten Amerisource distribution centers. The reason for Amerisource's interest in her wares was spelled out on invoices that she sent with the medicine: some bore the notation "special price."

Cavalieri even sold her medicine to Manhattan's fanciest pharmacy on Madison Avenue, Zitomer, which caters to some of the wealthiest New Yorkers and also sells fine cosmetics and rhinestone-encrusted dog collars. In the first six months of 2002, Zitomer bought more than $800,000 in pharmaceuticals from Cavalieri. Those drugs came from companies later implicated in the traffic in adulterated and counterfeit medicine. On August 14, 2002, Cavalieri arranged for three boxes of high-dose Procrit to be shipped to Zitomer, which the pharmacy needed on short notice. This time Michael Carlow's company, G&K, supplied the medicine.

To distributors and pharmacies seeking a bargain, Cavalieri's medicine was an obvious choice. To Cesar Arias and Gene Odin, her prices told another story.

Cavalieri said that in retrospect, she should have asked tougher questions. But her father was dying; she didn't scrutinize the medicine. At one point, she tried to cut off Eddie Mor, telling him, "I can't do this anymore. It isn't safe."

"Susan, we're making money," she recalled him saying. In truth, she had become a little unhinged by the amount of money she was making.

It was noon on August 21, 2002, when Gary Venema and Gene Odin arrived in her office, both wearing their black Horsemen shirts. "I'm going to take Carlow to jail for thirty-five years," Gary Venema told her. "I'm going to take Eddie Mor to jail." Then he flipped down a piece of paper—a subpoena for all her records.

"I'm so disappointed in you," said Odin, standing behind Venema.

Cavalieri was shocked that in their eyes, she had become one of the bad guys, tailed and photographed by detectives. She was a simple Midwestern girl who had never had a lawyer, who wouldn't have known a subpoena from a parking ticket, and who found herself making a fortune as a pharmaceutical wholesaler, a business that she'd really enjoyed. It was a tossup now as to what bothered her more: the lost moral high ground or the lost money since the FDA had seized her inventory and her defunct company had been named in a lawsuit. "It's just money lost, money gone, money owed," she complained. "I didn't go into this so that I would get bankrupt."

18. The Guitar Story

April 25, 2000
Nashville, Tennessee

NEIL SPENCE HAD REASON TO BE PLEASED. THIS MORNING, HIS company, Cardinal Health, had issued a press release announcing that the revenue from its pharmaceutical distribution efforts, of which he was a part, was up 23 percent. Not only did that sector show strong sales to customers, the press release stated, but "cost controls and productivity improvements" also had enabled Cardinal to reduce expenses to an all-time low.

Chairman and CEO Robert D. Walter boasted that "disciplined operations" had allowed the company to deliver excellent financial results, as it had done every quarter since 1998, either meeting or exceeding analysts' estimates.

Neil Spence headed the division of the company called NSS, or National Specialty Services, which handled the most fragile and costly blood products and cancer medicines. The customers he supplied were industry behemoths whose medicine reached millions. They included US Oncology, the nation's largest network of oncologists at fifty-nine major cancer centers; and Novation, one of the nation's largest purchasing groups, which supplied more than fifteen hundred hospitals with medicine and equipment.

For an executive of only thirty-five, Spence had sizeable power. Each purchasing decision he made could impact public health. But he was also under continuous pressure to boost profits. The deals and discounts he negotiated helped him to make his numbers, which in turn helped him to collect his end-of-the-year bonus, which comprised up to 30 percent of his income. The bargains he found in South Florida had become important to his bottom line. Michael Carlow, one of Spence's suppliers, was part of the winning formula.

The two men appeared to have little in common. Spence, an amiable Tennessean with straw-colored hair, wire-rimmed spectacles, and a soft, almost boyish face, projected an air of legitimacy. He was preppy, neatly dressed, and a draw at industry conferences, respected for his knowledge and influence. By contrast, Carlow was a convicted felon who traded pharmaceuticals in restaurant parking lots. On most days he looked disheveled in his sagging shorts and T-shirts. Nonetheless, Spence's purchases from at least four companies that Carlow controlled catapulted Carlow's medicine into the heart of the nation's supply, where it inevitably reached patients. In return, Carlow expressed his gratitude to Spence.

On April 24, 2000, the day before Cardinal released its good news, Spence had gone to his bank in the morning with a check for $10,460. It had been written to him personally from Carlow's BTC Wholesale, and it was not the only cash infusion that Carlow sent him. Spence deposited the check in his personal account at his bank in Nashville just at a moment of ebb in BTC's cash flow. The check, number 5256, bounced and was returned to Carlow's bank stamped UNCOLLECTED. Carlow, however, had not intended to play fast and loose with Spence. The fact that Spence bought his medicine put Carlow in his debt.

From January 1999 until June 2000, Spence's division bought more than $131,000 in blood products and other specialty items from Carlow's Medical Infusion Services, which he formally renamed as Quest Healthcare in May 2000. From January through June 2000 alone, NSS bought $1.45 million in medicine from Quest, even though the company had no license to sell pharmaceuticals until May 2000.

Once Quest became licensed in May, it did not stay open long. The company was closed down permanently two months later after Carlow's arrest for buying cancer medicine in a Miami parking lot. He pleaded guilty to purchasing the medicine from an unlicensed source—two Miami social workers. State officials appropriated Carlow's black van and put it to use for Medicaid fraud surveillance. He received a suspended sentence and probation and was barred from the pharmaceutical wholesale business. Undeterred, he immediately bought a similar van and set up a new business, BTC Wholesale, in Kissimmee.

Spence bought $150,000 in medicine from Carlow's new company between August and December 2001. He also bought from another Carlow shell company, MedRx, sometimes receiving up to three packages of medicine a day.

Meanwhile, the personal packages kept coming to Spence. On October 1, 2001, BTC sent a package to Spence's home in Gallatin. Two months later, on December 1, BTC's ledger specified that a "consulting fee" of $5,300 in cash had been given to Spence for "outside services."

Mark Novosel recalled Carlow saying that in spring 2001, he and Spence panicked because a package of cash shipped from BTC to Spence at his home apparently broke open in the rain and was discovered by the shipper. Carlow told Novosel that to explain the cash if need be, the two men had concocted a "guitar story" which went like this: Spence lived in Nashville, home of the handmade Gibson and Gruhn guitars so coveted by Carlow and other guitar collectors. On occasion, Spence picked up a guitar for his Florida vendor, paid for it in cash, and then when Carlow had a chance, he would send back payment, mailing money or checks directly to Spence at his home.

The story made some sense. At Gruhn Guitars, smack dab in Nashville's tourist section, travelers often shelled out cash for guitars like a Fender Telecaster Junior or a Gibson Les Paul that sometimes cost more than $100,000. "We get people from all over the world," said Cody Newman, a salesman at Gruhn Guitars. And paying in cash is "more common than you think."

Guitars were the excuse they planned to use in case they ever needed one, said Novosel with a guffaw. In the fall of 2004, FDA investigators began seeking information about Spence from Carlow's associates in South Florida, according to Novosel and others.

SPENCE'S PURCHASES FROM THE SECONDARY MARKET WERE NOT unusual within the company. Cardinal maintained an internal list of the secondary wholesalers from which its trading divisions bought and the amount of savings realized from these purchases. The document, a version of which bore the title "CTC top ten ADV margin"—which one company executive translated as Cardinal Trading Company's top ten alternate distribution vendors—ranked the wholesalers by how much they saved the company. In one version of the document, the Stone Group, which had saved Cardinal more than 50 percent on the cost of its drugs, was near the top of the list.

Thomas Blaylock, who preceded Spence as director of NSS, said that though he had never seen a document like that, it reflected the way Cardinal liked to analyze things: "to see who you're dependent upon," as he put it.

Publicly, Cardinal stated that its purchases from the secondary market accounted for as little as 2 percent of its inventory. This aggregate number did not reveal the percentage of certain drugs the company bought from the secondary market, or the buying patterns of particular divisions within the company.

While some Cardinal divisions bought more often from the manufacturer, the Bindley divisions that comprised Cardinal's trading arm bought almost exclusively from secondary sources and then transferred the medicine internally to other Cardinal divisions. For example, the Bindley Trading Company offered an online purchasing service for discounted medicine through its Web site "WAC Minus" (named to underscore that the medicine was priced below the wholesale acquisition cost).

Bindley's records from two separate warehouses, one in Orlando, Florida, and the other in Grapevine, Texas, from January 2001 to mid-May 2002, show different buying patterns. When acquiring five drugs vulnerable to diversion and counterfeiting—high-dose Procrit and Epogen, AIDS drugs Serostim and Combivir, and Zyprexa for schizophrenia—Bindley's Orlando warehouse bought some direct from the manufacturer and some from secondary sources.

By contrast, the Bindley warehouse in Grapevine bought four of those drugs exclusively from the secondary market and only 8.4 percent of one of the drugs, Combivir, from the manufacturer. None of the 23,738 boxes of high-dose Procrit purchased by the Grapevine warehouse came from the manufacturer. Half came from internal transfers from other Bindley divisions; over 14 percent, 3,506 boxes, came from the Phoenix wholesaler Dialysist West; 1,831 boxes came from Raymar, a secondary wholesaler in Miami; and the rest came from other secondary wholesalers or other trading divisions of Amerisource. None of its Zyprexa had come from the manufacturer. None of its high-dose Epogen or its Serostim came from manufacturers. Company-wide between January 9 and February 26, 2002, Cardinal purchased 1,473 boxes of Procrit from secondary wholesalers according to records the company provided to Florida's pharmaceutical bureau.

Despite these ongoing purchases from the secondary market, a spokesperson for Cardinal told the *Washington Post* that since 2001 the company had bought all cancer, injectable, and other drugs attractive to counterfeiters directly from manufacturers. When asked in 2004 about buying records showing the purchases continuing into 2002, a spokesperson clarified that the company had ceased buying these drugs from secondary wholesalers by the end of 2002.

OVER THE YEARS, A NUMBER OF EXECUTIVES AT THE NATION'S TOP wholesalers had been prosecuted for buying medicine from corrupt middlemen in exchange for kickbacks. These cases were often treated as isolated incidents. Yet those who worked at the nation's largest wholesalers were gatekeepers of the public health. Their buying decisions, when corrupted, brought adulterated medicine into millions of homes. The few investigators who managed to expose this systemic and underlying corruption often found themselves marginalized and their efforts derailed.

In the early 1980s, a registered pharmacist and senior agent at the Georgia Drugs and Narcotics Agency, Richard C. Allen, launched an investigation that came to be called "Pharmoney." Joined by the FBI, Allen probed the buying habits of Bindley Western Industries, Bergen Brunswig, and other large wholesale companies that would come to form the Big Three. His findings led in part to the creation of the 1988 Prescription Drug Marketing Act.

Allen uncovered wholesale companies using every imaginable ruse to buy discounted medicine: they paid off ambassadors of foreign countries to order drugs for export; bribed FDA inspectors to ignore violations; and obtained leaks from pharmaceutical companies on anticipated price increases so they could buy in advance. Some wholesalers even grabbed up the expired medicine of patients who had died at nursing homes, jumbling them together to be resold, recalled Gale McKenzie, the Assistant United States Attorney in Atlanta who prosecuted the Pharmoney cases.

At Bindley Western, two top executives told the FBI over weeks of interviews that the company bought widely from diverters, believing the middlemen somehow "cleansed" the transaction by allowing the firm to say it did not know where the drugs originated. Yet the company knew that 99.9 percent of all the pharmaceuticals bought at reduced prices from anyone other than drugmakers had been diverted or obtained fraudulently, according to one of the executives, Stephen Lee Asher, vice president of Bindley Western's drug company.

Asher, who had taught for a year at a Catholic school before training in management at a grocery chain, found himself crossing the line into illegal activity. Among the unsavory drug wholesalers he cultivated was Martin Thuna, who dealt exclusively in diverted medicine and cash-only transactions. Many of Thuna's goods, including $750,000 of the heartburn medicine Tagamet labeled for export to Nigeria, was stolen from trucks and airport cargo holds.

Asher told the FBI that the company regularly advanced Thuna up to $500,000 to obtain the pharmaceuticals and that

Thuna kicked back cash to Asher and others each time they bought medicine from him. Asher estimated that between 1981 and 1985, he and the company president, Jack Earl Laughner, received sixty payments from Thuna ranging from $5,000 to $50,000 each.

Asher told the FBI that he had passed cash to clients in hotel bathrooms and that Jack Laughner kept cash hidden behind the credenza in his office. All told, the men estimated that in 1984 Bindley's company purchased $25 million in diverted medicine, including boxes from a burned warehouse that Bindley's staff scrubbed with toothpaste and bleach so they would look new.

After two weeks of being deposed in a hotel room, Asher stated that he wanted to start serving time immediately to pay for his transgressions. But just months later he killed himself, leaving behind a note saying that he would have been sent away for life if the extent of his crimes had been uncovered. Laughner and two other Bindley executives pleaded guilty. As a company, Bindley Western also pleaded guilty to fraud for diverting pharmaceuticals. Thuna was convicted of hijacking truckloads of medicine and was sentenced to fifteen years in prison.

Despite the setback of Asher's suicide, Allen's investigation was on fire, with more than one hundred guilty pleas and about fifty more pending. And then one day in 1987, after four years of meticulous work and a half dozen typed confessions in his briefcase, Allen's investigation was stopped. Suddenly, the once-zealous prosecutor Gale McKenzie would not read any of Allen's investigative reports. The FBI pulled back its resources. His team got no answers and barely got excuses. Despite screaming matches, an FBI supervisor told his team, "I'm sorry, guys. We're going to close it down."

Speculation abounded over whether incompetence or weariness or interference was to blame. Allen and others came to believe that the U.S. Justice Department had instructed the United States Attorney in Atlanta to stop. McKenzie said that she did not recall Pharmoney being shut down and that some leads were handled by other districts.

Allen testified before Congress in 1990 that his investigation was halted "for what appears to be nothing more than a lack of willingness by the Department of Justice to prosecute. I cannot really say what happened, for I am but one lone state agent and as I have been told, I do not have the authority or ability to understand the decisions made by this country's major Federal law enforcement agencies."

The *why* still preoccupied Allen fifteen years later. Not to prosecute all that he had uncovered was bad enough, but to allow potential harm to patients to continue seemed unconscionable. In hearings leading up to the Prescription Drug Marketing Act, Allen had warned Congress: "The Tylenol poisoning cases are nothing but a drop in the bucket compared to what could happen with misbranded or adulterated drugs."

The Prescription Drug Marketing Act did little to change corporate behavior, as Allen and future investigators learned. In part this was because no one was enforcing the law, as Allen told the House Energy and Commerce Committee's subcommittee on oversight and investigations in 1990. The FDA's apparent refusal to fight diversion and to cooperate with state investigations and its failure to clarify key provisions of the PDMA had left states hopelessly confused over how to enforce the law, he said.

As though to prove Allen's point, a sweeping investigation of drug diversion in the mid-1990s turned up many of the same abuses that he had seen a decade earlier—even at Bindley

Western. This time, an FDA criminal investigator, J. Aaron Graham, penetrated the diversion market by posing undercover as a corrupt institutional pharmacist. His investigation, called Operation Grey Pill, quickly came to center on Bindley Western, where a corrupt management team in San Dimas, California, was selling discounted medicine to closed-door pharmacies that were illegally diverting it, in exchange for kickbacks.

Three Bindley managers from the San Dimas division were convicted on charges that included mail fraud, conspiracy to commit mail fraud, and conspiracy to transport stolen goods. In 2000, in an agreement with the government, Bindley Western again pleaded guilty, this time to one count of conspiracy and agreed to pay a $20 million fine. No officers at the corporate level of Bindley Western were ever charged. And in its settlement, Bindley Western stated that the scheme in San Dimas had occurred without the knowledge of any of its corporate officers.

"We were shocked to learn of the crimes committed by our former employees and deeply regret the embarrassment brought to the hundreds of other law-abiding and honest employees at the company," Bill Bindley was quoted as saying.

John Ransom, a financial analyst at Raymond James, defended Bindley at the time as "one of the most honorable, ethical guys in America."

Six months after the settlement, Bindley sold his company to Cardinal Health for $1.2 billion. He emerged from the sale with seven million shares of Cardinal stock and went on to become a philanthropist and civic leader. In September 2002, he gave $52.5 million to his alma mater, Purdue University, the school's largest gift ever from a private individual. The donation led to the creation of the university's Bindley Bioscience Center.

The purchase of Bindley made Cardinal the nation's largest drug wholesaler for a time. One out of every six pharmaceutical products dispensed to American patients traveled through its efficient nationwide network, the company advised in its promotional materials. In 2002, the last year that Spence and Carlow did business together, *Fortune* magazine ranked Cardinal as the nation's twenty-third largest company.

19. "They're Going to Die Anyway"

Summer 2002
Flagami
Miami, Florida

JOSÉ ANTONIO GRILLO TRUSTED ALMOST NO ONE. NOT BANKS OR business associates or those who worked for him. Secretive and controlling to the point of obsession, the forty-four-year-old body builder trusted his beautiful young wife least of all and kept her within arm's reach and out of other men's sight whenever possible.

He stood six feet tall and had fair skin, narrow brown eyes, and several hoop earrings. He usually wore a baseball cap pulled low over his face, his lips drawn habitually into a tight, expressionless line.

Grillo kept no bank accounts and stored his money—which at one point amounted to almost $1 million—in a duffel bag on his closet floor. He paid for almost everything with cash and money orders, almost never using credit cards. He changed cars and cell phones frequently, which made him difficult to follow or to reach. He initiated most phone calls and rarely gave out his last name. Most of those he dealt with—including the co-owners of Playpen South, Nicholas Just and Dr. Paul Perito—referred to him only as "Tony." Grillo was the source of the medicine hidden in the cooler of the strip club's back room.

Though he brought his wife, Leticia, almost everywhere, he rarely allowed her to see his business dealings. When she accompanied him to Just's garish home, with its statues of naked dancers on the lawn, she waited in the living room while the men conducted business on the back porch. On occasion, when she inquired about the huge amounts of money he brought home after being at Playpen South ($25,000 often being the smallest bundle), he became abusive, once throwing her down the stairs.

If inclined to trust anyone, it probably would have been his late father's wealthy friends with whom he socialized. Grillo's father and the other men, all Cuban nationals in exile, had served together in the legendary Brigade 2506, a paramilitary group that tried to unseat Castro during the Bay of Pigs invasion. The men were prominent in Cuban business circles. On the many occasions when Grillo and his wife dined at their homes, he treated them with a respect usually accorded family elders.

He had another reason to pay them respect, as he would later explain. They were his underwriters and had paid up front for the machines and raw materials of his current business. Consequently, as investors they had made millions, he claimed.

CARRYING A FIVE-GALLON PAINT CAN, GRILLO STEPPED FROM HIS Toyota Sequoia and made his way down a dusty lane and alongside a trailer, its makeshift stoop crowned with wind chimes, pinwheels, and small American flags. Trailer homes sat jammed together on either side of the narrow cul-de-sac. Corrugated iron sheets that stretched above the parked vehicles comprised a series of rooftops.

The neighborhood, Flagami, right beneath the flight path to Miami International Airport, was a checkerboard of vacant lots heaped with tires, outsized warehouses, and streets created by the trailers parked in rows. As usual when Grillo came down here, Leticia hung back some distance as he picked his way alongside the trailer to an improvised hut in the backyard.

The man who emerged to greet him, Silvino Cristobal Morales, had worked for Grillo's family for years. Formerly a welder on the Miami horse farm that Grillo's father had owned, Morales had recently been diagnosed with emphysema and could work only odd jobs. The job he did for Grillo allowed him to work at home on nights and weekends.

Grillo set his paint can on a table top. Inside, five hundred tiny vials clanked and rolled in frigid ice water. He had also brought a box of perfectly printed labels and cardboard planks to be folded into boxes.

Morales's wife fetched several bags from the refrigerator containing the work her husband had completed. Inside were five hundred more vials. Grillo examined them, holding several between his rough fingers. The new labels on them, for high-dose Epogen, appeared straight and clean and were affixed snugly. Morales had followed his directions perfectly: He had soaked the vials overnight in soapy water, removed the labels for low-dose Epogen, and reaffixed new ones—the 2,000 U/mL becoming 40,000 U/mL.

Morales's wife also brought another bag. Inside were small, perfectly constructed boxes with intricate lettering; Grillo was pleased to note the clean folds along the sides. Morales had done a fine job and had worked more quickly than Grillo's last helper, his elderly godmother. He peeled off some bills from the roll in his pocket, paying Morales twenty-five cents a vial.

The former welder believed that he was relabeling a solution for weightlifters to be marketed overseas.

Grillo loaded the new vials into a different paint can and placed ice packs on top in an almost compulsive effort to keep the medicine he counterfeited cold. The medicine inside would now bring in $225,000; he had increased its value by more than $214,000.

NOTHING EVER WENT PERFECTLY, OF COURSE, AND THERE WERE small glitches that forced Grillo to tinker with his scheme.

In August 2001, his printer in Hialeah had made so many small errors that Grillo had to shut down his operation for a month because the boxes and labels would not stand up to scrutiny. The text on some of the boxes had not been aligned properly. Some boxes didn't fold correctly because the perforations were in the wrong place, and some labels had been double stamped while others appeared blurry. The mistakes angered him, given that he paid the printer $25,000 for each run of materials.

After resolving the production errors, he again changed his cell phone number to make his phone calls that much harder to track, then resumed business.

ARMANDO RODRIGUEZ WAS A STUDY IN ANXIETY. THE DIMINUTIVE Cuban—known as "Dito" to friends—fidgeted continuously, always in motion except for his head, which he was unable to turn due to a bad neck surgery. His relationship with Grillo did little to relax him.

Three times a week, Grillo arrived at Rodriguez's home in Sunset to pick up the latest purchases from J&M Pharmacare, which was less than a mile away. These included as many as two hundred boxes at a time of the low-dose Epogen. While Dito believed that Grillo was counterfeiting medicine, he thought that all of it was for sale overseas, thereby making the consequences more distant and more palatable.

Grillo also seemed in control of every detail. He insisted that Dito buy the more expensive ten-pack vials, as opposed to the discounted pack of twenty-five, as he believed the adhesive on the labels of the ten-pack was easier to rub off once the vials had soaked overnight.

Once a month, Grillo directed him to buy a single box of the high-dose Epogen so he could check whether Amgen had changed the labeling and packaging or created any new security features. It helped that the plastic tops above the stoppers on both the low- and high-dose vials were an identical red. Dito knew that Grillo shredded all the leftover boxes and other paper from his work and used a mechanized folding machine to make perfect creases in the inserts he replicated (which warned patients of every possible side effect except the one he created—not getting better).

On one occasion, Grillo asked Dito as he was loading up the Sequoia, "Do you think they'll ever catch on?"

Earning $100 on each box of medicine he procured for Grillo, Dito had made $1.1 million in less than a year. He blew most of the proceeds on a dancer at a strip club for whom he bought a car and jewelry and rented a lavish beach house. As a result, life with his wife in their well-appointed home had become more tense than usual.

The last time Grillo went to Dito's house in early June 2002, he backed his Sequoia into the courtyard and, this time, carried two five-gallon paint cans inside. Dito wasn't home, but Grillo asked his wife whether he might store them in the garage. Later, Dito discovered a revolver and several handguns in the cans, which he interpreted as Grillo's way of telling him to keep his mouth shut. Days later, state investigators arrived at J&M Pharmacare asking questions.

In June, after inspectors confronted Maria Castro and Jesús Benitez at J&M, Dito and Grillo panicked. "Pero como puede ser?" Grillo asked. "Nadie sabia." *How could this be? Nobody knew.*

Shortly afterward Grillo vanished, as he did when things got hot, and Dito did not hear from him for a month. When he called Dito again from yet another cell phone, he mentioned that a business contact—"El Viejo"—wanted $620,000 back because some of the product he was selling had been identified as counterfeit and was seized. On occasion he mentioned "El Viejo" (the old guy) and "El Gringo" (the American), but he never named names.

In late August, Grillo resurfaced and called Dito once more, asking to meet. This time even Grillo was not being cautious enough. By then, Dito had secured a lawyer and was no longer working for Grillo. By the terms of his new deal, he was working for the Horsemen.

ON A STEAMING MIAMI AFTERNOON, HEAT RISING FROM THE pavement and life around the city reduced to a late-summer crawl, Grillo swung his Sequoia into the parking lot behind a Cuban restaurant, La Caretta. The Horsemen sat waiting several rows behind him. John Petri had his video camera rolling behind the darkened windows of his van.

Dito drove slowly into the parking lot, took a lazy swing past Grillo's Sequoia, and then pulled alongside it, the nose of his car in the opposite direction and the two drivers a handshake away through their open windows.

The men greeted each other in English and then shifted into the more comfortable Spanish. Jack Calvar, the only Spanish speaker linked to Dito's wire, listened and made occasional translations as the other investigators watched what amounted to a silent movie, Grillo's face appearing and disappearing in his side mirror.

"My partner, the big one, do you remember him?" Grillo asked. "I never heard from them again. He disappeared and I disappeared. I'm with someone else now."

"You told me you lost one," said Dito casually.

In a relaxed tone, Grillo explained that he had just recovered from the "big credit" that he had issued the man after some of the tainted product had been taken and now they were on good standing again. "And the guy wants to return to that but they insist on too much quantity and I'm afraid." As usual, Grillo mentioned no names.

Not mentioning those at J&M by name either, Dito explained that his previous source would no longer sell them large quantities.

"I knew that was going to blow up," said Grillo, mentioning yet another man as the cause of the breakdown. "The guy wanted too much, five hundred, one thousand, sometimes two thousand per week. I thought the other guys were exporting. They told me they were exporting."

"That was a story?" Dito asked.

"Yes," said Grillo.

The Horsemen could not be sure whether Grillo actually had believed that his counterfeits were being sold overseas. Instead, they had been sold within the United States and had reached both Maxine Blount and Tim Fagan.

The men also spoke about the design of the boxes that had vexed some of Grillo's associates at Playpen South the night they studied the boxes under the black light. "Did you resolve the problem with the seal?" Dito asked. "The one that you put it under the light?"

"Yes, yes, yes," Grillo said. "Now there's a new one that comes with a new seal also. The *Pro* [Procrit] comes with another seal. I haven't been able to get it . . . I'm still selling the other one."

Sounding restless to update his merchandise, Grillo continued: "Do you know anything about Combivir? Viracept? Trizivir? For an AIDS clinic. I have a new connection with a lab in Mexico."

Excited, Calvar quickly translated for the investigators that Grillo had just mentioned several new medicines, expensive ones for AIDS.

"I've heard of Viracept," said Dito. "Trizivir? Are these vials or pills?"

"They're pills," said Grillo.

"To do something with this?" Dito asked, sounding wary.

"Of course!" Grillo exclaimed, as though the opportunity was obvious. "We have to see because we can't buy at the price of everyone else."

"Well, if they come in tablets we can crush them up and put them in capsules," Dito suggested.

"Unless the capsules are the same but less," said Grillo, referring to a strategy of uplabeling similar to the one he employed now.

"Like if it's a five-milligram you label it as a ten-milligram," said Dito.

Grillo ignored his observation. "Can you get me one bottle of each?" He added, "Do you remember the vial with a yellow cap . . . Aranesp?" It was a new form of Epogen that Amgen was selling. He wanted one of those too.

AS THE HORSEMEN PREPARED FOR THE NEXT MEETING BETWEEN Grillo and Dito, Venema thought to himself, *Life is good.* They would procure the medicine for Dito to sell, obtain a court order to put a tracking device on Grillo's Sequoia, and prepare to watch Grillo in a hand-to-hand transaction. *Add one more party to the "Mr. Racketeering" indictment,* Venema thought.

Then Grillo vanished in plain sight. Just days after the meeting behind La Carreta, he changed his cell phone number, fell silent, and did not contact Dito again. And Dito had no way to contact him. Though Petri and Jones were able to find and follow Grillo, they had no more lucky breaks. They learned the minutiae of his personal habits, like buying food from the Wendy's drive-in and eating in the parking lot with his wife, in part so that male patrons inside the restaurant would not look at her, the investigators learned later. But their window into his scheme had closed. Somehow, he knew to avoid their informant.

After several months, the investigators did the only thing they could think of: They bumped Dito into Grillo after following him to a Burger King where he was eating with his wife and young daughter. At the back of the restaurant, Grillo acknowledged to Dito that he had gotten a warning to avoid him.

As the Horsemen tried frantically to trace the leak, their suspicion fell on the Miami-Dade Police Department, where a current of gossip, rumored corruption, political ambition, and long knives ran from the officers directly to Miami's streets. Petri spent sleepless nights at home staring up at the ceiling, reviewing every stray remark he'd uttered and every report he'd written, wondering who knew and who talked. The department's internal affairs division began to investigate.

The leak had damaged their case against Grillo immeasurably. Despite a taped conversation and an informant who could serve as a witness, the investigators had nothing on Grillo. They didn't have a single document linking him to the counterfeits: no sales records, invoices, memos, or bank records. His transactions had all been in cash. He had no criminal record and practically no employment history, except for having worked as a personal trainer. The man was potentially a mass poisoner, but the most damning case they had against him was for obtaining new license plates for his stolen Corvette.

BY EARLY 2002, COUNTERFEITING MEDICINE HAD BECOME ALL too easy. One could buy second-hand pill-making machines and other manufacturing devices on eBay, the Internet auction site. Printing technology had become so ubiquitous that any reasonably equipped printer could create accurate and convincing labels. Criminals in search of a niche were drawn to the liquid gold of the biotechnology revolution and the skyrocketing prices in the American market.

While counterfeiting had long been endemic in China, India, and certain African countries, where deaths occurred as a result of tainted medicine, the American market offered a unique incentive: Medicine here cost far more than anywhere else in the world. From 2000 to 2004, the FDA's criminal cases that involved counterfeiting increased almost tenfold, from six a year to fifty-eight. As of October 2004, ninety-one counterfeiting cases were open and active at the agency's Office of Criminal Investigations. The cases, too, showed a much greater increase in size, sophistication, and complexity, said Terry Vermillion, director of the FDA OCI. Each case involved cells of traffickers and counterfeiters likely operating in numerous states and potentially saturating the country with their bad medicine. One case alone involved counterfeit cholesterol medicine that may have reached 600,000 patients.

For years, those who engaged in diversion were able to view it as a gentleman's crime. One got rich buying and selling medicine that came with a cover story, no matter how feeble. While this secondary market was corrupt and porous, no one outside it was pumping counterfeit medicine into it. During the Pharmoney investigation of the mid-1980s, Richard

Allen had uncovered gross abuses of drug diversion but little evidence of counterfeiting. Of the approximately 150 people he charged, almost all had college degrees, and most had advanced medical or business degrees.

That changed as the drug war's ever-harsher penalties led narcotics traffickers into pharmaceutical wholesaling, which they found to be more profitable and lower risk. "The psychopaths got involved and ruined it for everybody," complained Mark Novosel, who viewed himself as an old-school diverter, never intending to hurt anyone.

The rapidly expanding network of middlemen, the uneven patchwork of state enforcement, and the lack of comprehensive federal regulations made it relatively easy to slip convincing counterfeits into the stream of commerce. And the wrong people had figured this out. Hard-core criminals who already moved contraband like narcotics and weapons through underground distribution networks added pharmaceuticals to their inventory.

One glance at the photographs of the Operation Stone Cold suspects reflected this change. Though almost all were licensed medicine distributors, the men looked dangerous. They were scowling, scarred, and drug addicted, with long rap sheets of crimes that ranged from drug smuggling and truck hijacking to assault and kidnapping.

The Horsemen began to encounter terrified witnesses and heavily armed enforcers, one a 280-pound bodybuilder who strode toward Randy Jones in a parking lot with clenched fists until Jones drew his gun and forced him face down onto the asphalt. Another wholesaler, James Robert Suozzo, who owned the Georgia shell company Premier Medical Distributors, was an eighth-grade dropout and long-time heroin addict who'd been arrested for cocaine possession, kidnapping, and aggravated battery.

The new counterfeiters and traffickers surprised the old diverters with their remorselessness. Grillo's own brother, José L. Grillo, who had run nursing homes and been convicted of drug smuggling, had died of cancer at age forty-four. Another counterfeiter, Eddy Gorrin, who was arrested by the FDA and pleaded guilty after attempting to sell vials of Miami tap water relabeled as Procrit, was himself an insulin-dependent diabetic. During one conversation with an undercover agent, his associate rationalized their scheme by explaining that the patients taking their medicine were "going to die anyway."

FROM A DISTANCE, OPERATION STONE COLD APPEARED TO BE going well. The five investigators continued to make headway against an entrenched, hydra-headed, and deadly enterprise to launder and counterfeit the nation's prescription medicine. Nine months after the first break-in at Marty Bradley's Miami warehouse, fifty-five of Florida's 477 in-state drug wholesalers were either subjects or targets of the investigation, their particulars taped on the wall of FDLE's small, windowless conference room.

By September 2002, at the request of the new prosecutor Robert Penezic, Venema drafted a far-reaching wish list of "bad boys" to be prosecuted, regardless of the hurdles involved. At the top sat Grillo, Michael Carlow, and Carlos Luis. Down at the bottom of the list sat Maria Castro and Jesús Benitez of J&M Pharmacare.

By year's end, the Horsemen had served sixty-four subpoenas, conducted almost a dozen search warrants, seized more than $14 million in bad medicine and cash, and had negotiated the return of over $3 million in assets to the state. As these checks came in, the men brought them to the money people in their various departments, making them the equivalent of rock stars in a lean budget year. Venema alone had written 146 detailed reports on their activities and findings so far.

If the Horsemen ever lagged, Venema would exhort, "Never let a day go by without annoying or aggravating a bad guy or an attorney!" Or he would explain that the Bible had some inaccuracies: "The meek will not inherit the earth," he would say. "The meek will inherit flame throwers and cruise missiles!" On another occasion, he stressed that while the FDA started "years ahead of us we are showing them how to do it— we are leading them by the hand. Don't stop, this thing is working!!!!" He would sign off his e-mails as the "Starfleet commander" or "from the bridge of the Enterprise."

As the volume of counterfeit medicine reaching patients increased, the media also began to call about the Horsemen's investigation: not just the Fort Lauderdale *Sun Sentinel* and the *Miami Herald* and local Channel 4, but the *Washington Post*, *60 Minutes*, and *Primetime Live*. Names like "Dan Rather" began to travel up the command chain. Everyone seemed to agree instinctively that it would be best not to let Venema speak freely to anyone.

They all gathered excitedly for a journalist who had come from New York City to interview them and talked of incredible progress and imminent arrests.

But in truth, by the fall of 2002 things were not going well at all. Despite the extensive corruption they had uncovered, Arias's bureau continued issuing wholesale licenses without verifying the criminal backgrounds of the applicants, effectively pouring more sludge into the funnel that the Horsemen were trying to clean up at the other end.

Venema, new to the workings of the health department but now dependent, in part, on its regulatory initiative, became furious with the sluggish pace. "We are breaking our butts down here, search warrants, surveillances, taped statements, undercover calls, living in our cars," he vented in a status report to the group. He complained that one bureaucrat there "needs to find the freakin' gearshift and move it out of neutral."

To Arias, Jerry Hill's interest in Operation Stone Cold still appeared minimal. He had said nothing to his inspectors about their work or the millions in adulterated medicine they had seized. In fact, since the investigation began, he had called Arias only once: to find out whether he knew anybody who could help get his best friend's daughter into a dormitory at the University of Miami. "The guy makes $92,000 a year and he doesn't know what this case is about," Arias complained.

Others cared too much, the case's sheer size and its promise of media attention bringing out micromanagers and obstructionists everywhere. Penezic found himself in treacherous political water the moment he took over from Stephanie Feldman. While Penezic lacked Feldman's innate toughness, he was smart and capable and had immersed himself in the evidence, working late hours to pound out subpoenas and search warrants for the men. He despised the politics and game playing of state government. He had won the Horsemen's trust. Yet his honest and even blunt style, which some took for naïveté, made him an easy target.

In August he grabbed Venema for a whispered meeting in the lobby of his office. Penezic had heard that he was about to get kicked off the case. The Horsemen were stunned: Just as they were negotiating with killer defense attorneys over cooperation agreements, they would again be left with no prosecutor? There seemed to be only two possible explanations: Either the statewide prosecutor's office did not want the case to go forward, or it did not care whether it did.

Foreseeing a nightmare, Mike Mann lobbied successfully to keep Penezic on the case. But two weeks later, Mann himself got promoted to head a money-laundering strike force. He had been the perfect supervisor, a deft political fighter on behalf of his men who knew when to get out of the way, and now he would be going too. Venema griped that by reputation at least, his replacement was a micromanager who needed a six-page "op plan" to go knock on someone's door.

Venema finally lost all semblance of composure when the accountant at FDLE called him, dissatisfied with a forfeiture check for $3,500 he had brought her. It needed to be made out to the Florida Department of Law Enforcement, she scolded, not FDLE. After days with little sleep, he bellowed over his cell phone, "Well you're not going to like the one I have in my hand. It's for $180,000 and it's made out to FDLE!" These were giant sums for a public safety agency facing a budget shortfall. Mike Mann, who happened to be passing by in the hall and had not left yet, plucked the phone from Venema's hand and soothed the accountant.

The dynamic at the Miami-Dade Police Department was even worse, as the commander of a rival unit vied with Petri and Jones to get a piece of the case and a portion of the proceeds. As Petri and Jones saw it, the men from the other unit generally showed up hours after they did to begin their surveillance and left hours earlier. On a day when the two investigators had spent twenty-four hours straight "sitting up" on a suspect and then securing his home, after a judge finally signed their search warrant, a sergeant from the competing unit drove out to tell Petri that it was only fair that his unit get half the proceeds.

"We're doing this every day, all day," Petri told him, smoldering.

Shortly afterward, the police director issued a memorandum stating that all pharmaceutical investigations should be turned over to the competing unit. "We got more problems internally with our guys inside than with the bad guys on the street," Petri said.

The Horsemen had problems with the bad guys too. On October 1, a well-planned sting ended in disaster when two hundred boxes of adulterated Procrit from a Carlow-controlled company in Maryland vanished just as investigators closed in. Arias and Calvar had seen the packages of medicine against the front wall of a medical supply store, where they had wandered in pretending to shop for a wheelchair. In the ten minutes between that sighting and the hasty departure of the woman behind the counter, who locked up the store moments after the two men left, the drugs vanished without a trace.

The inspectors fanned out, stopping the cars of everyone involved in the medicine's purchase, including the woman from the store. No drugs. A lawyer for one of Michael Carlow's deputies told them the drugs had been thrown away. But they could not be sure if the medicine had been resold and, if so, whether it posed a danger to patients. Unhappily, Venema wrote a four-page, single-spaced memo to his new micro-managing boss to explain how the medicine got away.

Just as the lost Procrit cast a pall over the group, Grillo changed his phone number and slipped away from their informant.

Worst of all was the chorus growing in the highest ranks of the statewide prosecutor's office that Michael Carlow and his associates had committed offenses that were not worth prosecuting. Frantic phone calls went out after the statewide prosecutor Melanie Hines told FDLE commissioner Tim Moore that she saw more misdemeanors than racketeering. Her comment reflected the view of several senior prosecutors that, taken separately, Carlow's offenses did not amount to a

hill of beans, if one strictly interpreted the law. He had passed
on phony documents and obscured the origin of medicine and
had bought and sold without a proper license. These were offenses under the state's weak health statute that amounted to
fines and probation. Using these to make a racketeering case
was a legal adventure, to say the least, the last thing Hines or
any other ambitious prosecutor wanted to embark on.

Despite the proceeds, almost everyone else wanted out
from the case, which was starting to look like a huge, complicated career-wrecker. And though Mann's boss, Tim Moore,
was getting lip service from the other agencies, he sensed that
each was formulating an out under the popular state theory:
Big Cases, Big Problems. Little Cases, Little Problems. And then
60 Minutes came along.

ON DECEMBER 17, FIVE DAYS BEFORE A *60 MINUTES* SEGMENT ON
the problems in Florida and the nation's drug supply was
scheduled to air, Governor Jeb Bush swept into a conference
room and those seated around the table stood. Arias's stomach was turning over as readily as a hand-cranked flour mill.
He had no reason or right to be there, he figured, among all
these department chiefs. The commissioner of FDLE, the
statewide prosecutor, the secretary of the Agency for Health
Care Administration, which oversaw the Medicaid program,
and their various aides and associates were present. Penezic
was there too, as were several of Arias's supervisors and
Michael Mann, who was still involved in briefings about the
case, though no longer supervising it. Bush sat down at the
head of the table directly next to Arias.

No one was particularly happy to be there at a meeting in
which they would deliver bad news. After all, they were about
to inform the governor that Florida appeared to be contami-

nating the nation's medicine supply and the state's Medicaid program was paying for it.

As Mike Mann began his presentation on Operation Stone Cold, Arias, the only Horseman in the room, grew calmer. "No one knows this stuff better than I do," he thought to reassure himself. He watched as Mann projected photographs on the wall of the shell company BTC Wholesale in Kissimmee and the undercover buy at AD Pharmaceuticals. Arias tried to watch Bush as he watched Mann, in an effort to read the governor's thoughts. The room was so quiet you could hear people swallowing.

Mann explained that the task force had identified eight to nine different cells of suspects who were trafficking in counterfeit and adulterated medicine. The suspects had opened shell corporations in Texas, New Hampshire, New Mexico, California, North Carolina, Tennessee, Utah, Georgia, and elsewhere to disguise the trail of their diverted, stolen, and counterfeit drugs, Mann explained. These shell companies obtained medicine by stealing it, acquiring it through fraud, or buying it from Medicaid patients. While the drugs appeared to come from all over the country, they left South Florida only after being sold to big national or regional wholesalers. The investigators also believed that they had identified at least one counterfeiter responsible for creating subpotent medicine.

They estimated that fraud accounted for 20 percent of the $1.8 billion that the state's Medicaid program spent on medicine, paying for drugs that had been dispensed more than once and sold back and forth in this gray market. Rhonda Medows, the secretary for the Agency for Health Care Administration, which oversaw the state's Medicaid program, reacted angrily. "How did you arrive at that number?" she demanded, dismissive of Mann's claim.

"Are you sure you mean that?" Bush asked. "Because I could balance my budget then."

Mann explained that the criminal cells they were investigating had collectively moved $200 million through bank accounts in a one-year period, all from the proceeds of diverted and counterfeit medicine, most of which Medicaid had paid for at least once. Since they were only investigating a portion of the cells engaged in similar activities, the 20-percent estimate seemed reasonable.

"Is he right, Rob?" Melanie Hines asked the young prosecutor.

"No he's not," whispered Penezic. "I'd say the number was higher."

"Where is this happening mostly?" Bush asked.

"South Florida," Mann said, those two words making the scenario he painted sadly plausible.

Some of the wholesalers handling the medicine were narcotics traffickers with felony convictions who had actually obtained state licenses, he explained. The weakness of the law, statute 499, made such offenses hard to prosecute in any meaningful way.

Then it was Arias's turn to present the two identical vials of Epogen. "Governor, here's one real and one fake vial. Can you tell the difference?" he asked. He felt so nervous that his hands were shaking as he held up the vials for the governor, who leaned in for a closer look.

Bush studied the two vials and broke the silence after a moment. "This is ripe for terrorism," he said. He then asked Hines how long it would take to convene a grand jury, which was needed for fact finding as a precursor to changing state laws.

Hines, a holdover from the last administration who was hoping to be reappointed, suddenly announced that she could convene a grand jury in little over a month and have everyone indicted six weeks after that.

"Twelve weeks from today?" the governor asked her, incredulous.

Everyone knew the accelerated timetable she was proposing could not possibly take place. Penezic was even shaking his head as the governor looked around the table. Commissioner Moore didn't care whether Hines's projection was fiction; he'd been gunning for a thirty-day estimate so that the investigation didn't bog down in the inertia of state government.

Having seen "No" on everyone's faces but not having heard it, Bush then said, "Okay, I'm going to take it under advisement," and he swept out of the room.

20. They Know
We Know

February 2003
Fort Lauderdale, Florida

FIVE DAYS BEFORE CHRISTMAS, GOVERNOR JEB BUSH DIRECTED the Florida Supreme Court to impanel a grand jury. Once he did that, it took only weeks for a group of ordinary Floridians to understand what Cesar Arias and Gene Odin had been trying to explain for fifteen years.

The eighteen men and women who comprised the seventeenth statewide grand jury met in a fifth-floor courtroom at the Broward County Courthouse in Fort Lauderdale. They heard testimony from the investigators, from industry representatives, state government officials, and cooperating defendants. They almost immediately grasped the significance of the evidence presented—namely, that a devastating question mark now hovered over every trip to the drugstore.

On February 27, they issued a searing report on Florida's drug supply and the grave danger it posed for patients. The report, forty-seven pages long, spared no one. It lambasted the health department for failing to enforce the law, the wholesale industry for its "willful blindness" in purchasing tainted drugs, and the traffickers for raking in profits that rivaled those of the narcotics trade.

The grand jurors brushed aside all industry spin and applied their common sense. "No one in their right mind would buy" an open bottle of medicine off the drugstore shelf, they stated. "But that is what many Floridians are unwittingly doing as a result of lax enforcement of the pedigree papers requirement."

For the Horsemen, the report was nothing short of miraculous. They had never imagined such a rapid or unequivocal statement from the grand jurors, who Odin had earlier dismissed as "a bunch of yahoos who don't know a drug from a cosmetic."

Their report stated that "a sizeable portion" of Florida's drug wholesalers are "uneducated, inexperienced, ill-informed rank amateurs with no pharmaceutical experience, many with criminal records." Though not trustworthy enough to perform "the most trivial tasks, yet through their hands pass some of the most expensive and delicate life saving drugs that exist today. No one has to go to their warehouses to buy their tainted product," the grand jurors concluded, "for eventually they show up in our hospitals, clinics and pharmacy shelves."

They found the industry's claims that verifying pedigree papers would be an overwhelming burden to be "weak and unpersuasive." "In the absence of any good faith effort by the industry to substantiate their claims, we have no choice but to disbelieve it," the grand jurors concluded. If pedigree papers were not useful, they continued, it was only because wholesalers failed to verify the information in them, for fear of losing purchasing opportunities.

The grand jurors also took aim at Arias's bureau and the deliberative process that had led to the Ad Hoc Committee on Pedigree Papers. After meeting for seven months, the committee issued a watered-down set of recommendations in October 2002. The grand jurors were not impressed. They wrote: "We fail to understand why, almost ten years after the passage of a state law, and seven years after promulgating its own rule, a state agency would have to embark on a nearly year long debate with the industry it regulates as to when, whether and how to enforce state law." But that is exactly what happened, they said, when the health department backed away from its "expressed intention" in November 2001 to enforce the pedigree law.

The Ad Hoc Committee had emerged with a recommendation requiring full pedigrees for only thirty of the most vulnerable drugs, even though more than 140 different drugs had been diverted or counterfeited over the years. "Criminals can change the drugs they divert and counterfeit much faster than the state can amend its rule to add or delete drugs from the list," the grand jurors noted.

To fix these problems, they recommended a complete industry cleanup: fingerprints, criminal background checks (as opposed to the current honor system), and a $100,000 security deposit for wholesalers—a hefty increase from the current $200. They also proposed dramatic increases in the penalties for falsifying pedigree papers and medicine itself. They recommended that the act of counterfeiting be a first-degree felony and that any counterfeiting that results in death should be a capital offense, subject to the death penalty.

The report had a magical effect on state legislators, who voted unanimously just weeks later to pass new legislation. As Representative Ed Homan, one of the legislation's sponsors, told a reporter at the *Orlando Business Journal*, "If everyone read that report, they would totally lose confidence in our health care system."

The Horsemen, who had no confidence left to lose, were ecstatic. A powerful ray of sunshine had exposed the problem and the entrenched interests that contributed to it. They were certain that arrests and a clean sweep of the management in Arias's bureau would follow. Every external indicator signaled the need for change. The lights were green, the levers on go.

DESPITE THIS, ARIAS AND ODIN FELT THAT IT WAS BUSINESS AS usual at the state's pharmaceutical bureau. The men and women in Tallahassee still poked along, their stately pace unchanged as they moved through the day, seemingly untroubled by a sense of moral urgency.

Arias learned that his bureau had released *all* of his personal information, including his social security number, home address, telephone number, and wife's name, to a Miami lawyer known for representing narcotics traffickers. The man had requested Arias's file under the state's public records laws, but that personal information was supposed to have been redacted. Had Arias told his wife, she almost certainly would have demanded they move. So he kept it to himself, even as he considered suing his department.

Instead, he abruptly stopped writing the daily activity reports and ordered Odin to stop too. He was afraid the department might release the reports to drug diverters who wanted to know how he spent his time—or who wanted to have him killed. But what bothered Arias even more than the careless and dangerous leak of his personal information was the sense he got that the status quo prevailed.

"At every other agency, their main concern is the public health. They're outraged," Arias observed. "But at our department, where it's our job, they don't care." There was no obvious explanation for the lassitude, though the Horsemen conjectured that the problems stemmed from corruption, stupidity, sloth, or all three.

Where they saw failure, Jerry Hill saw success. In his view, his staff had clearly demonstrated its teamwork and had succeeded in protecting the public from adulterated drugs. The bureau's efforts had had a galvanizing effect on state legislators, he said, which was no easy task.

But the Horsemen viewed the bureau as part of the problem, not the solution. "You could get rid of half of them and accomplish more," Arias fumed, slumped over a Cuban coffee, his hair unruly and his nails ragged. "You look at their desks, there's nothing on them. It's sad. It boils my blood to think that Jerry Hill still has a job. If this whole thing goes down and he still has his job, that's the biggest crime of all."

In perfect agreement, Odin chimed in: "Even if they want to do the right thing, it's for all the wrong reasons."

AS ARIAS AND ODIN POURED OVER BOXES, PILL BOTTLES, AND vials, they found themselves in an Escher-like world. Everything was suspect and nothing certain. Medicine that looked good turned out to be bad and visa versa. This uncertainty, with black bleeding into white, permeated every aspect of the investigation. Even those who appeared to be victims of the diverted-medicine trade became suspects as well. The government smelled "blood in the water," said Marty Bradley at BioMed Plus, still recovering from the burglaries at his warehouse. They were "trying to nail everybody."

By the end of 2002, Bradley found himself under federal investigation. He suspected that a former business associate, who pleaded guilty to racketeering and Medicaid fraud in 2001, had been whispering about him to the feds. In a surprise raid in December 2002, federal agents swarmed Bradley's warehouse, his home, his parents' home, and other facilities his company operated in Georgia and Tennessee.

They were looking for evidence that he had relabeled medicine from overseas for sale in the American market. Bradley told them that he had done no such thing. After hours of searching, the agents left and the raids appeared to turn up little of significance.

As the investigations spread, the bad medicine kept coming. On April 10, 2003, Arias stuffed his Buick to the roof with boxes of adulterated medicine. These contained Lipitor, a cholesterol drug, and Celebrex, for arthritis, that they had seized the day before in Jensen Beach.

From what they knew, the medicine had been smuggled from England through a Louisiana port and purchased illegally by Florida wholesalers, including Michael Carlow. While nothing about the medicine looked right, the question of what, exactly, was wrong with it would become a fraught and complex question. Unable to actually lock the Buick's malfunctioning passenger door, Arias left it open, despite the valuable cargo of some $300,000 worth of drugs, as he met with Odin.

Inside a Dunkin' Donuts at a strip mall in Weston, the two medicine detectives slumped over a table and rummaged in a plastic bag that sat between their coffee cups. It was full of tiny Procrit vials that belonged to an elderly woman in New York. Her daughter suspected they were counterfeit.

Some FDA investigators in New York examined the woman's medicine and concluded it was clean, but the two inspectors wanted to see for themselves. Arias lined up the vials on the table, sank down to eye level, and studied them to see if they were even and if the labels lined up.

"You can see them better if you turn them upside down," Odin reminded Arias impatiently. One by one, Arias turned the vials onto the stumpy bottle-cap ends, exposing a view of the bottoms of the rubber caps. The two men peered over the bottles, their chins at table level, as the patrons at the next table stared.

"They look good. Get the wand," Arias told Odin, who then went to the parking lot.

Odin returned with what looked like a doctor's implement, the kind of tool you'd poke down someone's throat or in an ear. In their low-tech, shoe-leather world, this black instrument was their cruise missile. The device lit up the codes or "chemical tags" that manufacturers embedded in packaging to help identify and authenticate their products.

Odin waved the wand over one Procrit box and gazed into the little screen, seeing a telltale marker. "It's good," he said.

With some relief, Arias turned one of the vials right side up and rotated the metal cap on the lid. "It turns," he said. "The bad guy's lids don't turn." He plunked the vials, each worth more than $1,000 when full, back into the plastic bag, knowing that his conclusion was not definitive and could not be.

The drugs belonged to Shuchu Hung, who lived in Flushing, Queens. She had been receiving Procrit injections for a rare blood disease. At first, the weekly treatments at her doctor's office left her feeling energetic. Then, when she picked up the medicine from a pharmacy and injected it at home, it had no effect, even after she upped the dose at her doctor's suggestion.

Shortly afterward, she was admitted to the hospital for a seemingly unrelated intestinal infection. One of her daughters, a pharmacist trained in Taiwan, saw an FDA advisory about counterfeit Procrit. The lot number and expiration date matched her mother's prescription.

It was the same lot as the counterfeit Procrit made by Eddy Gorrin, who had filled hundreds of vials with bacteria-laced tap water. Shuchu had picked them up from a Flushing pharmacy on the same day FDA agents arrested Gorrin in a Miami parking lot.

The FDA believed that it had seized all Gorrin's medicine. But Hung's case pointed up the uncertainty confronting those who investigated bad medicine. Only Gorrin knew for sure how much medicine he made, when he sold it, and to whom. Just because the box was authentic didn't mean Hung's medicine was. Even if the box was counterfeit, the medicine might be good. Though the medicine in the bottles appeared good, not all of Shuchu's medicine was in the bag. Also, she had already used the medicine, so the bottles were largely empty.

Yet Shuchu's symptoms were consistent with the counterfeited batch, down to her intestinal infection, which could have resulted from the *E. coli* in Gorrin's vials. When Shuchu first suspected her medicine might be fake, she lost hope and felt betrayed. Then, when she learned that the medicine may have been real after all, she felt even less hope. If the real stuff couldn't help her, what could?

IN APRIL, THE WINDING MAZE OF PEDIGREE RECORDS, COUNTER-feit lot numbers, and aliases led the Horsemen to another of Michael Carlow's cooks. His name was David Ebanks, and on a bright afternoon in Miami's Liberty City, Venema, Arias, Petri, and Jones surrounded him on a weedy sidewalk next to an abandoned lot, having just arrested him for using a forged driver's license to open a bank account. They wanted his cooperation. But Ebanks was sulky and silent. Dressed in linen pants and sandals, he stood with his head hanging down and his handcuffs cinched tight.

Steve Zimmerman, a Miami-Dade detective now working with the Horsemen, was also there. He had just switched from homicide to surveillance on the advice of his doctor, because of his hypertension. Unlike Petri and Jones, Zimmerman did not see the beauty of sitting motionless in a car for twelve hours at a time until one became an invisible part of the automotive landscape. But his outsider status as someone who retained an accent from his days in Brooklyn, New York, recommended him to the investigators.

Ebanks was a social worker with a penchant for prostitutes and fat cigars. The Horsemen believed that he collected and sold medicine for Michael Carlow using false identities, post office box addresses, shell companies, and obscure bank accounts. They also suspected he'd stolen the name of a patient in a drug treatment program where he'd worked as a counselor, used the man's social security card, and opened bank accounts through which he'd laundered millions of dollars. It was Ebanks who'd brought John Bullock, the recovering alcoholic, to Houston, Texas, to open up JB Pharmaceuticals, another one of Carlow's shell companies.

Three weeks earlier, on April 9, after weeks of surveillance, Petri, Jones, and Zimmerman pulled over Ebanks late at night and found a loaded handgun, two counterfeit driver's licenses, and seven bottles of HIV medicine in the back of his car. Ebanks had refused to cooperate. Based on that traffic stop, they had an array of felonies to charge him with. Venema decided to approach the project the way they used to work homicides in Hialeah: keep on arresting Ebanks, jailing him, and towing his car, one charge at a time, causing so much disruption to his life that his resistance would finally crumble. Today, John Petri had radioed Venema and Arias, who were eating dolphin sandwiches in Fort Lauderdale, to say they had Ebanks again. On arriving, Venema plucked the business card of *Craig Brand Esq.*—the lawyer who represented many small secondary wholesalers—from Ebanks's shirt pocket.

"Tell Craig I said hello," Venema guffawed.

The Horsemen surfed through Ebanks's cell phone and took note of a recent call to Brand.

Ebanks ventured, "Officer, if you're going to arrest me . . ."

"Oh I am," interrupted Venema, now turning his attention to Ebanks's wallet. "So what have you and Fabian been up to?" he asked, referring to Fabian Díaz, another Carlow associate.

Ebanks denied knowing Díaz, even as Venema pulled a car insurance card in Díaz's name from Ebanks's wallet.

Meanwhile, Ebanks's cell phone rang steadily. "Is business that hopping?" asked Venema.

Zimmerman, who had the native gift of the homicide detective, an ability to talk to anything, even a rock, until it said, "I did it," answered Ebanks's phone and began a cordial conversation with the person on the other line. It was a schizophrenic patient calling Ebanks from a local psychiatric ward, offering to sell his medicine Zyprexa, a costly favorite of diverters. Zimmerman made an appointment to meet with the patient the next day.

Exultant, Venema sang out "Roses are red/ Violets are blue/ I'm schizophrenic/ And so am I."

Petri explained to Ebanks that for using false documents to open a bank account he could be charged with racketeering in the future if he didn't cooperate.

Ebanks mumbled, "You mind if I scratch my nose?" He leaned over Zimmerman's car door and rubbed his nose against the top.

"You know," said Venema as though thinking it over, "I don't mind ruining your entire life."

"Your handcuffs okay?" Zimmerman asked, and Ebanks shook his head miserably. "Too tight?"

"Good, I like it that way," said Venema as they loaded
Ebanks into the police car and slammed the door. "Turn up
the heater," he told Petri jokingly, the car sauna being another
tried-and-true Hialeah technique.

As Venema watched Ebanks disappear down the street, he
smiled broadly. "The beauty is, they know," he said. "We know
they know. They know we know they know. We know they
know we know they know." He had made it clear to the bad
guys that he was coming, and he did not want them to forget
him anytime soon.

AFTER THE GRAND JURY REPORT CAME OUT, JERRY HILL AND HIS
managers at the Bureau of Statewide Pharmaceutical Services
in Tallahassee hailed the document as a positive step forward
that would educate the public and strengthen their hand.
They launched an effort to overhaul existing legislation and
turned to lobbyists for the wholesale industry to help them
draft the changes. The series of sit-downs with the industry,
which the local press got wind of, came to be known as "cooperation meetings."

The group that met on a regular basis included Hill,
Compliance Officer Sandra Stovall, and two lobbyists—one
representing the largest wholesalers' trade group, the Healthcare Distribution Management Association (HDMA), and
the other working for Salvatore Ricciardi's group of secondary wholesalers, the Pharmaceutical Distributors' Association
(PDA). The group's efforts became so interconnected that the
lobbyist for the PDA, Ross McSwain, openly referred to their
work product as the "DOH/industry bill."

As the draft went back and forth, the lobbyists focused on
several sticking points that displeased their members. One
was the proposal that wholesalers who failed to pass along a

pedigree with medicine they sold would face a third-degree felony charge, not just a misdemeanor as in the past. Mc-Swain wanted the word "knowingly" inserted into the text so that no one would be charged with a felony simply because he made an "innocent isolated mistake."

The other sticking point was the criterion that defined a *primary* wholesaler, an authorized distributor that would be exempt from passing along a full pedigree with every drug it sold. Wanting to increase the documentation of a medicine's origin, the regulators proposed making it harder to qualify as a primary wholesaler, a designation that would require 250 direct purchasing accounts with manufacturers, as opposed to the current fifty. On seeing this change for the first time, HDMA lobbyist Bonnie Basham sent an e-mail to Hill saying she was "shocked to see dramatic changes in the 'final' draft which you appear to have made without consultation with either Ross or myself."

She continued, "You have, by making these changes, now subjected large/regional/full-line wholesalers to the same scrutiny and additional licensure requirements you have placed on small secondary wholesalers. Why?" As to Hill's explanation that he faced an internal deadline for handing the draft to his legislative committee, Basham hammered, "What is the rush? I thought we were working as a 'team.' This is not how 'teams' operate."

Little more than three hours later, Hill wrote back saying, "I did not intend to upset the application/licensure process for major wholesalers. So to satisfy your concerns, we will add the old language back. . . ." He had just made it easier for midsized wholesalers to evade the pedigree requirements.

And so it went. The lobbyists pushed to delay the new regulations until 2007.

As e-mails circulated, FDLE Commissioner Tim Moore

wrote to the new statewide prosecutor, Peter Williams, "What's up with the 2007 date for the pedigree papers. Have you agreed to this?"

Williams wrote back that the lobbyist Ross McSwain had put in that provision, while acknowledging that the date was a "detail" to be worked out.

The final bill was a compromise that pleased almost no one. The grand jury had recommended that each drug sold should have pedigree papers tracing its origin from the manufacturer—nothing more than what Florida law had required since 1994. But the final bill required only that the largest companies provide such papers for thirty vulnerable drugs; the smallest wholesalers would be required to provide the documentation with all drugs they sold to other wholesalers. State senator Walter "Skip" Campbell, one of the bill's sponsors, called the legislation a Band-Aid. Though he had not known about the "cooperation meetings," they sounded highly irregular to him given the department's mandate to protect public health and not to satisfy the industry.

Representative Ed Homan, a Republican and an orthopedic surgeon who had sponsored the bill in the House, explained that the bill was a necessary compromise so that the legitimate wholesalers affected didn't try to torpedo it down the road. "The people that were not at the table were the people that were corrupting the system," he explained, adding, "The only people you cannot worry about are the criminals 'cause they're not going to show up."

But neither he nor anyone else in Florida government actually knew who lobbyist Ross McSwain represented, because Sal Ricciardi's group had not disclosed its members. "We haven't had a list," the health department's lawyer, Robert Daniti, acknowledged. Ricciardi's deputy, Bruce Krichmar, explained this by saying, "Given the climate today, there's not an upside to getting our names known."

As for the HDMA, widely viewed as representing the industry's "legitimate" wholesalers, almost a dozen of their members were being actively investigated, and several would be involved in a subsequent counterfeiting incident that would trigger the most widespread recall of medicine in American history.

Jerry Hill explained that he had been bound by what was feasible, perhaps not by what was ideal. "I recognize that our field operation sometimes wanted operations in Tallahassee to take other actions," he said. "We talked about what we could do, how we could make changes, and part of that change was to take advantage of media opportunities and other people who were partners," namely, the industry. "Those were some of the kinds of things we did in an effort to get where we wanted to go."

ON JUNE 13, THE HORSEMEN DONNED SUITS AND TIES AND GATH-ered at the Broward County Courthouse to watch Governor Bush sign the 2003 Prescription Drug Protection Act. Despite its shortcomings, the law did impose heavy new restrictions on wholesalers and required criminal background checks to get a license, doing away with the state's honor system. It created serious criminal penalties for those who trafficked in adulterated drugs or falsified the documents intended to authenticate them. In the glow of the signing ceremony, the investigators actually found themselves hopeful about the state's commitment to solving the problem. The legislation made Florida a leader in the nation's minimal reform efforts.

The Horsemen smiled broadly for an official photograph with Governor Bush, all of them present except Randy Jones, who never attended an official event if he could avoid it. Afterward, Jones joined them for lunch to celebrate the legislation and his retirement, which he had postponed for two years to work the case.

Before leaving the bill signing, Arias made a point of thanking the governor for his support. "You guys scared the hell out of me," Bush replied jovially, referring to the horror stories that had emerged at his briefing and in the grand jury's report.

Not missing a beat, Odin replied, "That was our intention, governor."

21. Inspector Arias Goes to Washington

June 9, 2003
Washington, D.C.

DEEP INSIDE THE CATACOMBS OF THE RAYBURN BUILDING ON Capitol Hill, a congressional researcher, Chris Knauer, got off the phone with security officials from Pfizer Inc., contemplating a more frightening and insidious threat than the one he had spent years documenting. In the past month, counterfeit versions of Lipitor, Pfizer's anti-cholesterol medicine, had been distributed all over the country and had possibly reached 600,000 patients. A wholesaler in Kansas City, Missouri, recalled more than eighteen million tablets, the legitimate and the counterfeit intermingled—one of the largest recalls of medicine in American history.

The incident had clearly exposed the vulnerabilities of America's drug supply, as Pfizer's top security official would later tell a federal task force.

Knauer had long studied the growing danger of medicine from unregulated markets overseas. But Pfizer officials had just told him that the counterfeit Lipitor had not reached patients through Internet orders or overseas purchases. It emanated from a Miami wholesale company and had reached patients through their pharmacies.

Knauer had recently returned from Miami, where he had visited one of the nation's fourteen international mail facilities.

Each week more than 150,000 packages of drugs arrived from overseas. By law, each package had to be inspected and either released to its American buyer or returned to sender. But in Miami, the manpower to do this lay with two part-time FDA inspectors who between them didn't work a full, five-day week. The backlogged packages awaiting inspection overflowed mail carts, spilled off shelves, and formed mountains on the floor.

Several years ago, the FDA adopted a "personal use" policy that allowed individuals to buy pharmaceuticals from abroad, but prohibited the importation of foreign drugs for commercial use. Many of the packages Knauer saw in Miami appeared to be illegal bulk purchases. Even the drugs that Americans thought they were ordering from Canada came from everywhere: India, China, Brazil. In a dangerous bureaucratic blunder, some FDA officials had recently approved the release of a huge shipment of Viagra from the Miami facility, even as the drugs were being investigated as counterfeit.

The situation was a mess. If Congress further loosened the restrictions on importation as they were considering, the sluices could open and all hell would break loose, the potential for terrorism staggering. But the Lipitor problem meant something entirely different—that even if the country's drug supply could be secured against foreign threats, it remained wide open to dangers at home.

Within three days of his phone call with Pfizer officials, Knauer and two of his colleagues boarded a plane to Florida. This time, Knauer made an appointment to meet with those who helped to reform the state's law, including Robert Penezic and Cesar Arias. He planned to recruit them to come to Washington and testify at a Congressional hearing about the problems they had seen.

———

FOR FOURTEEN MONTHS, MICHAEL CARLOW'S TRASH HAD SERVED as an early warning system of contaminated medicine about to spread through the nation's distribution system. True to form, evidence of the counterfeit Lipitor had shown up there first.

On February 22, Venema had extracted a curious invoice. It showed that a California company had shipped medicine to a repackager in Lexington, Nebraska. The shipping bill, more than $15,000 for next-day delivery, had been sent to one of Carlow's shell companies in Maryland. The buyer of the drugs was a wholesaler in Kansas City, Missouri, Albers Medical Distributors. Venema did not know the exact significance, but figured it was nothing good. "Where Carlow goes, nothing grows," he liked to say.

Rummaging amid cigarette butts and infant diapers, Venema had collected enough documents to piece together a remarkably detailed picture of Carlow's profits. He learned that Carlow's business enterprise had made $54 million in sales. In about eight months, he had cleared profits of at least $2.3 million, which he funneled into accounts in his wife's name.

Venema also pieced together a map of Carlow's new companies, the network of his front men, and the deals he was brokering both in and out of the country. He could even see Carlow's reaction to the investigation. One trash pull yielded a lawyer's bill for a consultation about the *60 Minutes* segment on counterfeit medicine that had aired in December 2002. Michael Mann had been interviewed and had mentioned "upcoming arrests."

To Venema's delight, he unearthed evidence of cash flow problems that would make it that much more difficult for Carlow to make bail should that day ever come. A notice from the IRS regarding a $290,000 debt stated that unless Carlow made some provision to pay, the agency would "be forced to believe" that Carlow was engaged in a "criminal act of evasion of income taxes."

He also retrieved a cordial letter that Carlow had sent to a Bahamas wholesaler:

> *It was a pure delight to speak to you. I will truly make an effort to bring myself and Noah to Freeport. We will enjoy the weather, the hospitality, the food and the casinos. By the way, I want to buy 100,000 Celebrex and 50,000 Viagra. We agreed at $450,000. While my wholesale pharmacy license is carried in the state of New Hampshire, please address all correspondence to the above address.*
>
> *Michael Carlow, Accucare*

The address was his home in Weston, Florida.

The letter revealed Carlow as the head of Accucare, a company he claimed his wife, Candace, ran. Venema believed that Noah was Noah Salcedo, a buyer for Albers Medical Distributors, the large regional wholesaler in Kansas City, Missouri. It appeared that Salcedo and Carlow were headed out of the country on a buying spree.

Albers Medical Distributors belonged to Douglas Albers, a pharmacist and popular civic leader named Ernst & Young's Entrepreneur of the Year for the Kansas City region in 2001. Students from the Kansas City School of Pharmacy rotated through his pharmacy as part of their training.

Albers's company had been in the FDA's sights for some time. A parade of counterfeit medicine, including Serostim, Procrit, and the AIDS medicines Ziagen and Zerit, had traveled through the company at the same time that Albers's sales had ballooned from $1 million to $47 million in a single year.

Since 2000, the Missouri Board of Pharmacy had repeatedly cited his company for buying from unlicensed out-of-state distributors.

Venema had collected evidence that Albers Medical Distributors should have known that it was buying diverted medicine from Carlow. The company had returned hundreds of boxes of medicine to Carlow's companies as too sticky and ragged to be resold, yet continued the buying relationship without interruption.

In March, with the relationship between the FDA and FDLE healing, FDA agents invited Venema to Kansas City to help serve a search warrant at Albers's warehouse. Records there showed that the company had bought from a rogue's gallery of South Florida wholesalers. Investigators seized a note from Doug Albers to his two medicine brokers, Noah Salcedo and Paul Krieger, indicating that he would dock their pay until they had made up for the cost of adulterated Procrit that the FDA had recently seized. They also found a confidential proposal for a business partnership between Albers and Cardinal Health.

Venema returned to Florida and fished a document from Carlow's trash showing that in just three weeks Carlow had sold more than $11 million in medicine to Albers, a quantum leap over sales from earlier months, which had run from $2 to $3 million. Venema shared this information with astonished FDA agents. The volume of adulterated medicine was bigger than any of them had imagined. From a single month of sales to Albers, Carlow made a profit of $700,000.

IN LATE FEBRUARY 2003, THE DRUGMAKER PFIZER STARTED GET-ting calls from patients and pharmacists all over the country about unusually thick and bitter-tasting Lipitor tablets that dissolved rapidly and left a burning sensation in their mouths.

Just weeks later, some regulatory inspectors from the FDA walked into the warehouse of a medicine repacker outside Chicago and saw workers taking Lipitor out of five-thousand-pill containers with Brazilian labels and repackaging them in smaller ninety-pill bottles. The inspectors, who looked only for regulatory and not criminal violations, did not catch the inherent falsity: five-thousand-pill containers did not exist in Brazil or in any market other than the United States. Someone here was reassembling the Lipitor from packaging that had originated at home and did not exist abroad.

The counterfeit Lipitor was all over the country. It contained some authentic ingredients, but also a range of other chemicals that, while not necessarily harmful, contained little that was salutary. The crime was different from the Procrit and Epogen counterfeiting, which tainted an astronomically expensive medicine taken by relatively few patients. Whoever did this had targeted a big-volume seller: Eleven million Americans took Lipitor. Pfizer sold almost $7 billion of it a year in the United States alone.

On May 23, the FDA announced that Albers was recalling 300,000 doses of Lipitor that posed a "potentially significant risk to consumers." All had been repackaged by Med-Pro in Lexington, Nebraska, the company mentioned on the invoice in Carlow's trash. The alerts kept coming over the next month: more lot numbers, more bad Lipitor. The Rite Aid pharmacy chain asked 200,000 customers who bought Lipitor between April 3 and May 23 to stop taking the medicine. The massive recall spread anxiety across the globe. In the Far East, the health ministry of the kingdom of Brunei even issued a statement to reassure consumers that their supply had not been affected.

Within weeks, a trail of evidence would lead FDA agents

and the Horsemen, now working in concert, to a convicted cocaine dealer in Miami named Julio Cesar Cruz. They suspected him of masterminding the Lipitor counterfeiting and having developed the scheme with associates he met while serving time in federal prison.

ALMOST FOUR MONTHS HAD PASSED SINCE THE STATEWIDE PROSecutor promised swift indictments in Operation Stone Cold. Since then, Melanie Ann Hines had been replaced, no one had been arrested, and Venema could not find an outlet for his restless anger. As much as he threatened and ranted and paced, the bureaucrats regarded him with indifference. So he figured it was time to do something crazy—which ended up being something stupid. He called Doug Albers in Kansas City to let him know that Carlow's business and his business were next.

"Your boy Noah [Salcedo] is on a plane to the Bahamas with Michael Carlow," Venema needled, showing off the knowledge gleaned from Carlow's trash.

"Wait a second, he's not my boy," Albers insisted angrily. "I'm not associated with him anymore."

Venema swept past the remark: "This stuff is all bad. You're putting people's lives at risk."

"I'm a pharmacist," Albers insisted. "I would never do that."

"You talk to Noah," Venema continued, "and tell him I'm coming for him."

After hanging up, Venema cackled, "He wanted to have a root canal and a rectal exam more than he wanted to talk to me." He was feeling cocky at that moment, but would quickly come to regret his call.

———

SEVERAL DAYS AFTER TALKING WITH ALBERS, VENEMA STEPPED out of his truck under a bright moon, sneaked past a row of dark, silent mansions, and zeroed in on the familiar bags, which contained the cigarette ashes from Carlow's study. He grabbed them silently and sneaked back to his truck. It was 4 A.M. He could not even recall what *rested* felt like, but each time the adrenaline and excitement over what he might find far outweighed the missing sleep.

Now, the bleary-eyed detective pulled into a gas station and ripped open the night's haul. He stared down, too shocked to do anything until he began laughing and could not stop. The next morning he called Odin, still laughing. They had always joked that when Carlow got out of jail in thirty years, he'd buy a paper shredder powered by an F-16. "The unthinkable's happened," he told Odin. "The worst thing that has ever happened."

"What?"

"He got a fucking shredder!" Venema bellowed. "And not just any shredder. A Cadillac." The laser paper in the bags had been shredded so finely, there wasn't a scrap large enough to hold a punctuation mark. Carlow must have figured out that there was only one way for Venema to learn so much about his trip to the Bahamas.

JOHN PETRI HELD THE SECOND ANNUAL "OPERATION STONE Cold" party at his house, the night before Arias was headed to Washington, D.C. Venema declared the event top secret and off limits to micromanaging bosses and other "enemies" of the Horsemen. All the men brought their wives. Michael Mann and Robert Penezic also came. The party had a Hawaiian theme, in which everyone came dressed in a flowered shirt. Petri videotaped the guests as they entered and his wife, Gloria, looped a plastic lei around their necks.

Randy Jones's wife, Dee, also a police officer, made her famous rum-soaked coffee cake. Petri, a natural master of ceremonies, had prepared a plaque for his partner Randy, who was finally retiring after thirty-five years on the job, along with an awards ceremony and a photomontage of great Operation Stone Cold moments (including Venema and Odin asleep, side by side, at a conference, their chins sunk onto their chests).

On the one hand, there was a great deal to celebrate. They had changed state law. Tomorrow, Penezic, Arias, and Arias's supervisor, Gregg Jones, would depart for Washington, D.C., to testify before Congress. Yet a sense of unease hung over the party. Carlow was free, and even Venema had stopped using the word "imminent" to describe the arrests he'd imagined so many times. "We're not even doing it to make a big splash anymore," he explained with a new tone of defeat. "We're just doing it for our little group."

Penezic had arrived late to the party, looking stressed. In his office, the testimony he was to give before Congress had become a group writing project. His remarks were massaged to thank everyone, detail a "spirit of cooperation," and avoid angering the drugmakers or the Big Three. In the ongoing turmoil surrounding the case at the statewide prosecutor's office he had recently been back-benched, and another prosecutor far more skeptical of the evidence against Carlow had been put in charge. To Penezic, it seemed like Operation Stone Cold was going off the rails.

"He wanted to resign several times," Arias said of Penezic. "He's so dejected." With a woeful air and dark circles beneath his eyes, Penezic looked as though he had seen to the bottom of something terrible, and it was not just the degradation of the drug supply. Like Arias, he remained uncertain as to whether the hearings in Washington would be a whitewash or a showdown.

ARIAS'S FATHER, CESAR SILVANO ARIAS, HAD SPENT THE LAST YEAR rebuilding from memory the small Cuban village where the family had lived decades earlier. Using an architectural software program on his home computer, he re-created Central Alava in its entirety, including every tree, park bench, and building. As a civil engineer, he knew the village in a way that few people did, and the program allowed him to view his creation from a vast distance or from inside its tiniest detail.

Arias's notion of a pure drug supply was not unlike his seventy-nine-year-old father's memory of a free and prosperous Cuba. It was something that he had imagined and built in his mind over and over, from its subterranean infrastructure to its global effects. When Governor Bush actually convened a grand jury just three days after being briefed, a new legal edifice began to rise quickly around the abuses Arias knew so well. Justice was no longer just around the corner in the movie version he'd imagined. The construction of it had begun, and his trip to Washington was akin to erecting foundation walls.

Sixteen years to the day that he began work as a Florida drug inspector, Arias watched the monuments of the nation's capital come into view as his plane touched down at Reagan National Airport. Once outside, he fought the throngs of commuters to find a taxi to the hotel where his mother and father were waiting. They had flown up separately from Miami to witness the testimony of their eldest son.

As the taxi sped past the Pentagon and the Lincoln Memorial, Arias marveled at how far two powerless pharmacy inspectors and three retirement-age cops had come. His appearance tomorrow before the House Energy and Commerce subcommittee on Oversight and Investigations would be on behalf of the Horsemen, five unknowns who might actually change the way the nation's medicine was distributed.

As he entered the hotel, his euphoria stopped short. His cell phone rang and it was his bureau director, Jerry Hill, calling for only the third time in more than a year. Hill explained that his superiors had reviewed Arias's proposed testimony and did not want him to mention the problem of Medicaid fraud. He needed to report to Governor Bush's Washington, D.C., office, where staffers would help rewrite his remarks. The call knotted Arias's stomach. Unless he was prepared to quit, he had no choice but to follow the bureaucratic directive and conceal what he knew to be a significant prelude to, and cost of, the counterfeiting problem.

He slumped in the hotel restaurant and raised Odin on his walkie-talkie. "The politics have started," he declared. "I should have declined to go. How can I sit there and be quiet about this? I don't know why they dragged me out here if they don't want to hear the truth."

"If you're going to be silenced," Odin advised, "then you need to quit."

"If I have a job tomorrow, I'll let you know," Arias said. Then he raised Venema and explained the predicament. "What do I do now, coach? Stick to my principles and get fired?"

Venema's voice crackled back over the phone: "This is what separates the men from the boys," he said. "It's time to make a decision."

As a group, the Horsemen believed that public disclosure was essential to righting the wrongs they had uncovered. But *the truth,* as they were learning, was subject to interpretation and political expediency, and once expressed drew opponents from the shadows. Speaking the truth was not necessarily the end point they had once imagined, but a place to start winning converts to their cause. And so Arias weighed the benefits of the unvarnished truth—which would mean his certain firing. Or some shade of it—which would help him to keep his job.

That evening, Arias's mother, Maria, waited for him in the hotel lobby. Dressed for dinner in her best white suit and jewelry fit for a first lady, she was furious to learn of the state's efforts to censor her son's testimony. Arias had been instructed to eliminate all mention of "Medicaid recipients," who may sell their drugs for cash, and instead use the phrase "street brokers" to describe those who divert medicine. As Arias understood it, state officials did not want to disclose at a budget-crunch moment that a significant portion of the federal money they received was being devoured by fraud. It was a restriction he decided to live with.

But Arias's proud Cuban mother, who saw shades of Fidel Castro in any infringement of personal liberty, advised her son to make a principled resignation. "Tell them, 'I'm not a puppet,'" she declared bitterly in the lobby of the Marriott.

"Mom thinks it's corruption," Arias explained to Penezic, who had joined them for the evening. Arias's frustration receded somewhat as they walked the streets of the capital, feeling a sense of awe.

Though his mother continued to lament what she viewed as a cover-up, Arias assured her that the problem was too big and too dirty to be hidden any more. "It isn't a wash," he said over dinner. "They couldn't wash it."

———

SHORTLY BEFORE 10 A.M. THE CROWD IN THE CORRIDOR OF THE Rayburn building filtered into an oak-paneled room covered with long red curtains. The hearing, entitled "A System Overwhelmed: The Avalanche of Imported, Counterfeit and Unapproved Drugs into the U.S.," was about to begin.

The event was instantly different than Arias had imagined. Though the gallery was full, only a few congressmen, including the chairman, James C. Greenwood, occupied seats at the dais.

Greenwood began by describing the American system as one in which poor-quality foreign drugs and domestic counterfeits easily penetrated the most nominal of firewalls. He held up two identical boxes of Serostim—one real, one fake, saying, "There is no way on God's earth you could tell these two products apart." He added that the citizens of Florida could not presently know if their drugs were legitimate. Nor could anyone else, Arias thought to himself from his spot in the front row.

The first panel of witnesses, one U.S. Customs representative and two FDA officials, took their places at a table piled with boxes of seized, misbranded, and counterfeit drugs. It quickly became clear that they had been called to answer congressional accusations that they had not done enough to combat foreign drugs pouring into the country. William K. Hubbard, the FDA's associate commissioner for policy and planning, calmly explained that the FDA had no intention of frisking senior citizens as they returned from purchasing medicine in Canada or Mexico, unless Congress wanted to expand the agency's statutory authority.

The circular dance between the FDA and the lawmakers continued, each side wanting the other to incur the political liabilities of restricting affordable medicine for senior citizens. "Nothing seems to change here," Congressman Bart Stupak of Michigan grumbled.

Thumbing through Penezic's printed testimony, which he had yet to give, Congresswoman Jan Schakowsky of Michigan turned to John Taylor III, the FDA's associate commissioner for regulatory affairs, and said, "There are cocaine traffickers entering into drug wholesaling. What are we going to do about the growing criminal element now involved in the trafficking of prescription drugs?"

Taylor confirmed that his agency had seen "a wide array of people get involved in counterfeiting," and that the agency's first priority was to protect the public health.

"What is it that will assure people that when they go to Walgreens or CVS, they're not going to get counterfeit product?" Schakowsky persisted. She went on to answer her own question: As long as the cost of drugs remained sky high, criminal elements would come in, she said.

Congressman Bobby L. Rush of Illinois then asked Hubbard how many Americans had died because of counterfeit medicine. "We believe that's an unknowable thing," Hubbard said, explaining that patients who receive subpotent drugs don't necessarily keel over and die. They just don't get better.

After the lunch break, Penezic, Arias, and Gregg Jones took their seats at the witness table. The dais was nearly empty. Only one legislator, Congressman Greenwood, who had to be there to chair the panel, was seated. Penezic had two false starts with his microphone off before getting underway. His rewritten testimony began with his praising almost everyone in Florida state government:

> *The citizens of the state of Florida and the United States are safer today due to the collective efforts of a broad spectrum of representatives in the state including: Governor Jeb Bush, Attorney General Charlie Crist, State Senators Durell Peaden and Walter "Skip"*

Campbell, State Representative Ed Homan, Department
of Law Enforcement Prosecutor Pete Williams, and the
members of the Seventeenth Statewide Grand Jury. This
overall process that led to the new state law was truly
government at its best, because all parties recognized
the problem and worked diligently and tirelessly to solve
it. I think it is especially important to recognize the
contributions of Governor Bush and Attorney General
Crist for their leadership and vision.

With these formalities behind him, he became eloquent, explaining that unless the federal government mandated a
strong audit trail for drugs, no one in the country would be
safe because bad drugs (and bad wholesalers) will just move
from state to state.

Gregg Jones was strong and clear in his testimony. He described a secondary wholesale market "riddled with corruption." Because the law requiring that all drugs come with
transparent audit trails has not been enforced, Jones told the
legislators, "Your vision of a pedigree revealing the true origin of a drug has never been realized."

As Arias began his testimony, which had become strange
and stilted with revision, he was clearly nervous. He described
many sources of diverted drugs including those "purchased
from street brokers." He went on to say that the "street broker is an unregulated individual, who has no clue how to
handle or store these temperature-sensitive products." These
"brokers" funnel huge amounts of diverted drugs from the
streets of South Florida back into the drug distribution system, he said, "not just in Florida but throughout the nation."

The words were not his own, of course, and so they did
not plainly explain the problem: that "cooks" working for
drug traffickers like Carlow parked their Hummers outside of

AIDS clinics and gave Medicaid recipients petty cash in exchange for their newly prescribed drugs.

The rest of his testimony was stronger, as he explained
that every time counterfeit drugs entered the mainstream distribution system, it was through a dishonest wholesaler. "Once
the drugs enter the system they can end up in any pharmacy
in the nation," he said. "That is why there is no patient in the
nation that can know with 100 percent certainty that the drugs
they are getting are what they are purported to be—or if they
are—that they have not been in the trunk of someone's car, or
sitting in a hot warehouse or a crack house in South Florida."

One of the Congressmen appeared confused. "What is a
street broker?" he asked. Arias nervously twisted a paper clip,
his face appearing on a big-screen television that overhung the
dais and showed the faces of the witnesses, who had their
backs to the spectators. He struggled to explain the role of
these brokers without using the word "Medicaid."

When Greenwood asked Penezic if the Florida legislation
had been compromised at all, the prosecutor responded haltingly, "I was not in Tallahassee when the compromises were
being made." Regaining his footing, he added, "The industry
has an argument that a full pedigree can't really happen now."
He added that the legislation was a great first step that would
make consumers safer than they were. Consciously or not,
both Arias and Penezic had chosen to keep their jobs.

Greenwood concluded the hearing with gratifying praise
for his three witnesses from Florida: "Thank you for being on
the cutting edge of this issue and for your fine, fine work."

22. The Ultimate Box Case

May 23, 2003
Plantation, Florida

OVER LUNCH AT THE QUARTERDECK RESTAURANT, MICHAEL CARlow extolled the virtues of Costa Rica. He spoke of nice real estate, interesting "retail" opportunities, and the "hot women" he had enjoyed there recently.

"Smoking hot," his lunch companion and business partner Steven "Doc" Ivester interjected.

Carlow did not mention another benefit of Costa Rica, one that must have occurred to him during the state's slow-motion eighteen-month pursuit. The steamy Central American country would be a good place to disappear. Later, he would tell Ivester that he would go to Costa Rica if the investigation in South Florida came any closer to him.

From a distance, the two men might have been mistaken for good friends, ogling women, talking business, and even sharing personal difficulties. Carlow divulged that he found his young wife, Candace, "very immature," while Ivester confided that he went to therapy. "You have this really tangled personal life that's intense, very tangled," Carlow observed. "It's like a bowl of spaghetti that's been drying out."

But Ivester hated Carlow. In fact, the wire tucked beneath his shirt was recording the medicine trafficker's every word. Over the two-hour lunch, Ivester kept leading the conversation back to Carlow's previous schemes and his future plans, and how he managed to amass $4.1 million in the four years since filing for bankruptcy. At one moment, Carlow spelled out his business ethics, stating, "I do not put friends, neighbors, acquaintances into any deals that I am not in myself." At another, he noted approvingly of a woman in the restaurant, "That's a tight pussy there."

As the conversation shifted between business and women, with Carlow doling out advice on both matters, Ivester knew more about his lies and betrayals than Carlow could have guessed.

In 1998, Ivester, a technology inventor and entrepreneur, launched a company called Navigator PC, to develop advanced navigational devices for the United States Navy. Carlow, whom he had not known previously, became an early investor, pledging $500,000, though he gave the money in dribs and drabs.

At first, Ivester tolerated Carlow's peculiar behavior. One day through the office's thin sheetrock walls, he heard Carlow offer a particularly beefy secretary $25 if she would show him her panties. As those who spent any time with Carlow knew, he liked big women, particularly those others deemed fat and ugly.

Another day, a janitor at the Navigator office told Ivester about some men taking photographs of a car in the parking lot. It was the Horsemen photographing Carlow's car. When Ivester asked his new partner whether he was under investigation, Carlow blew up, screaming, "You don't fucking know me. I'm going to ruin you."

Shortly afterward, Ivester found Carlow hugging his girlfriend, who was slim and attractive. Carlow had begun a campaign of seduction that ultimately divided the couple. Ivester believed it was Carlow's revenge for his asking about the investigation.

But Ivester knew something about revenge too. "I'm not a badass but I'm not dumb," he said. At the right moment, he had a friend reach out to state law enforcement to offer his services. When Gary Venema learned of Ivester's offer in late May, it was as though Christmas had come early in the form of a savvy and jealous entrepreneur with an impeccable motive for revenge.

Two days later, Ivester was at the Quarterdeck listening as Carlow—cocky and disheveled as ever—provided a veritable map of his criminal activities. He described expanding his pharmaceutical business with a new shell company in Kansas City, World Pharma, run by his former banker and confidante Jeanie McIntyre. But the home of this new endeavor was now in question. Kansas City had been the site of a "pretty big explosion . . . with a . . . big buyer of mine," Carlow said, referring to Albers, who "got raided by the FBI and the FDA . . . and the ripple effect kind of like spread to the four corners of the earth."

He also explained that McIntyre would become his new bookkeeper, replacing his mother-in-law, Marilyn Atkins, who Carlow said he'd recently terminated for "piss-poor record keeping" and being "in-fucking-competent."

He even described a whole separate business: smuggling knockoff designer clothes. One thousand cargo containers from Karachi had just arrived in Los Angeles. The clothing scheme had much in common with his pharmaceutical efforts. He planned to "uplabel" the clothing from no-name to designer. This disclosure explained the large wire transfers to Karachi that Venema had seen in Carlow's bank records.

He went on to talk about Costa Rica and the weekend he

spent there at a club that catered to American men. But his lurid description of the weekend with a twenty-year-old named Danielle mattered little to Ivester or to the Horsemen. What grabbed their attention was Carlow's almost incidental remark that he might go back to Costa Rica again "this coming week."

JOHN PETRI FEARED MICHAEL CARLOW WOULD FLEE. FROM HIS pickup with blacked-out windows, the sergeant observed the telltale signs he knew so well from twenty years of watching people. There was a buzz at Carlow's Weston mansion, with people coming and going. Recently, Carlow's wife had brought suitcases to her mother's house. The two women had driven to the courthouse, where they pulled records on the deeds to their property, Petri learned after following them inside.

Petri and Randy Jones started watching Carlow in twelve-hour shifts because their department had refused to assign more officers, despite the risk that Carlow might flee. They "sat up" on his house from a neighbor's driveway and followed him along South Florida's roads.

For a man so hunted, Carlow appeared utterly blasé. He slouched along in rumpled shorts and a T-shirt, cell phone glued to one ear like any middle-aged dad tracking down his kids. He continued to meet associates at his favorite outdoor café and unwittingly led the Horsemen to a new warehouse, where they snapped photos of him taking out a key to the door.

Carlow's conversation with Ivester had sparked a frenzy among the investigators. While the prosecutors dithered, Carlow appeared to be planning his getaway. After eighteen months of advance warning, he was free to go to Costa Rica at the hour of his choosing and was also free to never come back.

IF CARLOW NEEDED ANOTHER INCENTIVE TO LEAVE, HE GOT IT on May 26, three days after his lunch date with Ivester. A Fort Lauderdale newspaper, the *Sun Sentinel,* ran a front-page series, FORMER CONVICTS TRY A SAFER VENTURE: PHARMACEUTICALS. The article, which quoted liberally from Venema's search warrants, described Carlow as a "major wholesaler selling millions of dollars worth of questionable medications out of his $1.3 million home." The paper ran the mug shot from his earlier arrest on the front page.

The article quoted attorney Craig Brand, saying that his client planned to file a lawsuit against the state for persecuting him unfairly: "It's easy to throw stones," he said. "I look forward to seeing their proof."

After the article came out, Carlow told Ivester that his lawyer believed the investigators didn't have anything on him. "If they had, they wouldn't have printed an article, so they blew their wad," Carlow said.

But his mellow demeanor did little to reassure the Horsemen, particularly after Ivester unearthed a document that Carlow had left in his Navigator offices. It was entitled "Michael Carlow Offshore Wealth Preservation Planning Business Structure Diagram" and listed various offshore accounts, essentially providing a template for a life on the lam. Petri told Penezic that if Carlow headed for an airport, he planned to handcuff him and "hope someone shows up with some charges."

IF MICHAEL CARLOW HAD BEEN CAUGHT SELLING CRACK COCAINE at a Miami intersection, he would have been arrested instantly and faced serious prison time. Instead, he was suspected of selling more than $54 million in adulterated medicine to wholesalers nationwide, tainting the country's drug supply, and potentially killing patients. And almost no one in Florida government could figure out how to stop him.

Weeks had rolled into months of inter-agency bickering, as officials with the Statewide Prosecutor's Office, the Attorney General, the state Health Department, and the Florida Department of Law Enforcement argued over what laws, exactly, Carlow had broken and who had jurisdiction to prosecute them.

In truth, the most senior state prosecutors were afraid to proceed. They liked "3x5" cases, those in which the evidence fit on a file card that small. Conversely, they hated "box cases," in which the evidence arrived in a box. Carlow was the ultimate "box case."

Carlow's talk of Costa Rica and the *Sun Sentinel* articles increased both the panic and chaos of the debate. Some of the investigators argued that if they reconvened the grand jury to issue racketeering indictments, the lengthy process might allow Carlow to flee. Penezic argued that it was better to arrest him immediately on a misdemeanor. That way, if he fled afterward, at least he would be marked as a fugitive with an outstanding warrant.

Those who witnessed the ongoing disagreements became convinced that the dispute was not necessarily about how to approach the case, but whether to approach it at all. Skeptical prosecutors had made Penezic assign the investigators fruitless tasks, which the lawyers claimed were necessary as a prelude to prosecution. One such task: determine how long it would take to remove a pharmacy label using the paint remover Goo Gone and how long it would take to reglue and dry it. The politics in the statewide prosecutor's office had become so noxious that Odin declared his "renewed respect" for the statewide pharmaceutical bureau.

———

THE HORSEMEN LOOKED BEAT. VENEMA'S WATCH STRAP HAD broken and his watch dangled loose like a bangle around his wrist. His eyes were puffy, his face creased and poorly shaven. His pants were only half-zipped. He wolfed down meals, sauce dribbling down his chin, walkie-talkie squawking. He had gained twenty pounds since the start of the investigation.

Arias, too, was a mess. Exhausted, he laughed so often that he actually drooled, wiping away saliva with the back of a hand. The pocket of his favorite shirt had ripped and the pens he clipped to it swung back and forth on the detached fabric like a pendulum.

The mood was such that Venema actually basked in the hatred of his enemies and was delighted when Craig Brand requested his internal affairs file, which the lawyer would find disappointingly slim. Brand had already subpoenaed John Petri to give a deposition as he tried to build a case that the police had harassed his client David Ebanks, the social worker they had arrested on four separate occasions. Venema swore that if he got a subpoena from Brand, he would wipe his ass with it.

PENEZIC'S BOSS, THE NEW STATEWIDE PROSECUTOR PETE Williams, determined suddenly in late May 2003 that his office had no jurisdiction over state health statutes and wanted to hand off the Carlow case to a local Broward County prosecutor.

The problem with this strategy appeared obvious to the Horsemen: What prosecutor in his right mind would sign up for a case so complex, politically fraught, and legally booby-trapped that another prosecutor was literally giving it away?

In late May, Venema attended a meeting with the Broward County prosecutor and brought along Gene Odin, hoping, with his irrepressible optimism, that it would put the arrests back on track. Afterward, Odin slipped into Arias's car in the FDLE parking lot to report on what had happened during the thirty-minute meeting, and the two men sat there as a torrent of afternoon rain fell from a black sky.

"They're arguing over one word in there," Odin said.

"What's the word?"

"Guilty."

Odin, punchy with frustration, let loose a recital of almost every indignity and absurdity that had beset them. First, he said, the Broward prosecutor honed in on the obvious: Why had Williams only realized now that he lacked jurisdiction, when his office had more than a year to ponder every possible charge?

When Williams tried to explain their sudden concern that Carlow might flee, the county prosecutor looked incredulous and suggested that they put a tracking device on Carlow's car. The investigators' bosses had already chewed endlessly on this idea, finally rejecting it because Carlow had so many cars and drove slowly enough to make following him a relatively safe activity.

Appearing annoyed, the local prosecutor ended the meeting by saying that since they'd had eighteen months to work on the case, he wanted at least a day to review the evidence.

"Here's the whole story in a nutshell," said Odin from the back seat of Arias's cluttered Buick, where he'd jumped in to get out of the rain. "Together we have thirty-one years' experience. We've been warning about this stuff for a decade. . . . Then *60 Minutes* runs a twelve-minute segment and the governor forms a grand jury. Everyone's like, 'Holy shit. We have to look into this.' These people have read all Gary's reports. But when they read it in the *Sun Sentinel,* now they're scrambling as if they had no heads.

"They act as though Michael Carlow suddenly found out what he did: 'Holy mackerel, the cat's out of the bag, the crooks just found out. Who told the crooks what they were doing?' It's like a Chinese fire drill. You say, 'Charge,' and everyone goes in different directions."

Arias laughed so hard, he doubled up at the steering wheel.

BY NOW EVERYONE HAD JOINED IN THE EFFORT TO WATCH CARlow. Even Arias revved his Buick behind their suspect's forest-green Explorer. Pharmacy inspectors always used to lose suspects at tollbooths, because their department refused to purchase the electronic pass that would allow them to pay automatically. Arias had spent the $50 himself.

On an early June morning, Petri, Arias, and Jones followed Carlow along the Sawgrass Expressway into a complex of modest rental homes where Jeanie McIntyre—the woman who had been his banker, was now a girlfriend, and on paper planned to be the president of World Pharma in Kansas City—lived.

As they covered the entrances to the complex, Arias gleefully read aloud an editorial in the *Sun Sentinel*, which blasted the health department for its failure to prevent adulterated medicine from reaching patients.

"Oh I love it," he declared. "How long can you do nothing and get away with it?"

After an hour, Carlow emerged from McIntyre's house and was headed to see attorney Craig Brand. The investigators followed at a discreet distance.

In Brand's parking lot, the weather went dark and an ibis, the last bird to leave before a hurricane, marched about the grass median. Suddenly, Petri's walkie-talkie crackled. His office wanted him to come in directly and report to his superior.

The matter sounded urgent, if not dire. Pale with anger, Petri said, "I might be retiring earlier than expected."

Petri had a guess as to the problem: The well-connected commander of the rival police unit was still wrangling to claim the case as his own and may have succeeded.

Hours later, when Petri finally reemerged from his closed-door session, the walkie-talkies lit up. He faced a new problem—this one involving the leak to Grillo that had exposed their informant. Petri suspected that the leak had emanated from within his department and had broached the possibility with a sergeant he knew well. Word of his suspicions had traveled through back channels all the way to the director's office. As a result, his supervisors accused him of voicing concerns outside the chain of command. They gave him a deadline of noon the next day to officially report his suspicions to the internal affairs department.

"Heard you need some Zantac," said Arias.

"You got anything stronger?" Petri asked. "One of those things on *Animal Kingdom* they use to tranquilize elephants?"

ENSCONCED BEHIND HIS DESK AT FDLE, PLOWING THROUGH BANK records and organizing the group's evidence, Venema could have used some tranquilizers too.

In April, his son Kevin had been sent to Iraq as part of a maintenance battalion for the Fourth Infantry Division, and the light in Sandy's beautiful blue eyes, which Kevin shared, had gone out. Every day, she pinned a yellow ribbon to her blouse, her lips drawn and unable to form a smile. Ribbons adorned the trees in their neighborhood.

As weeks passed without any word from Kevin, Venema dealt with his anxiety by throwing himself even deeper into the investigation, if that was possible. During his sixteen-hour days at the office, he fielded every imaginable complaint from his new supervisor: His desk was too messy. His evidence was in disarray. He had filed for too much overtime in May (twelve hours total for the entire month).

Even with no arrests foreseeable, officials were already fighting over who would get to stand at the podium during the press conference. The director of one state agency made Venema break away from crucial work and drive an hour to his office to discuss the matter.

But what really drove Venema crazy was the thought that another agency entirely might end up arresting Carlow first. Venema had given Customs agents the tip about the smuggled clothing and related wire transfers to Karachi. To his surprise, the agency mobilized instantly for what it hoped might be a vast smuggling case.

The whole thing practically made Venema want to cry. After selling adulterated medicine throughout America, "he's going to be nabbed on an *apparel* case," Venema hollered into his walkie-talkie, as he walked into a waterfront bar in Fort Lauderdale to meet the other investigators.

Though he almost never drank, he approached the bar. It had not been an easy few days. First, Carlow had actually driven to an airport, creating panic in Venema's office. Phone calls flew. Petri stood by to arrest him. As it turned out, Carlow was headed to Los Angeles to inspect the delivery of smuggled clothing. Venema alerted Customs, which quickly stationed a surveillance team at the Los Angeles airport and followed him until he boarded a flight back to Florida.

As the days ticked by still with no arrests, Venema's mood grew dark. It's no wonder he needs blood pressure medication, his wife Sandy reflected over dinner.

———

SINCE THE *SUN SENTINEL* ARTICLES, NERVOUS CUSTOMERS AT THE Walgreens where Arias's former colleague Robert Loudis now worked as a pharmacist had begun to ask him whether the drugstore bought medicine directly from the manufacturers. Loudis knew that he had no honest way to reassure them. Instead he would say that all their medicine came directly from an enormous Walgreens warehouse in Jupiter, Florida, that was so state-of-the-art it was staffed by robots. What he didn't discuss was where the drugs had been before that. After all, neither he nor anyone knew for sure, so why go into it, he figured.

By now, the nationwide director of security for Walgreens had written to Venema, asking how the company could protect itself against adulterated medicine coming from secondary sources.

IN EARLY JUNE, ARIAS AND ODIN SAW A POTENTIAL SOLUTION TO the problem of their boss, Jerry Hill. They headed to a local Hooters after work to discuss it.

They had learned that the health department's inspector general and auditor general were coming to investigate their bureau. They were told that a health department employee in Jacksonville had shot a co-worker and that the visiting inspectors wanted to make sure no one else had an anger-management problem. Hoping this might signal the beginning of the end for the managers of their bureau, Arias and Odin prepared for the visit by creating a list of the bureau's failings in the last decade. The list included everything from a failure to inform the state's Secretary of Health of grave public health dangers to not supplying the inspectors with thermometers.

As it turned out, the inspector general's arrival had little to do with the Jacksonville shootings. When Odin and Arias asked the inspector general why she had shown up, she responded, "We go anywhere the Secretary of Health sends us." Perhaps their complaints had been heard after all.

IN EARLY JULY THE STATEWIDE PROSECUTORS' OFFICE NOTIFIED grand jury members that they were being recalled, starting on July 14. Venema was slated to testify for three days straight and expected to emerge with indictments.

In June the Horsemen's bosses finally got religion. Red lights turned green. Convinced that Carlow was ready to flee, Venema's supervisors now wanted to arrest him for something.

While Penezic had no explanation for the sudden change, Michael Mann believed the relentless pressure from FDLE had finally forced the statewide prosecutor's office to get serious. Top officials at FDLE had demanded that the prosecutors either lead, follow, or get out of the way. The grand jury report, the media reports, and the Horsemen's own hardheadedness had added to the cumulative pressure. And the prosecutors had finally decided to lead by recalling the grand jury.

FOR ARIAS AND ODIN, THE ARREST OF MICHAEL CARLOW HAD ALways been one of three goals. The second was to change Florida law, which had been accomplished. The last was to remove the leadership of their agency.

To that end, Gene Odin had kept exceedingly busy pouring poison into the inspector general's ear. Arias concluded fondly, "Gene is doing his best work."

Part Four

Part Four

23. The Rosetta Stone

July 20, 2003
Windmill Ranch Estates
Weston, Florida

LATE AT NIGHT, GARY VENEMA FLIPPED OPEN HIS BADGE AT THE Windmill Ranch security gate. "FDLE," he announced.

The security guard, a slight man who did not recognize him, looked skeptical.

"Who are you going to see?" he asked, peering out of his guardhouse and studying the badge.

"A resident. Can't tell you who he is," Venema shot back.

The guard waved him uneasily into the gated community that until recently boasted Dolphins quarterback Dan Marino as its most famous resident. Carlow lived just a few doors down from Marino's old place.

In the moonlight, Venema saw right away that the Carlows were home. Two vans were parked in their semicircular driveway and the yellow nose of Candace Carlow's Viper peeked out from the side garage. Carlow usually worked late, the lights in his home office burning as he chain-smoked and drank coffee. Tonight the house was dark and light reflected off the pool in the back.

The detective took a lazy swing past the house, satisfied that the Carlows would still be home by first light. In his truck he had copies of nineteen arrest warrants and two indictments. The first indictment was for Carlow and seventeen associates, among them his wife; his brother-in-law, Thomas Atkins; his mother-in-law, Marilyn Atkins; his supposed cooks, Fabian Díaz and Henry García; the silent social worker, David Ebanks; and the diverter who sold medicine from his construction company, José L. Benitez. The ninety-five-page indictment listed thirty-two charges that included racketeering and grand theft.

The second indictment targeted José Grillo. He faced one count of running an organized scheme to defraud patients by mislabeling thousands of boxes of Epogen, and eighteen counts of violating statute 499 for purchasing, possessing, and intending to distribute prescription medicine unlawfully.

"I've been waiting my whole life to do something like this," Venema reflected back on the highway, "just so I could say I did."

When he returned home to Davie, he found Sandy busily assembling a care package for their youngest son, Kevin, now stationed in Tikrit. Canned vegetables, Chunky Soup, cereal, and his other favorite snacks were piled on the velour La-Z-Boy.

Venema settled down in the dining room to organize for the next morning, his favorite TV show, HBO's *The Wire*, humming in the background. He reviewed some of the paperwork that had taken over the dining room table and sat in boxes stacked in the living room. He then set out his black jeans and FDLE raid jacket with neon yellow letters.

Meanwhile, Sandy set the coffee maker and hunted down a travel coffee mug on the good assumption that her husband would need it when he awoke at 4 A.M.

Farther north, in Boynton Beach, Gene Odin couldn't sleep at all and got out of bed at 3:30 A.M. He had been waiting fifteen years for this day. Down south in Miami, Cesar Arias had turned off his walkie-talkie the day before and took his wife out for her birthday, perhaps the first time in more than a year that he had gone incommunicado.

John Petri stayed out till midnight, driving by the addresses of their suspects to see if they were home. For him, the anticipation was bittersweet: the politics in his office would force him off the case after tomorrow.

Venema actually slept a few hours. By 4:30 A.M. he was out the door, the flag and the fishpond lit up by the ground lights on his lawn, a sliver of moon still in the sky.

VENEMA'S DEPARTMENT HAD DECIDED TO WAIT UNTIL DAWN TO make sure there were no mistakes and that Carlow and the others could clearly read the investigators' field jackets. The dangers they faced were twofold: Someone threatened with arrest might strike up a gun battle. But more likely, Carlow would hire lawyers to bury them in procedural complaints. Sunlight, as well as transparency, seemed like good things.

At 5 A.M. Windmill Ranch was steamy and silent. A neighbor up early fetched the newspaper from his driveway. A few sprinkler systems blew on. Birds flew over the palm trees, planted down the lanes in regular intervals.

Venema parked his truck on a perpendicular street with a clear view of the house, Carlow's Explorer and his wife's van still nose-to-nose in the driveway. And then he waited, the moon slowly abating into a languid broiler of a dawn.

By first light, a line of unmarked cars with darkened windows rolled slowly and silently toward Carlow's home. Other units moved into place behind the house. And then two marked police cars, lights turning silently, joined the caravan.

The sound of car doors opening and slamming shut echoed in the silent neighborhood. Agents with guns drawn crawled up an embankment behind the mansion, covering it from both sides. It took Venema only a few seconds to reach the door and start pounding. To start with, he announced a search warrant. Carlow appeared in a pair of shorts, surveyed the line of idling cars, and said casually, "Come on in."

Then he sat down at his kitchen table and shook out a cigarette from his pack.

AT THE SAME MOMENT, EIGHT OTHER LAW ENFORCEMENT TEAMS fanned out across South Florida. In Lake Worth, Gene Odin confronted Candace Carlow's mother, Marilyn, whom he had known for years from when Carlow had a legitimate state license.

"You got a real fine opportunity to talk. You can help yourself," he said, adding, "You can't help Michael. He's going in for thirty years."

"You have some nerve," she responded, making it clear there would be no further discussion. "You have some nerve!"

She was still yelling as officers marched her into a waiting squad car.

Back in the house, Marilyn Atkins's husband remarked dryly, "We know that Michael walks close to the line." Armed with a search warrant, Odin scooped up a pile of documents that had not yet been shredded, including records of wire transfers.

DOWN IN SOUTH MIAMI, CESAR ARIAS ARRIVED WITH A TEAM OF police officers at the home of Julio Cesar Cruz, the convicted cocaine dealer who had masterminded the Lipitor counterfeiting and whose money had moved through a Carlow shell company.

Cruz's wife, a sales rep for a large pharmaceutical company, began to cry. Wearing only boxer shorts and an electronic monitoring bracelet around his ankle—he was still on parole for his cocaine trafficking sentence—Cruz said to her, "What are you crying about? This is nothing." Casually he said, "Don't worry about it. Call the attorney."

As he was led to a waiting squad car, he asked, "What's the bond?" When he heard $4 million, he shouted frantically to his wife, "Call the attorney! Call the attorney!"

In Doral near Miami, José Grillo looked stunned to find himself under arrest. At police headquarters, Jack Calvar and Cesar Arias sat him down and explained that they knew all about the eleven thousand boxes of Epogen that he had relabeled and that if they could prove that one person died from the medicine, he would be charged with first-degree murder.

"The guy I work for is very upset about this," said Calvar. "I'd hate to be in your shoes."

Grillo looked horribly upset and his lower lip began to quiver. He said despondently, "I'll never get out."

DESPITE THE COMMOTION AROUND HIM, CARLOW KEPT HIS COOL as Venema separated him from his wife and confronted Candace in a separate room. The detective showed her a diagram with her husband's picture in the center and the photographs and names of seventeen others ringed around him in a concentric circle. All of them—including her mother and brother—were being arrested simultaneously.

Her eyes widened, but she told Venema, "I don't want to talk to you about anything."

In the other room, Carlow asked Venema what his wife has been charged with. "Racketeering," he said. "Her bond is

$1.5 million." He did not yet tell Carlow the amount of his own bond: $7 million.

The telephone rang incessantly, the caller ID showing the home of Candace's brother, Thomas Atkins, in Kissimmee. No one picked it up.

"Can I make some coffee?" Carlow asked coolly.

"You're not going to be here that long," one of the officers responded.

OUTSIDE, CARLOW'S NEIGHBORS WOKE UP SLOWLY TO THE arrest. Drivers headed for work slowed but did not stop. As Carlow's baby-sitter arrived with Weston's daily army of nannies, maids, and dog walkers, one officer reassured a passerby, "Don't worry. We're taking a police action. Your neighborhood is safe." Officers emerged from Carlow's house sporadically, toting a computer monitor, a shredder, and boxes of files. A search of his van turned up his passport on the seat.

Annetta Epstein and her teenage daughter, dressed for their morning yoga class, peered over their hedge. "They're arresting Mike," Epstein said in astonishment.

"He's been arrested before," said her daughter, home for the summer from the elite Miss Porter's boarding school in Farmington, Connecticut.

"No," Epstein chided, explaining that a police officer had only asked at one point to use their driveway as a lookout.

A friend had mentioned an article in the *Sun Sentinel* that described Michael Carlow as being under investigation, but Epstein hadn't paid much attention. "I don't get the paper," she said. "If it's not on Fox News, I miss it."

Now she watched, amazed, as officers led Michael Carlow *and* his wife, Candace, both handcuffed, from the house and into the back of a squad car. A TV crew jumped from a van, the reporters scrambling up the driveway and yelling questions. A chopper from another TV news station flew overhead.

Epstein sank to the stairs, angry that the police would humiliate the Carlows in front of their children. "I hope this turns out to be a big mistake," she said doubtfully, describing Carlow's many virtues.

He had been so friendly, often out doing yard work, chatting by the mailboxes, his house always open for a visit. Their children played together. Her husband, Dr. David Epstein, an interventional radiologist, loved his company. The two men were in a poker group where they played for small stakes. Carlow's wife, Candace, was an animal lover, with a stable of horses at their nearby ranch house. "She's like that horse-whisperer type of person," the daughter offered.

Epstein now instructed her daughter to go check on Carlow's kids. Two children, a boxer, and a Spanish-speaking nanny holding the couple's infant peered out the door. All seemed remarkably calm for the trauma they had undergone and claimed not to need any help.

By noon, twelve of the nineteen indicted were in custody. Exhausted, grubby, exhilarated, the Horsemen went home, showered, and donned suits for the press conference.

AT THE PRESS CONFERENCE THAT AFTERNOON IN FORT LAU-derdale, only four men stood at the podium: Daryl McLaughlin, FDLE's new interim commissioner; Dr. John O. Agwunobi, Florida's health secretary; Peter Williams, the statewide prosecutor; and Charlie Crist, the state attorney general. Before the Florida press corps, the agency directors praised each other for their hard work and inter-agency cooperation.

Agwunobi praised the "collaborative model" of state and local agencies working with the pharmaceutical industry to safeguard the public health. He offered "special thanks" to the industry for recommending changes implemented by the legislature. He did not mention that the industry's recommendations watered down the changes proposed by his own prosecutors and the grand jurors.

Williams, the statewide prosecutor who tried to fob off the case onto a local prosecutor just a month earlier, described his agency's investigative efforts, including "hour after hour of listening to the heart-wrenching story of victims." The investigators, shunted off to the side of the room, knew that no such sessions ever occurred. Such was the display of self-promotion that even Michael Mann looked stunned. All the while, Venema stared straight ahead, a smile on his face. Nothing could dim his sense of accomplishment.

Agwunobi, the afternoon's sound-bite leader, told the reporters that he planned to "drive out of Florida those parasites that would hurt our most vulnerable citizens." Their activity, he said, "preys upon a fundamental trust of our health-care community that what is in the bottle is what it says on the label."

After the press conference ended and the officials made their way warily among the crowd, Williams explained to one reporter that reforming a "powerful industry" consists only of "what you can get the industry to do." He then introduced a prosecutor from his office, Oscar Gelpi, who had recently taken over the case from Penezic, to the Attorney General as the "man who made it all happen." Penezic, overhearing this comment, made a mental note to start his job search.

Arias stood back observing the officials at work, looking both amused and relaxed for the first time in months. "What are you going to do? These are the [oxen] we have to plow with," he said, invoking an old but very apt Cuban saying.

Despite all this, Gene Odin was in heaven, holding court

with the reporters and particularly schmoozing one pretty blonde newspaper writer. He relayed his best stories of the investigation as he patted her knee.

In the elevator afterward, on their way to margaritas and a retelling of the day's events, the Horsemen began laughing. "If I'd had false teeth, they would have fallen out right then," said Petri, recounting Agwunobi's praise of the pharmaceutical industry. Venema mimed teeth chattering across the floor.

AT THE EMORY UNIVERSITY LAW SCHOOL, WHERE SHE WORKED as a legal writing instructor, Stephanie Feldman celebrated the arrests, which she thought of as the "midway point of a very long dream." She felt the success in her bones when Venema called and she did a victory lap around her desk, arms raised and hands pressed into fists.

MARTY BRADLEY AT BIOMED PLUS FELT ENORMOUS RELIEF WHEN he saw the indictment against Carlow. More than two-and-a-half years after the burglaries at his warehouse, investigators had finally linked Carlow to the crimes. The indictment did not specify who, exactly, had broken into the warehouse, but it charged Carlow and his cook Fabian Díaz with grand theft for obtaining and using the stolen medicine.

AMONG THE DOCUMENTS SEIZED FROM CARLOW'S HOME, THE Horsemen found records of checks to a Florida company, Fortissima Secretarial Services, with payments of about $2,500 every few weeks. They had never heard of the listed CEO, Gina Catapano. But the address of the company—in a small cul-de-sac of rental homes—was familiar. They had followed Michael Carlow there on several occasions.

Two days after Carlow's arrest, Gene Odin and Steve Zimmerman headed to the address in the late afternoon, not expecting to stay long. Two hours later he and Zimmerman were still at Catapano's kitchen table, two enormous cats crawling around them. She told the investigators that she was Carlow's personal secretary, girl Friday, and his lover for the past eight years, a relationship that continued through his last three marriages.

Catapano explained that she had created invoices for every batch of drugs that Carlow had sold into the supply chain. The investigators could hardly believe what they were hearing: It was exactly the evidence that the Horsemen had been looking for all along.

Odin asked, "Where are the records?"

Catapano explained that she no longer had them. Just a few hours before the investigators arrived, she received a coded e-mail from a Carlow associate, who told her: "Per Mike's instructions I am to get with you immediately and get whatever it is that you have . . . I am in Coral Springs right now at 12:50."

So Catapano brought the box to the woman who sent the e-mail. She was a friend of Carlow's who worked from a warehouse in Coral Springs and was trying to launch a new design of a garment bag on the Home Shopping Network. Carlow had invested in her efforts.

Odin and Zimmerman immediately cut short their visit. It was 6:30 P.M. They stepped outside and raised Venema on his walkie-talkie.

Venema's response: "Nobody sleeps tonight until we get those records!"

———

DRIVING NINETY MILES PER HOUR, JOHN PETRI GUNNED HIS truck toward Coral Springs, a community northwest of Fort Lauderdale. He was the first to reach the warehouse, next door to another warehouse where he had followed Carlow once before. The place was locked up.

Venema, meanwhile, doing ninety-five miles per hour, raced to get a judge's signature on a search warrant, only to be pulled over for speeding. Venema and boss, John Vecchio, who was tailing him, flashed their badges and explained their frantic speed.

By the time the investigators located the landlord, found a locksmith, obtained the search warrant, and hunted for the woman who supposedly received the box, four hours had passed. By 10:30 P.M., with the Horsemen assembled in front of the warehouse and ready to torch the doors, the woman arrived with several family members and explained that she no longer had the box. She had given it that afternoon to Jean McIntyre, Carlow's banker, associate, and sometime girlfriend.

Again they leaped into their cars and raced to McIntyre's home in a caravan that included Odin, Arias, Petri, Zimmerman, Venema, and Vecchio. McIntyre's teenage son answered the door and told Venema that his mother had headed to the home of Candace Carlow, who had made bail.

Their hearts sank. "You call her up and tell her to get back here *immediately*," Venema instructed the teen. By then, McIntyre's husband was up and had come outside to wait with them.

By 11:30 P.M. two cars rolled up carrying McIntyre, a big, imposing woman, and her brother, an attorney.

"Where are the records?" Vecchio demanded.

As McIntyre hedged, he said, "Either you produce those records now or you're going to jail as a co-conspirator—tonight."

At her brother's urging, McIntyre told the investigators that the box was at her sister's house.

For the third time that night they leaped into their cars with McIntyre in tow. With a grim expression, McIntyre marched into her sister's house. After several minutes, she emerged with a full black file box and a white plastic bag containing more documents.

THE HORSEMEN FILED INTO A DUNKIN' DONUTS WELL PAST 1 A.M., cheering and high-fiving each other as they plunked down their booty. A young couple and a few senior citizens looked on as the men pulled out one document after another and passed each one around the table. Venema whooped and howled. Each record was a missing link that helped to lay bare a vast pharmaceutical fraud.

"It's like the Rosetta Stone," said Odin. "It's the entire code."

MICHAEL CARLOW APPEARED BUOYANT AS EVER AT HIS BAIL hearing on July 28. He entered the courtroom in a jail-issue jumpsuit, waving, blowing kisses, and giving thumbs-up. He greeted his friends packing the courtroom.

Candace Carlow, however, looked haggard and distraught. Her mood visibly deteriorated as the hearing progressed. As all three prosecution witnesses spoke not only of Carlow's alleged pharmaceutical misdeeds, but also detailed his extramarital affairs, Carlow turned to his wife and mouthed, "Are you okay?" His secretary and his banker testified that they had both slept with him. His former business partner, Steven Ivester, testified that Carlow had seduced his girlfriend.

"I can honestly say we beat out the Jerry Springer show," John Petri later observed.

The judge ultimately reduced Carlow's bond from $7 million to just under $3 million. Carlow posted bail and enjoyed his freedom for a day—until the bail bondsmen learned that the Carlows had already defaulted on the mortgage that Candace Carlow had offered as collateral for her own bail. The bail bondsman apprehended Michael Carlow and brought him back to jail.

24. A Wink and a Nod in Las Vegas

October 2003
Las Vegas, Nevada

THREE MONTHS AFTER MICHAEL CARLOW'S ARREST, JUST AS A federal task force released a report on America's vulnerability to counterfeit medicine, two groups with radically different agendas convened amid the desert's glittering hotels and casinos to determine who would control the nation's drug distribution system. Both groups met in Las Vegas during the same week, just a few blocks apart.

In a conference room at the Las Vegas Chamber of Commerce, the Nevada State Board of Pharmacy convened a public hearing for two wholesalers accused of conspiring to buy diverted medicine that proved to be counterfeit. Such proceedings might have yielded a slap on the wrist in the past. But this case had mushroomed into a full-blown trial with implications far beyond the two companies involved.

Nevada authorities spared no expense as they endeavored to show how counterfeit drugs moved easily through the system's loopholes to reach unsuspecting patients. They invited witnesses from around the country, including pharmaceutical executives, an AIDS patient who had received counterfeit medicine, and Gary Venema. The investigator planned to testify about the Florida companies that sold tainted drugs to the Las Vegas wholesalers.

Across town at the Monte Carlo Resort & Casino, the nation's secondary drug wholesalers gathered for their first-ever national conference, a summit on their industry's survival. They represented more than fifty companies from Kentucky, Florida, California, and elsewhere. They came to talk about the investigations, the creeping regulations, the bad press, the crackdown in Florida, and the lacerating effects of counterfeit medicine moving through their companies. They hoped to restore their image as the small businessmen, rather than the pariahs, of the supply chain. And to do so, they needed to bring a credible public face and more disclosure to their secretive industry.

Their plenary session was entitled "Secondary Wholesalers: Why You're Good Guys Too." Conference organizer Robb Miller, a savvy Las Vegas wholesaler with wire-rimmed spectacles and a goatee, climbed the podium to deliver his positive message. "We are lawful citizens," he began. "We provide many services. We move billions of pharmaceuticals across the country each year. We provide inventory management services. We help new manufacturers."

Expounding on his theme of usefulness, he added, "I would say we are essential to national security." He suggested the small wholesalers would be the front line of defense if terrorists attacked the computer systems of the Big Three. "How will they maintain their orders? Where would the emergency response teams get their drugs?" This was the image makeover he wanted: one that equated small wholesalers with homeland security.

"Without us, loved ones would be lost," he said. "We are in every city across the country. We could be called on."

His colleagues remained divided about the best way to survive the onslaught of negativity. Most believed nothing good could come from a PR offensive and that any public mention of their names could lead larger wholesalers to cut them off or regulators to probe their businesses. Already they had risked exposure just by coming to a gathering that the FDA and the larger wholesalers' trade group, the HDMA, had canceled their registrations to attend.

In a further effort to influence his colleagues, Miller invited fourteen of the wholesalers to a private club that night on the top floor of the Mandalay Bay Resort & Casino. The Foundation Room, with its guarded entrance, separate elevator, and absence of signs, was a perfect place for a more candid discussion. Miller gathered the men on the balcony for drinks as they awaited their table. Some of his guests were members of the secretive trade group run by Sal Ricciardi in Boca Raton, the Pharmaceutical Distributors' Association, which refused to publish its membership list. Tonight, Miller planned to make his pitch to wrest the leadership of the trade group from Ricciardi.

With a sweeping view of the Las Vegas strip, a blaze of electricity that ended abruptly in blackness at the undeveloped desert, the wholesalers smoked cigars, drank Scotch, and traded stories about how it felt to be treated like criminals.

"Before, it was just fine because it was just a money thing, nobody was getting hurt," said a short man, hair in disarray, who used to work as a techie for a major news network before moving to Las Vegas and finding his way into the pharmaceutical business.

A well-dressed Miami wholesaler who had been on the short end of Gary Venema's wrath joined in. "You have guys come in with badges and guns," said the Miami wholesaler. "They have no idea what business is. All they do is look at an invoice and say, 'That's too much money.'" He concluded, "Well, fuck you."

Fresh from the Florida dustup, he explained that the wholesalers there had been forced to give up everything, with no concessions from the state. He recalled that first pedigree meeting in Tallahassee eighteen months earlier and the speech by that young prosecutor. "She referred to the industry as contaminated rats," he said. "'Rats' was definitely in there. 'Dirty rats' is what she said."

Lance Packer, one of the Las Vegas wholesalers on trial that week, sipped his cocktail and joined the discussion. Packer and his father-in-law faced the prospect that their licenses would be revoked. Now among friends, the thirty-seven-year-old businessman excitedly proclaimed his innocence.

When their table was called, the men headed to a private dining room with its own balcony and a magnificent view of the city's neon forest of casinos. Fresh drinks and appetizers arrived. The talk grew more voluble.

The conversation turned to meetings that Ricciardi's group, the PDA, had held with the FDA months ago, regarding the most contested part of the federal Prescription Drug Marketing Act. The final rule, as it was known, would have required wholesalers nationwide to pass on complete pedigrees with their drugs. In January, the group managed to get yet another stay of the rule, the fifth since it had become law. "How could they *not* issue us a stay?" one of the men asked.

Each guest seated around the table occupied a different rung on the wholesale ladder, their operations reflecting varying shades of the gray market. While the men viewed themselves as operating legal businesses, they understood that their tactics were of interest to law enforcement and probably best left unpublicized. Most also seemed to agree that a real scourge facing their industry was a man named Gary Venema—a "cocksucker," the Miami wholesaler declared with venom as he described how the FDLE agent had barged into his warehouse and threatened to take away his business.

"You *are* a secondary wholesaler," Ricciardi's deputy chimed in.

The Miami man laughed and joined in the satire. "I can't help myself. I'm a low-life scumbag diverter."

By now everybody was warm with liquor and ready for more cigars. But Miller stood and handed out copies of his eighteen-page proposal to revamp the Pharmaceutical Distributors' Association and replace Ricciardi as its leader. Miller argued that the changes would help preserve what they had worked so hard to build.

With bemused detachment, his dinner companions rifled through the pages. The formality of Miller's pitch—including his suggestions of a mission statement, funding forecast, weekly bulletins, and political action committee—flowed against the tenor of the dinner, which was dark, private, and full of shorthand talk. The men did not appear ready to come out of the shadows.

Later on the balcony with a cognac in hand, Miller put a foot up on the balcony rail, aired out the sides of his Versace suit jacket, and pulled at a cigar. He appeared unbowed by the chilly reception. "I have a romantic notion of justice and I think that what's happening is unjust," he said.

The Miami wholesaler agreed. "It makes no difference whether you sell paper clips or Procrit, you have a right to exist," he said. "I think it's wrong that you punish everyone for the bad acts of a few."

As the evening devolved, so did the sanctimony. Long after midnight they went their separate ways, each returning to his own little trade war with regulators and each other, not knowing that their nemesis, the "cocksucker" Gary Venema, was in town, too.

———

TWO NIGHTS EARLIER, VENEMA HAD BEEN TOO EXCITED TO SLEEP
when he arrived in Las Vegas. He walked the strip of mega-
hotels until 1 A.M., barely noticing the garish casinos, the
dancing jet sprays from fake lagoons, and the amusement park
rides that towered above the neon billboards for show girls.
Instead, he was thinking about the Horsemen and the surprising ripple effects of their work. Breathing in the desert air, he
declared to a companion, "Finally, I feel that I have accomplished something of relevance in my life."

Since Michael Carlow's arrest, the Horsemen had become
a more integral part of a growing cross-country network. In
Georgia, Connecticut, New York, Florida, California, and
Nevada too, drug security executives, state investigators, regulators, and patients had begun to work together. These were
individuals who really understood the issues involved and who
were obsessed with safeguarding the nation's medicine. They
attended the same hearings and registered the same complaints. They were dissatisfied with the FDA's sluggish pace
and its tepid explanations of problems. They viewed the drugmakers as complicit through their secrecy and pricing schemes,
which they believed encouraged diversion. They believed the
Big Three wholesalers had made a devil's bargain in their
quest for discounted medicine. And they wanted to see the entire system rewired so that patients came before profits. While
most had never met each other, they were committed to fighting what they saw as a sellout of America's patients.

The nation's drug supply had long been governed by an
unspoken covenant: Americans pay the highest prices in the
world in exchange for the highest manufacturing standards
and a guarantee of purity. When drugmakers emphasized that
foreign medicine—which could be riddled with impurities—
threatened the integrity of our own supply, they reinforced

the connection between higher prices and greater safety at home. Consumers who went to Canada or Mexico in search of cheaper drugs were seen as abandoning the essential protections of our blue-chip supply.

But sky-high prices had not made us safer. Instead they made America the go-to market for counterfeiters and the prescription business a lucrative draw for criminals—as even an internal report prepared for drugmakers by their own research organization, the Pharmaceutical Security Institute, pointed out.

The drugmakers fought bitterly against importing cheaper medicine from abroad, calling it a grave threat to public health. Yet their internal report stated that the United States experienced fifty-four counterfeiting incidents in 2003—more than any other nation in the world. Each incident posed a serious potential threat to thousands of patients, as the Lipitor counterfeiting and Grillo's uplabeling of Epogen had shown. The report stated that increased reporting of incidents and a free press in the United States were only partial explanations for the larger numbers. The size of the nation's prescription market, the high demand, and the increasing pressure to find lower-cost medicine had helped create a market for counterfeiters—and made it likely the number of incidents would increase, the report stated.

Sure enough, by the first half of 2004, the number had risen. The United States had experienced twenty-eight counterfeiting incidents, behind only Colombia, with thirty incidents during the first six months of the year, according to the internal report. India was third, with nineteen incidents.

The Pharmaceutical Security Institute report showed the hidden costs buried deep in the nation's drug distribution system. Even patients who got medicine from pharmacies or hospitals had no way of knowing whether their medicine had been recycled or sent on dangerous journeys that left it sub-potent, adulterated, or outright counterfeit.

"In the final analysis, this is a fraud against the consumer," said Oscar Gelpi, the assistant statewide prosecutor who took over Operation Stone Cold. "What kind of moron pays Rolex prices for a knockoff?" Americans, he said, "are paying top dollar for what they think are legitimate drugs."

Before the Horsemen came along, state prosecutors and drug inspectors had only weak laws to use against those who tampered with medicine. But the Florida task force had convinced the legislature that adulterating medicine was as bad as trafficking in cocaine. At least cocaine addicts knew they were buying a harmful product. But cancer patients were unwitting victims and might never know the difference once they dispensed the evidence of tainted medicine into their veins. Remarkably, despite all the obstacles, the Horsemen's efforts had made Florida's drug supply the most strictly guarded in the nation.

Other state pharmacy boards looked to emulate Florida's 2003 law. Drugmakers, too, began to tighten their networks, threatening to cut off large distributors who bought their products from the secondary market.

By the end of 2003, everyone, it seemed, had a five-point plan or a task force to improve the safety of the drug supply, and invitations poured in for the Horsemen to speak out about what they had discovered. Venema flew to Delaware, Boston, and Chicago to present his findings. He and a CIA agent were the featured speakers at a meeting of the Pharmaceutical Biotechnology Security Council, a working group of drug security experts that met every few months around the country. When Venema went to testify in Nevada, he landed at a critical moment in the state's battle for control of its drug supply.

THE NEVADA PHARMACY BOARD'S GENERAL COUNSEL, LOUIS LING, a youthful reformer, and its executive director, Keith Macdonald, a patrician silver-haired pharmacist, had been fighting the state's drug distributors for thirteen years. By 2000, the two men realized that many of the distributors were phantom businesses with few employees. One company operated from a U-Haul storage shed. Another had its inventory delivered to a pizza parlor. The men pushed through regulations in 2001, requiring Nevada companies to employ real people in real offices and keep real business hours. As a result, the number of in-state wholesalers dropped from fifty to eight. Those who remained detested Ling and Macdonald and waged a battle against them that became bitterly personal.

In January 2003, Ling had received a call from Gary Venema about a Nevada drug wholesaler called Dutchess Business Services. Venema explained that Dutchess had been buying medicine from Crystal Coast, a Miami wholesale company suspected of selling counterfeit AIDS medicine. Eager to help, Ling sent a detective to Dutchess, where an angry Lance Packer announced that he wouldn't provide any "fucking records" and that he was tired of the state trying to put him out of business. It took Ling seven weeks to get the records. As soon as he saw them, he suspected that Dutchess had bought and sold a huge quantity of counterfeit Serostim.

As Ling investigated Dutchess, Packer, his father-in-law Paul DeBree, and others were investigating him. In March, five wholesalers—including Dutchess and Robb Miller's company, Caladon Trading—sued Ling and Macdonald personally, claiming that the state officials had interfered with, and slandered, the wholesalers' businesses. The companies had been cut off as suppliers for Cardinal Health, they alleged, after Ling raised questions about their buying practices with Cardinal executives. The lawsuit alleged that Ling and Macdonald aimed to drive *all* secondary wholesalers from Nevada. As proof, it quoted Ling as having told their lawyer that he would "continue to piss in your clients' soup until they get out of Dodge."

Five months later, in August, Ling and Macdonald filed a fifty-six-page complaint against two of the companies, Dutchess and Legend Pharmaceuticals. It alleged that the two companies—run jointly by Lance Packer and Paul DeBree—conspired to profit by purchasing diverted and deeply discounted medicine from companies not licensed in Nevada. It accused them of using pedigrees that perpetuated the false representation that their suppliers were authorized distributors, and of continuing to buy from them for more than a year after their medicine proved counterfeit.

The complaint sought to challenge in advance the defense wholesalers had used over the years: that they were innocent companies misled by corrupt suppliers. The complaint portrayed willful blindness as part of a conspiracy among wholesalers and alleged that if they didn't know the medicine they were buying was bad, they should have. And that is where Venema came in. He would testify as to how convicted felons in Florida used the thinnest of pretenses to make their dangerous drugs appear good—pretenses the Las Vegas wholesalers accepted without question.

ON THE MORNING OF OCTOBER 13, A PANEL OF BESPECTACLED pharmacists gathered in a conference room at the Chamber of Commerce to hear the case against Dutchess and Legend. Venema, who had arrived in black jeans and a polo shirt with a lightning bolt on it that said "Phi Zappa Crappa," surveyed the roomful of people in sober business attire. He muttered, "I should have worn a suit."

Ling and Macdonald were missing, but not by choice. They were huddled with their own lawyer in a courtroom across town, fighting an eleventh-hour defense motion seeking to bar both men from setting foot in the hearing. Ling was the state's prosecutor on the case. Another lawyer, Mary Boetsch, stood by to present the case in his stead.

Lance Packer sat amid the few spectators in the conference room. His father-in-law, Paul DeBree, a diminutive man in loafers and sportswear, sat primly near several of his lawyers. Together, the two men ran Dutchess, DeBree as its president and CEO and Packer as its day-to-day manager. Legend, a new and separate company established by DeBree, shared the same office and interchangeable staff. DeBree was an old hand at regulatory battles and even criminal investigation: After his last wholesale company became implicated in the diversion scheme perpetrated by Bindley Western in the late 1990s, another of his sons-in-law who helped run the company pleaded guilty to conspiracy.

As the hearing began, DeBree and Packer's lawyers presented legal objections to the proceedings that the pharmacy board's president, Larry Pinson, listened to with strained patience and overruled. He then directed prosecutor Mary Boetsch to call her first witness.

Rick Roberts took his seat at the witness table. An adjunct communications professor at the University of San Francisco, Roberts was an AIDS patient whose prescriptions included Serostim, an injectable growth hormone that reversed the sudden weight loss suffered by those with HIV. A month's supply of the drug cost $7,000.

He testified that in late 2000, he began to feel a stinging sensation when injecting his Serostim. When he went to his drugstore and inquired, the pharmacist told him, "You should go home and check. You may have gotten some of the fake stuff."

Roberts took the bus home, his head swimming. "When I got home I convinced myself that I must have misunderstood," he said. "It didn't make any sense to me."

As he took out all his vials and compared them, he noticed tiny variations. On one, the ink appeared slightly darker. On another, the cap was a slightly different blue.

He went online and found an article describing how counterfeit Serostim had been found in seven states and the FDA had launched a criminal investigation. One of his lot numbers matched that of a counterfeit lot.

Roberts testified that once he called the manufacturer, an executive there had urged him to send all the counterfeit medicine back and the company would replace it free of charge. But Roberts—seeing clearly that the vials remained his only evidence that he'd been the victim of a crime—decided that he wouldn't give them up until he could learn what had been in them. "I wanted to know what I had injected," Roberts told the board. He didn't know what the solution contained and he feared that it could kill him.

The counterfeits turned out to contain a far less expensive fertility hormone, human chorionic gonadotropin, which the counterfeiter had likely chosen because its packaging resembled that of Serostim. Another lot turned out to contain a pediatric dose of growth hormone not approved in the United States. Roberts had injected both.

As Roberts was about to end his testimony, the wholesalers' principal lawyer, Steve Gibson, swept into the hearing room. A district court judge had agreed to temporarily bar

Ling and Macdonald from the hearing. Gibson began a new debate over whether to continue the hearing at all.

As the witnesses and board members adjourned for lunch, they encountered Ling and Macdonald outside, looking dejected. Now under a court order, the two men kept their distance, not wanting to jeopardize the outcome of the hearing.

Over lunch they talked about the frustration of taking on the industry's well-paid legal team while their own defense came from a tired and overwhelmed state lawyer who represented the men in addition to his usual duties.

"I can't afford to fight Steve Gibson at $300 an hour," said Macdonald, staring into a glass of iced tea as he contemplated losing a lawsuit that could strip him of his home and other assets.

"We're getting our butts kicked," said Ling.

BY THE SECOND DAY OF TESTIMONY, GIBSON'S EXHAUSTIVE CROSS-examinations had slowed the proceedings to a crawl. Several witnesses had already given up and flown home, as the board acknowledged that it would likely have to convene at a future date to complete the hearing. Venema, this time wearing a suit, went off in disgust to see the Hoover Dam.

Pamela Williamson-Joyce, the vice president of regulatory and quality assurance for Serono Inc., testified that as a result of the counterfeit Serostim, the company had put a unique hologram on the boxes the vials came in, shrunk its customer list, and required all authorized distributors to accept being audited at any time. As well, each box bore a unique identifier, a number that would never be repeated.

On cross-examination, Gibson asked why the company didn't do more sooner.

Williamson-Joyce snapped back that he was talking about "the concept of the ideal. It is not humanly possible for any professional organization to eradicate what someone may choose to do in their basement."

While Williamson-Joyce went on to assert that counterfeits are not a "commonality," only Serono knew for sure how many patients had received counterfeit Serostim. She balked at revealing the number.

"More than ten?" Gibson asked.

"Yes."

"More than one hundred?"

"I don't know," she said.

"More than fifty?"

"Honestly, I'd have to go back and check." Only three patients who'd received it, including Roberts, had come to public attention. Two had settled confidentially with Serono and the major distributors. Roberts had not yet sued anyone because the counterfeiter had never been caught and he didn't know whom to blame.

All the procedural dueling absorbed three days of testimony. Venema took the stand briefly. He rattled off his credentials, including his years as a homicide detective. This caused a detective for the pharmacy board to nod approvingly, as though glad to have a murder investigator working this case. Venema did not get far in his testimony and the board resolved to resume the matter in January 2004.

Venema got back to Florida just in time for the arrests of Maria Castro and Jesús Benitez, the owners of the Miami pharmacy J&M Pharmacare. They pleaded not guilty to charges of organized fraud and the unauthorized sale of prescription drugs.

———

BY THE TIME VENEMA RETURNED TO LAS VEGAS IN JANUARY, A major stress in his life had been lifted: His son Kevin had returned from Iraq unharmed and was honored with a bronze star for leadership. The case against the Las Vegas wholesalers actually had grown stronger. The corrupt Florida wholesaler who had supplied them with counterfeit Serostim, Bill Walker, pleaded guilty to seventeen felony counts of racketeering, fraudulent use of personal identification, and violations of the health statute 499. Ling and Macdonald had won dismissal of the restraining order against them and were back on the case.

On the stand again, Venema described the criminal networks they found running through the pharmaceutical trade.

He detailed the investigation of Crystal Coast in Miami and the arrests of its principal officers, Bill and Elinor Walker, who had sold the counterfeit Serostim to Dutchess and Legend. A former narcotics trafficker and Norwegian national who was wanted by Interpol, Walker, who used an array of aliases but claimed that his real name was Per Loyning, operated six different wholesale companies. One was called Rekcus (Sucker spelled backward).

Venema detailed the crimes of the Walkers, the winding path that their medicine took through a series of shell companies, and how he found memos between them and DeBree arranging pharmaceutical buys.

This time, lawyer Steve Gibson embarked on his cross-examination of Venema, eager to establish that his clients were no worse than anyone else in the secondary market. He asked Venema whether he had encountered a "stretching of the rules" in the way wholesalers used their status as authorized distributors to disguise the true origin of the medicine they sold. "What I've encountered is more cheating" than stretching the rules, said Venema. "It's out and out lies."

He then testified about the unstated "Don't ask, nobody

has to tell" style of the wholesale market, in which buyers don't ask where medicine came from because they don't want to know. Even the Big Three participated in this system, he said, buying at least two 2 percent of their medicine from the secondary market, as long as the paper looked legitimate. "I personally think they are out of their minds," he said.

Gibson then asked: Had the Big Three conspired with the secondary wholesalers to move cheap, substandard medicine into the supply chain?

"I'm not saying it is a conspiracy on a grassy knoll with the guys who blew away Kennedy, but there are shenanigans going on," he said.

By revealing the seamy source of DeBree and Packer's medicine, Venema proved to be a damaging witness. But he was not nearly as damaging a witness as DeBree himself. Under questioning, the soft-spoken DeBree offered the most vivid and honest admission to date about the profit motive of the secondary market and the exploitation of the loophole that allowed authorized distributors to erase a drug's real pedigree.

DeBree explained that in doing business with Crystal Coast, his due diligence consisted of checking to see if the company had a license. "As far as priority is concerned," he said, "it has nothing to do with the quality of the product." Even after the manufacturer circulated a warning about the existence of counterfeit Serostim and two wholesalers in succession returned the Serostim, DeBree testified that he resold the drugs to a third company and never confronted Walker.

"It never entered your mind that he sold you fake drugs and it never occurred to you that he was a crook?" asked Boetsch.

"I was obsessed about economics and to be made whole," he responded.

As Louis Ling took over the questioning, DeBree explained that he had wanted to exploit the loophole of the authorized distributor, namely that "it washes" the pedigree, he said, allowing him not to list the trail of companies through which the drug had passed.

Though Crystal Coast claimed it was an authorized distributor, it refused to give DeBree any proof of this claim. When asked why he kept buying medicine from Crystal Coast anyway, DeBree paused and the room fell silent. "That's a good question," he said, finally offering, "Economic advantage?"

IN CLOSING STATEMENTS, THE PROSECUTORS TOLD THE BOARD that under Nevada and federal law, DeBree and Packer did not have to know the medicine was counterfeit to be held accountable for it.

DeBree made no effort to know because he was focused on money, not patient safety, Boetsch said, adding that there was "a nod and a wink and a smirk to the side between all these wholesalers.

"They say, 'They didn't know.' That doesn't matter. It doesn't save them." She continued, "He doesn't have to know it's counterfeit. He has to make sure it's not."

Gibson then argued that the prosecution had not provided "an iota" of evidence of a conspiracy between DeBree and Walker. Admitting that his own client's testimony gave him queasy feelings in his stomach, he concluded that DeBree had been conned by Walker, just like the Florida regulators who licensed Crystal Coast. "Did he exercise the best business judgment all the time? No." Gibson then asked for the board's mercy.

It took about thirty minutes for the board members to reach a verdict. They found the two companies guilty on all counts, revoked their licenses immediately, and fined Dutchess $1 million and Legend almost $400,000.

Outside, on the corner of Tropicana and Sands, Venema celebrated. On the day he served a search warrant on Bill Walker, his boss at FDLE had been furious, complaining that Venema should stay focused only on Michael Carlow and stop chasing bad medicine all over South Florida. But without his scattered pursuit, which had led to Walker and his customers in Las Vegas, DeBree and Packer might still have their licenses. To Venema, this signaled a revolution born in Florida beneath his hands. *The Horsemen are few*, he thought to himself, *but they ride like a thousand.*

25. The Education of Kevin Fagan

Winter 2004
Long Island, New York

More than two years after Tim Fagan was injected with counterfeit Epogen, he entered a nearby college in Westchester, still dependent on medicine he was terrified of taking. His mother, Jeanne, had gotten her Ph.D. in education and the family called her "Doctor Mom."

Kevin Fagan remained fixated on the same unanswered questions: Where had Tim's medicine come from? How had it entered the distribution system?

WHERE HAS OUR MEDICINE BEEN?

By the end of 2004, no one was any closer to answering those questions for any consumer who picked up a prescription. Drugstores knew only the last wholesalers that sold them medicine. Pedigree papers documented a drug's trail only from the last authorized distributor. Manufacturers said goodbye to their medicine at the loading docks—and from there, quality control moved into the hands of thousands of wholesalers regulated by a patchwork of weak state and federal laws.

As the battle over the hazards and cost savings of importing foreign medicine raged, counterfeit medicine continued to seep into our own porous distribution system. Federal agents turned up a trove of bad medicine in the nation's pharmacies: Celebrex for arthritis that had been made of calcium; Zyprexa to treat schizophrenia containing aspirin, vitamins, twigs, and even stones; counterfeit versions of Risperdal and Seroquil for mood disorders, the AIDS medicine Kaletra, and the arthritis medicine Bextra. A fiber mesh for repairing hernias that had been counterfeited and was not sterile reached numerous hospitals and was implanted in patients. A North Carolina man suffered repeated infections and severe complications from the fake mesh.

The bad medicine kept coming. The arrests in Florida were followed by arrests in New Jersey, New York, Tennessee, Georgia, and still more in Florida. The FBI, the FDA, and state investigators across the country probed illicit diversion networks, corrupt pricing schemes, and hidden warehouses from Florida to Costa Rica with stockpiles of counterfeit medicine bound for the American market. Michael Carlow's operation was a single open door to the distribution system. Countless doors remained open.

TO CONGRESSMAN STEVE ISRAEL, A YOUNG DEMOCRAT WHO REP-resented the Long Island district where the Fagans live, the problem seemed obvious enough. The companies that profited from the drug supply should not be policing it too. In response to what happened to Tim Fagan, Israel introduced new legislation, the Counterfeit Drug Enforcement Act, to give the FDA more authority to fight counterfeits.

The act, HR 3297, would require comprehensive pedigrees for all drugs, give the FDA authority to perform drug recalls, and strengthen criminal penalties for counterfeiting and diversion. It would do what Congress had envisioned in 1988 when it passed the Prescription Drug Marketing Act: make visible the Byzantine paths through which our medicine traveled.

Kevin Fagan poured his anger and uncertainty into building support for the legislation. He e-mailed many of the drugmakers' top executives. He called most of the congressmen on the House Energy and Commerce committee. He wrote to the FDA commissioner, the Surgeon General, and top officials at national organizations, including the American Pharmacists' Association. He also wrote to state pharmacy boards, asking about their policies and proposals for change. The more empty responses he received, the more committed he became. By the end of 2004, he had contacted more than seventy officials.

On his days off, he drove to the Long Island and Westchester offices of various congressmen. With no appointment, he simply walked in with packages of newspaper clippings about Tim's ordeal and asked for a meeting. In August 2004, Kevin, Jeanne, their ten-year-old daughter, and Tim set out for White Plains to the office of Congresswoman Nita Lowey. They waited in a cluttered conference room until Lowey bustled in, a briefing sheet in her hand that she had not yet read. As she studied it, her mouth dropped open.

"That's outrageous," she said. "I'd be happy to co-sponsor Steve Israel's bill."

"We're just an ordinary family," Kevin explained. "I never heard of counterfeit medicine. The doctor was totally perplexed."

Lowey asked about the corporations involved: "What do they say?"

As Kevin described the runaround he had gotten, Lowey looked sincerely amazed. "I will call Steven Israel and see how I can be helpful," she said.

Though Kevin looked pleased as he left her office, he did not know that the politics of Washington, D.C., had already blown past him. The FDA had granted the secondary whole-salers another stay of the pedigree rule—this time until the end of 2006. That decision, along with several others, made the possibility of a more transparent supply chain unlikely anytime soon.

Where had Tim's medicine come from?

The map of where Tim's medicine traveled was not unearthed by FDA regulators or by inquiries from Amgen, CVS, or AmerisourceBergen. They didn't know where it had been or how it got there and they had no way to know. No one had drawn the map, planned the route, or had the tools to document it.

What was known surfaced because of the Horsemen's investigation, the combined efforts of FDA agents, the information supplied by cooperating witnesses, and lawsuits between wholesalers. Some of the best information about Tim's medicine came from the man accused of tainting it. At the Dade County Stockade, José Grillo spoke briefly with investigators and began to name names: those he claimed set him up in business and those he said bought his medicine. It was a glimmer of information. Grillo fell silent again.

The counterfeit Epogen that reached Tim from lots P002970 and P001091 had begun as low-dose Epogen, sold in large quantities by the big distributors Cardinal and Amerisource to the small Miami pharmacy J&M Pharmacare.

Instead of selling it to patients, the pharmacy sold more than 110,000 vials of it to the middleman Armando Rodriguez, who had ordered it for José Grillo.

Grillo then packed the vials into a paint can and carried them to the dusty trailer of Silvino Morales, the former iron welder. The setting was as far from sterile and controlled as imaginable. Working from a hut in his backyard, Morales soaked the vials, rubbed off the low-dose labels, and glued on the fake high-dose ones. Even before Grillo sold any, the drug had already moved through four sets of hands.

He did not limit himself to counterfeiting Epogen. Using the same method, he also uplabeled the Procrit that reached Maxine Blount, the breast cancer patient in Harvester, Missouri.

The nation's drugs had become almost impossible to trace because those who profited from their sale didn't want them to be traced. The companies buying and selling had not only agreed to, but actively lobbied for, a veiled distribution system with few records and minimal inquiry. The regulations governing pedigree papers were so weak, limited, and unenforceable, said Terry Vermillion, at the FDA's office of criminal investigations, that "you can satisfy the requirement by writing it on a bar napkin."

But investigators including Vermillion still considered the pedigree papers a vital tool to track a medicine's origin. A false pedigree functioned as evidence, a fingerprint, a façade that investigators could knock down. The need for more transparency was so obvious that even the drugmakers reversed their longheld objections and in 2003 urged the FDA to implement comprehensive pedigree papers for all drugs immediately.

But the FDA made clear that it would not mandate a solution or impose any standard on the powerful wholesale industry. Instead, the agency would encourage the companies to move toward the use of technology that was still being developed: bar coding and radio frequency identification that could help track a drug's origin electronically, without the burdensome paperwork.

By September 2003, the drugmakers' anger at the FDA had reached a new height when their security officials and others in the industry gathered at a hotel in Washington, D.C., for a three-day conference on the integrity of America's drug supply. There, they dropped the long-held image of purity and impenetrability. Various speakers described what they'd seen: medicine with the wrong ingredients, without active ingredients, with insufficient ingredients; medicine that had been repackaged and recycled with false expiration dates; medicine made with bacteria-laced water.

Openly, they expressed fear that America was veering toward a third-world drug supply. All present knew what that implied, because they had battled the horrors of it in other parts of the world: microbial resistance caused by subpotent medicine; aspirin tainted with strychnine; children's cough syrup made with deadly antifreeze; dirty reused vaccines for infants; pills containing leaded highway paint. But all that paled in comparison to the big fear, which they talked about candidly: a terror attack on the drug supply. Just as Al Qaeda had used America's own flight schools and planes against us, those gathered could envision terrorists easily moving tainted pharmaceuticals through the nation's distribution system and obtaining state licenses to distribute medicine as easily as the narcotics traffickers had.

When the FDA's director of pharmacy affairs, Tom McGinnis, addressed the conference and said that he hoped an electronic pedigree could be phased in within five years, an angry line formed at the microphone.

"For him to say to go another five years is criminal," fumed J. Aaron Graham, a former FDA special agent who was now the vice president of corporate security at Purdue Pharma. "Oh my god. How about educating the FDA?"

> *Once the vials of Epogen had been successfully uplabeled, José Grillo brought them to his customers, who included the owners of Playpen South. In a dismal back room at the strip club, he sold the vials to Paul Perito and Nicholas Just, who stored them in a beer cooler.*
>
> *Once Grillo left, Carlos Luis arrived to buy the drugs. He had his customers already lined up. One was Eddie Mor, who on paper owned a wholesale company in Texas but actually sold the medicine out of his Davie home. In one two-month period, Mor bought more than $2 million worth of medicine from Luis.*
>
> *Mor, in turn, sold the medicine to the blonde pharmacist Susan Cavalieri, who had a legitimate state license to distribute medicine. In this way, the medicine labeled as high-dose Epogen from lot P002970 picked up a seemingly legitimate pedigree. Along this route, Tim Fagan's medicine moved through seven sets of hands before even reaching a company properly licensed to distribute it.*
>
> *Cavalieri sold the medicine to the Arizona wholesaler Dialysist West, which sold it to AmerisourceBergen.*

In February 2004, an FDA task force released its final report on the problem of drug counterfeiting. Those looking for forceful direction, strengthened regulations, or immediate solutions were disappointed. The report emphasized due diligence by those in the supply chain, the voluntary use of cost-effective technology, and better education of the public. It also emphasized the need to reduce the "regulatory burdens" for "stakeholders"—which included the middlemen.

On Long Island, Kevin Fagan was furious. He wrote to FDA commissioner Mark McClellan, saying that he found the voluntary standards "unfathomable," and accused him of allowing companies to place profits before safety. "I look to national leaders such as yourselves to ensure our safety by holding those that choose to turn a blind eye in the name of profits accountable for their actions," he wrote, adding, "You have surely failed in that mission."

Though Kevin didn't know it, another mission, a political one, had intervened. Before the FDA released this report, a far more candid and hard-hitting draft of the FDA report had circulated among government officials the previous September. That draft report blamed drugmakers' discounted pricing for widespread diversion and wholesalers' purchase of medicine that had no clear origin for allowing dangerous medicines into the nation's supply. Numerous transactions and a lack of comprehensive pedigrees had made the nation's medicine almost impossible to trace and therefore more susceptible to counterfeiting, the draft stated. It also criticized the FDA for using what little power it did have ineffectively. "There are no comprehensive efforts to educate the public about the threat of counterfeit drugs," the draft noted, "how to identify them and how to minimize the risk of receiving" them.

The early report saw great promise in radio frequency technology to track a drug's origin. But until such technology could be implemented, it recommended that manufacturers circumvent wholesalers when distributing drugs at high risk of being counterfeited. It also urged the immediate implementation of full pedigree requirements.

The public saw few of these recommendations. A rewritten interim report issued publicly one month later, in October, bore little resemblance to the draft. It minimized the problem of diversion in the secondary market, with previously neutral or critical sentences revised to legitimize and praise the role of secondary wholesalers. While the September draft explained that there were "many reasons why wholesalers sell to one another," the October report stated, "There are many reasons why sales from one wholesaler to another may benefit consumers."

Instead of blaming wholesalers, the report now blamed "illicit nationwide networks" that capitalized on the "inadequate due diligence" of those in the distribution chain—a problem that could be solved by the industry's adoption of voluntary standards. The report concluded that pedigree paper requirements were stayed because of "valid concerns expressed by industry, trade associations and Congress. . . ." It stated that implementing the rule now could "impose substantial costs, at a time when access to affordable drugs is also a major policy concern."

As for how consumers could protect themselves, the report suggested that they not buy foreign drugs: "It's safest to purchase ONLY from US state-licensed pharmacies, where the FDA and state governments can assure the safety of drug manufacturing, packaging, distribution and labeling." Though the report had been undertaken precisely because the FDA and state governments could *not* assure the safety of the drugs in America's pharmacies, it almost guaranteed a status-quo approach to the distribution of medicine. It stated that the task force would continue to work with the distributors' trade groups "HDMA, PDA and others to gather information about current and best practices in pharmaceutical industry."

Sal Ricciardi, along with principals from his company Purity Wholesale Grocers and their spouses, had contributed over $200,000 to congressmen, senators, and several political action committees from 1999 to 2004. With the support of his Florida congressman Peter Deutsch, who sat on the House

Energy and Commerce Committee, Ricciardi had been able to meet with high-level FDA officials. With these efforts, Ricciardi had earned his group full credibility and a place at the negotiating table. While in testimony and speeches he claimed to informally represent thousands of companies, Ricciardi's trade group actually had only twenty-eight members. The FDA knew this because each company had signed a voluntary integrity pledge that Ricciardi had sent to the agency as proof of the PDA's commitment to safety.

The member companies, whose sales ranged from $1 million to $1.7 billion a year, had varying reputations. While some had no known history of regulatory trouble, a number of others had been enmeshed in past and ongoing investigations into drug diversion and counterfeiting. One had been sued for allegedly distributing counterfeit Lipitor. Another was cited in a probe by the California attorney general into price gouging in the sale of scarce flu vaccine. At least eight of the companies bought medicine that had moved through Michael Carlow's shell companies. At least four had done business with the now-defunct Nevada wholesalers Dutchess and Legend. Regardless, the pledges served to stave off stricter federal regulations for the time being.

"The Association took a leadership role in the crafting along with HDMA of guidelines for wholesalers to follow," said Anthony Young, the PDA's lawyer, who added, "The wholesalers that are observing the guidelines are responsible parts of this trade."

José Grillo's prices could not be beat. He sold his Epogen and Procrit for one-sixth the average wholesale cost. As Cesar Arias observed, the price told a story that almost everyone in the industry knew: The medicine was impure in some way.

Counterfeit Epogen from the two lots that reached Tim Fagan took a variety of paths through the drug supply. Grillo sold the medicine to several buyers, including a former narcotics trafficker who owned medical clinics. Some counterfeit medicine from both lots, P001091 and P002970, moved upward through several tributaries, including through two unlicensed middlemen to a licensed wholesaler in Gainesville, Georgia, Premier Medical Group.

The owner, James Suozzo, the eighth-grade dropout and heroin addict, ran the company from his Fort Lauderdale home. He sold more than twelve hundred boxes of Grillo's counterfeit Epogen to a Tennessee Company, CSG Distributors, which in turn sold all of it to Dialysist West, which sold it to AmerisourceBergen.

The cheaper the medicine to start, the more hands it was destined to move through as it rose up the distribution chain. The Epogen moved from felons and diverters to licensed and increasingly legitimate companies, with each one raising the price until it approached the market rate. At that point, a national distributor bought it, though still for less than from the manufacturer. The savings vanished by the time the medicine reached patients.

"Why on earth can you buy it for less from a flea than from a giant?" asked Oscar Gelpi, the third prosecutor assigned to Operation Stone Cold. "It's the industry's dirty little secret," he said. "They know it's dirty drugs just like the tobacco companies knew that cigarettes will kill you."

For years, the drugmakers had benefited from the interwoven systems of multi-tiered pricing and distribution. By unloading a huge volume of drugs to just a few of the biggest wholesalers, they were able to keep distribution costs low and show the large sales that boosted their stock price. The relationship between the wholesalers and the drugmakers was symbiotic. The drug companies were sponsors of the lavish annual conference of the wholesalers' biggest trade association, HDMA. Those who registered were welcomed with an Astra Zeneca tote bag filled with samples of over-the-counter medicine.

The system led to abuses. In August 2004, drugmaker Bristol-Myers Squibb agreed to pay a $150 million fine after the Securities and Exchange Commission found the company had offered distributors incentives to overpurchase its medicine, a practice called "channel stuffing" that allowed Bristol-Myers to overstate its sales by $1.5 billion. In turn, distributors stockpiled the medicine until the price went up or sold off the excess, creating a glut in the supply chain ripe for diversion. And diversion, though barely mentioned in the FDA's final task force report, was *the* gateway crime that opened the door to counterfeiting. "There's no one in OCI who would separate diversion from counterfeiting," said Vermillion at the FDA's office of criminal investigations. "There's never been a counterfeit drug enter into the United States market that didn't come from an illicit diversion network."

The best and perhaps only way to fight diversion was to tighten the distribution system. By mid-2003, drugmakers began to do just that. Pfizer, Johnson & Johnson, and Eli Lilly demanded that wholesalers and pharmacies purchase their products only from them or from their authorized distributors. Ortho Biotech announced that it would terminate any distributor caught buying their products from unauthorized wholesalers. To foil diverters, GlaxoSmithKline announced that the discounted drugs it sold to third-world countries would now be a different color than those for the American market. Serono took perhaps the most radical step, shipping its AIDS drug Serostim directly to pharmacies so that middlemen no longer touched it. It also began selling the drug at the same price to all buyers so that any Serostim being sold for less would be easily identified as diverted. Drugmakers also began to limit the amount to one month's supply that wholesalers could buy in advance, thereby making diversion and speculative buying more difficult.

A number of drugmakers redesigned their packaging, embedding overt and covert features into their products, from color-shifting ink on their labels to actual chemical markers in the medicine itself. Ortho Biotech began using tiny seals of color-shifting ink on its Procrit boxes. It added color-coded caps for the vials that identified the strength of the medicine, along with matching color-coded seals for the vials themselves. The seals could not be removed without destroying the vials, in theory making "uplabeling," what Grillo had done, impossible. Those in the industry knew that security measures needed to be continuously updated. Given eighteen to twenty-four months, counterfeiters could copy almost anything.

KEVIN FAGAN NOW LOOKED AT HIS SON'S MEDICINE LIKE SOMEone who had donned night-vision goggles. In the individual vials and pills, he saw a playing field tilted against the average consumer and toward the pharmaceutical interests. His family paid the world's highest prices for drugs, even though the drugmakers, the distributors, and the government could not guarantee their safety. For all the money the distribution system supposedly saved, his family had yet to see any of those savings. Everyone talked about the dangers of foreign drugs, but we had yet to protect our own supply.

As Fagan wrote to the Surgeon General, "so much attention is given to drug importation, and the risk of counterfeit
pharmaceuticals entering the country. What I find so frus-
trating, hypocritical and enraging," he wrote, was the "run
around" he got for a counterfeit that had come from our own
supply. The FDA and the pharmaceutical industry seemed to
"turn a deaf ear when the issue of domestic counterfeiting is
raised." After several months with no response, he wrote to
the Surgeon General again.

The Fagans' lawyer, Eric Turkewitz, had also become immersed in the issue. Though his practice had taken him to the
state capital before, he had never spoken publicly about policy issues. But on October 15, 2003, he joined more than three
hundred people in a crowded conference room in Washington, D.C., for a public meeting called by the FDA as part of
its anti–counterfeit drug initiative. The gathering looked
more like an industry trade show, as salesmen of new technology clogged the halls, handed out memorabilia, and hawked
equipment they touted as the twenty-first-century solutions to
counterfeiting. Amid all the big-money interests in the room,
Turkewitz was the only one who had come to speak on behalf
of a single person.

Here, the Horsemen's accomplishments became plainly
evident, as speaker after speaker either debunked the new
Florida law as overly restrictive and unnecessary or read aloud
from the statewide grand jury report to argue for stricter enforcement. By late afternoon the crowd had thinned, most of
the top officials had left, and it was finally Turkewitz's chance
to speak. The lawyer made his way to the microphone. The
room gradually quieted as he described what young Tim
Fagan had gone through and the need for accountability in
the pharmaceutical industry. Then he turned to the question
of technology.

"You've heard plenty about technology, some of which is ready and some of which is not," he said. "This has got nothing to do with technology. This has to do with the fact that our drugs should not be treated like a commodity. They should not be traded on open markets like gold or silver or pork belly futures. They're used for people with cancer and HIV or people with transplants and other life-threatening diseases and conditions."

Speaking slowly and deliberately, he continued, "The crimes exist because it's a crime of opportunity. And just as if you had one house with five doors and windows and another house with five to ten thousand doors and windows, you know which one is going to be more secure. Nobody can secure five to ten thousand doors and windows. You've got to close down those doors and windows, you can't let them in. If you let them in, there are people within the pharmaceutical industry who are going to take advantage of it, who are going to turn that blind eye.

"We're asking you to restrict the number of times that drugs can be sold to cripple that gray market," he said. "Because if it exists, it will be exploited. Where there's money to be made, people will go after it.

"On behalf of the Fagan family, I thank you," he concluded. "They wanted their voice to be heard."

BY WINTER 2004, KEVIN FAGAN CONCEDED THAT THE LEGISLA-tion his congressman had proposed was all but dead. His letter-writing campaign had fallen on deaf ears everywhere. He had not heard of the Horsemen and still knew almost nothing about where Tim's drugs had come from.

But the family's lawsuit offered a ray of hope. In July a New York judge ruled that the case against AmerisourceBergen and CVS could proceed, despite those companies' efforts to be excused from the case. In her ruling against Amerisource and CVS, the judge focused on the companies' participation in a gray-market system that was known to be dangerous. Amerisource may have been in the best position to prevent harm to Tim by avoiding such purchases, the judge ruled. She concluded that CVS also may not have done enough to preclude buying counterfeit medicine. The ruling released Amgen from liability, stating that it did not have a duty to anticipate and prevent the criminal acts of third parties.

In Missouri, the Blounts' lawsuit crept through the courts, narrowly surviving several adverse rulings that weakened the case against Amgen and Ortho Biotech.

Publicly, FDA officials said that patients had a greater chance of getting struck by lightning than of getting counterfeit medicine at their pharmacies. But no one actually knew how much counterfeit, adulterated, or subpotent medicine was in our drug supply. In April 2004, the two authors of the FDA's task force report, Ilisa Bernstein, the FDA commissioner's senior advisor for regulatory policy, and Dr. Paul M. Rudolf, senior advisor for medical and health-care policy, wrote an article for the *New England Journal of Medicine,* in which they said, "Although it is hard to obtain accurate data, the proportion of drugs in the U.S. marketplace that are counterfeit is believed to be small—less than 1 percent."

In an interview that month at the FDA's headquarters, the authors acknowledged that the available data was "soft" and that their estimate was more an effort to reassure the public than to analyze the number scientifically. "We don't believe there's currently a substantial amount of counterfeiting," said Bernstein. Rudolf added, "We were trying to walk the line between saying this was a serious potential issue but not scaring people so they would stop taking their drugs."

Few if any reliable estimates existed, since no one had tested drugs system wide. Even John M. Taylor III, the FDA's associate commissioner for regulatory affairs, described the effort to quantify the number as the "longest running fiction." Though the World Health Organization had estimated that 7 to 10 percent of the world's medicine was counterfeit, the numbers were "fuzzy," he said. Bernstein and Rudolf wrote in the *New England Journal of Medicine*, "Although there have been few reports of serious adverse events, the number of treatment failures attributable to counterfeit drugs in the United States is unknown." They acknowledged that with more than a billion pills sold in the country every year, "it is possible that millions may be counterfeit."

Even by the most conservative estimate of less than 1 percent, the number of counterfeits was potentially huge. In 2003, Americans spent $216 billion on filling 3.4 billion prescriptions. One percent of that was 34 million prescriptions, the counterfeits certainly concentrated within the categories of the most expensive medicine for the sickest patients who needed it the most.

In November 2004, the FDA announced with fanfare an initiative to protect the drug supply. The agency said it would promote the voluntary use of radio-frequency identification tags—a "bar code that barks," as Robin Koh, director of applications research at the Auto-ID Labs of the Massachusetts Institute of Technology, told the *New York Times*. The FDA gave the pharmaceutical industry a goal date of 2007 for introducing the technology. While the announcement garnered positive headlines, the agency had done little more than back the voluntary use of technology still being developed, just months after it had again set aside the simple solution of paper pedigrees.

Those waiting for FDA action were not surprised. "They're making the good the enemy of the perfect," Cesar Arias concluded. "They don't want to do the good, because they have this imaginary thing coming up." Instead of a high-tech magic bullet, said Congressman Israel, "good old-fashioned enforcement and tougher penalties" are needed. "This is a no-brainer, it's a simple deal," he said, adding of the Fagans, "This family has had no responsiveness from the federal government and now a steeper mountain to climb."

By the end of 2004, Kevin Fagan felt manipulated and lied to and saw clearly how financial interests had been stacked against the health and well-being of his family. What did it mean to have the best medicine when you could not trust its efficacy? What did it mean that without a shred of evidence or a study showing that their businesses would be devastated, the government had accommodated twenty-eight companies that did not want stricter regulations? Tim had very narrowly survived, in part because he had a great surgeon, a committed family, and the miracle of a compatible organ donated by a man who turned out to be a neighbor. All those stars had lined up for him, but at day's end he had been struck by lightning. Even if true, the Fagans took no solace from the government's assurance that only a small portion of the country's drugs are counterfeit.

Despite the many obstacles he faced, Kevin stubbornly clung to the belief that an individual could overcome profit motives and corporate indifference to help change a system. Like Gary Venema and Cesar Arias, he had a stubborn, obsessive personality that made him keep going in the face of resistance and almost certain defeat. Though Kevin likened his efforts to "climbing Mount Everest in a wheelchair," he said that only change for the better would make him stop.

Epilogue

BY 6 A.M. ON MAY 27, 2004, MORE THAN A DOZEN AGENTS FROM the FDA, FDLE, and Miami-Dade Police Department had gathered in the parking lot of a commuter train station near Coconut Grove wearing bulletproof vests and guns strapped to their legs.

As Cesar Arias and Gene Odin milled among the heavily armed men, several greeted the pharmacists warmly, even leaning in so Odin could hear them better. The team reviewed procedures and checked weapons. One agent donned gloves, in case they had to break down the door at the elegant home of Paul Perito, the urologist who co-owned the Playpen South strip club.

The police cruisers and unmarked cars drove silently past ornate gates and towering palm trees and stopped a few feet from Perito's house. A flatbed truck backed up to his gate and the six cops inside swarmed over it. Their pounding and yells of, "Police, answer the door!" broke the early-morning quiet. Two small children in pajamas swung open the front door, their mother a few steps behind. "Please put them in the bedroom," one of the officers told her.

Finally Perito stepped out. He was young and handsome, with an air of refinement, despite his rumpled hair and the track pants and undershirt he had thrown on in a hurry. He also wore plastic doctor's clogs. He was charged with twenty-two counts of racketeering, conspiracy, fraud, money laundering, selling counterfeit goods, and numerous violations of the health law statute 499 in connection with alleged sales of counterfeit medicine through his strip club.

As Odin snapped photographs, Arias watched coolly from the street. "Another satisfied customer of the Horsemen of the Apocalypse," Arias said.

To the south, in Homestead, a larger and more heavily armed group gathered to arrest Perito's business partner, Nicholas Just, who was believed to keep weapons at home. At the sprawling ranch house, Just's aging father answered the door, blinking at the sight of so many cops carrying guns. He pointed the way to the bedroom where Just, who had arrived home from the strip club only a few hours earlier, was sacked out, a small child sleeping alongside him. Behind Just's bed hung three huge posters of characters from *The Sopranos* in ornate frames.

Gary Venema reached down to shake Just awake and flashed his badge: "FDLE."

"FP&L?" asked the groggy Just, referring to the Florida Power & Light company. Venema indicated for him to be quiet, motioning to the sleeping child, and got Just to follow him into the hall. Then the investigator read the eleven charges against him, including racketeering, conspiracy to commit racketeering, and buying and selling counterfeit goods. "And no," Venema added, "I am not the Florida Power & Light company."

As the team finished searching the house, Odin and Arias pulled up outside. One of the federal agents who had known Arias for years looked at his Buick and asked, "Are you still driving that piece of garbage?"

THREE YEARS AFTER THE HORSEMEN HAD BEGUN WORKING TO-
gether, everything and nothing had changed. They had built
the foundation walls of a clean drug supply and shown other
states where to look, how to challenge abuses, and how to
build cases bigger than the nation's antiquated health laws allowed. Spurred on by the National Association of Boards of
Pharmacy, California followed Florida in tightening its regulation of wholesalers. A dozen other states, including Arizona,
Illinois, and New Jersey, introduced similar bills. But the in-
transigence of the federal government and its capitulation to
moneyed interests meant that the fight against bad medicine
would remain an uphill battle.

By the end of 2004, the Horsemen had arrested fifty-five
suspects—more than thirty of them on racketeering charges—
and seized $33 million in bad medicine and almost $3 million
in cash. Sixteen suspects had agreed to cooperate, most plead-
ing guilty to an array of charges.

The Horsemen had changed Florida law and testified before Congress. In the wake of the state's reforms, the number
of licensed drug wholesalers in Florida had dropped by almost half. And statewide Medicaid costs plunged for certain
categories of drugs that had been overprescribed, billed to
Medicaid, diverted from clinics, and prescribed again. Monthly
Medicaid payments for blood products like intravenous immune globulin fell from a height of $2 million in August 2002
to $160,000 by March 2004. Operation Stone Cold expanded
its reach, working to break up a ring making at least $50 mil-
lion a year selling painkillers over the Internet and another that
had submitted more than $700 million in fraudulent claims for
prosthetic limbs to the state and federal government.

The cases against the Horsemen's biggest targets plodded
through the legal system. Michael Carlow and José Grillo had
pleaded not guilty to the charges. Both men had been aban-

doned by confidantes and former associates, who now lined up to testify for the state.

Carlow remained in jail in Fort Lauderdale, awaiting trial in 2005. His wife, Candace, had sold the couple's pink ranch house with the horses, the proceeds instantly sucked up by an array of creditors including the IRS. She filed for bankruptcy in August 2004 and was also awaiting trial.

Carlow's former deputy, Mark Novosel, took a job buying and selling warehouse equipment while awaiting sentencing in his own case. He faced up to eight years in prison, but likely would serve as little as three because of his cooperation.

Among Carlow's other alleged co-conspirators, Fabian Díaz and José L. Benitez remained at large, with fugitive warrants out for their arrests.

At the Dade County Stockade, José Grillo had sunk into despair. He was placed on a suicide watch after twice trying to kill himself. His wife, Leticia, whom he had always controlled, broke away and began cooperating with investigators. In 2004, Grillo learned that he had been indicted a second time—on racketeering and other charges in connection with counterfeit medicine sales at Playpen South.

The Horsemen made another sweep in February 2004 and arrested Carlos Luis, Eddie Mor, Javier Rodriguez of L&L Distributors, and others allegedly connected with the counterfeit Epogen and Procrit sales. When Luis refused to come out of his house, the cops took the hinges off his front door and took him out, guns drawn.

The arrests in Operation Stone Cold had flowed from the initial insight of a young prosecutor who had managed her own chronic illness from childhood. Stephanie Feldman believed that those who lied, ran shell companies, and passed phony drug pedigrees were defrauding innocent patients whose lives depended on the effectiveness of their medicine.

The racketeering statutes had allowed her to target this illicit activity as a criminal conspiracy.

As the cases headed to trial, some of the defense attorneys prepared to argue that the government had overreached and had found conspiracies where none existed. "A lot of these people have nothing to do with each other," said one attorney, Edward O'Donnell, who represented Carlos Luis.

MARTY BRADLEY RECOVERED FROM HIS LOSSES AND, BY LATE 2004, had moved BioMed Plus to a much larger warehouse. In 2003, his company had reported sales of $185 million. He had gotten a contract from one of the Big Three distributors, McKesson, to be the company's principal supplier of specialty blood products. And he was pleased by what he viewed as the cleanup of his industry.

In 2004, Bradley hired a new consultant for his company: Neil Spence, the former vice president of NSS, who had a longstanding business relationship with Michael Carlow.

By late 2004, Bradley faced a resurgence of the previous trouble he'd had with government investigators, unrelated to his hiring of Spence. He found himself in the middle of an inquiry into whether his company had fraudulently acquired and resold prescription medicine and blood products, some of which had already been billed to Medicaid.

Turmoil ensued at Cardinal Health as the Securities and Exchange Commission and the U.S. Attorney's office for the Southern District of New York announced they were pursuing an investigation into how the company classified revenue from its pharmaceutical distribution business. In July 2004, the company's chief financial officer resigned, acknowledging that, "certain financial reporting practices and judgments that

occurred during my tenure as CFO have come under scrutiny."
The company also faced a lawsuit from shareholders alleging
that its officers manipulated the accounting and inflated the
company's revenue.

At both Cardinal Health and AmerisourceBergen, margins from their pharmaceutical distribution businesses shrank,
in part because of decreased purchases of medicine from the
secondary market.

Meanwhile, the ongoing investigation into the Lipitor
counterfeiting produced indictments, a flurry of lawsuits, and
a detailed picture of how Julio Cruz and his associates had
gamed a poorly guarded system. In Kansas City, Missouri, a
federal grand jury indicted two executives at wholesale companies in California and Illinois for allegedly taking secret
commission payments from Albers Medical Distributors after
they got their companies to buy Lipitor and other medication.

Cruz led investigators to a filthy Miami warehouse filled
with $15 million worth of counterfeit, diverted, and illegally
re-imported medicine. Investigators found pill-making machines and two million tablets of counterfeit Lipitor. Cruz and
his associates had obtained the raw ingredients, dyes, and
punches to make the drugs by forging a letter from Pfizer to a
Missouri engineering firm requesting equipment they claimed
was for a new manufacturing plant in Costa Rica. In January
2005, Cruz pleaded guilty to charges of conspiring to sell, and
selling, counterfeit and illegally imported Lipitor.

AS A TEAM, OPERATION STONE COLD WON A DAVIS PRODUCTIVITY
Award, which honors Florida civil servants who save the state
money through hard work and innovation. It was the second
year that Arias and Odin had won. At the awards luncheon,

Petri spoke for the group and accepted a check for $1,500—to be split twenty-one ways among all those listed as nominees, including the Horsemen's higher-ups.

To his dismay, Petri continued to be shut out of the ongoing work of Operation Stone Cold by a competing unit at the Miami-Dade Police Department.

Despite Arias and Odin's best efforts, little seemed to change at the Bureau of Statewide Pharmaceutical Services. If anything, things got worse. By the end of 2004, there were only six drug inspectors statewide, three less than when the Florida grand jury report had lambasted the bureau for being understaffed. Since then, two inspectors had resigned in disgust. A third inspector was forced to resign. Arias suspected it was because he had spoken candidly to the health department's inspector general. But the Tallahassee managers cited daily activity reports that did not match the inspector's gas receipts and tollbooth records in several instances. In the Tampa and Orlando regions, only one inspector remained to regulate hundreds of businesses, including warehouses for each of the Big Three. The management said it hoped to fill three inspector positions by the end of 2004. By the end of January 2005, one of the positions had been filled.

In September 2003, before the staff decreased, six of the department's eight drug inspectors wrote a three-page letter to the state's deputy secretary for health. Arias, Odin, and the others accused the bureau's senior managers of idly sitting by as the illegal distribution of tainted drugs grew progressively worse. The inspectors wrote that their managers took no action, displayed no leadership, "or worse yet in many instances impeded any progress to be made in protecting the public from counterfeit, stolen, diverted, misbranded and/or adulterated drugs." They charged that the "do-nothing policy" of bureau chief Jerry Hill and others had "put the public at risk" of receiving bad medicine.

The letter followed a critical report by the health department's inspector general, who observed that the bureau had no functioning computer system to track the cases the inspectors developed and that it took months for inspection reports to reach those who needed to approve or terminate licenses. The report also noted low morale, high turnover, and perceptions of favoritism. Jerry Hill told the inspector general that he held meetings, had an open-door policy, and maintained a suggestion box.

A second inspector-general report released in October 2004 found that it sometimes took the bureau almost two years to notify businesses of violations found by its inspectors. The report recommended that the bureau speed up, streamline, and strengthen its case management system, which the bureau's managers agreed to do.

Hill did not seem troubled by the inspector general's findings. "These management reviews are used to provide leadership with opportunities to continue our departmental journeys towards excellence," he wrote in response to questions from a reporter. He continued seeking ways to improve morale by providing raises to drug inspectors, improving technology, and visiting field offices more frequently. He said, "I recognize that employee satisfaction is linked to productivity."

At seventy-five, Gene Odin finally had his eye on retirement. Arias had considered quitting his job many times during the past decade, but never more seriously than now. The bureau's Tallahassee office had placed an advertisement in the state government listings for a manager in the Miami field office—Arias's job. Arias's boss told him there would be no change in his pay, but that Arias would no longer supervise

the Miami office. Instead, he would report to the new supervisor, who would not be a pharmacist and would earn $20,000
to $50,000 a year. To do this, the managers had converted an
open position for an inspector to that of a bureaucrat.

Arias began a job search. "I'm going to go home and sleep
like a baby—wake up every half-an-hour and cry," he said.

Jerry Hill remained bureau chief and received credit for
being a reformer. In November 2004, Hill was appointed to the
National Drug Advisory Coalition, a committee responsible
for creating a list of medicines susceptible to counterfeiting. "I
am pleased that the National Board of Pharmacy not only rec-
ognizes Jerry's expertise and contribution to the state of
Florida but has asked him to lend his knowledge in helping to
set national pharmacy standards to ensure patient safety,"
Florida's health secretary John Agwunobi announced. At the
Jackson Memorial Hospital pharmacy, where the small theft
by Sydney Jones had originally brought the Horsemen together, the refrigerator that contained cancer medicine had
been secured with a bicycle lock.

The Horsemen remained as cohesive as ever, through
good and bad. Petri continued his annual Operation Stone
Cold Christmas parties. The group had regular dinner parties
and even took vacations together. In late 2004, all five men and
their wives headed to Amelia Island on Florida's northeast
coast for a long weekend. They stayed by the ocean, rode
horses along the beach, and at night had a cookout, spreading
a tarp across the sand. Odin lay on his back, staring up at the
stars.

For him, the case and the resulting friendships had been
the true bonus of a lifetime. He had never dreamed he would
be part of a team like this one. He knew they had failed in
some of their goals and that they had exposed only a small
sliver of a systemic problem. Nonetheless, he believed that

bashert, the Yiddish word for destiny, had brought them together. "If you're a fatalist to any degree, you have to say 'wow,'" he concluded.

Arias, stretched out on a beach chair, thought, "It doesn't get any better this." Through serendipity, he felt that he had met some of the finest people in his life. Petri and Jones agreed that the weekend together represented for them the case's greatest accomplishment. On a nearby beach chair, Venema let his mind drift back to the first time that Arias came to his house and sat on his back porch, laying out how all those dangerous doses moved from Florida into the nation's supply. The case had been an incredible learning experience. The problem was so huge and out of control that it filled his imagination and almost required him to be a maniac, at which he excelled. He closed his eyes. With just a little rest—he had drafted two search warrants over the weekend—he would be fired up to do more.

Afterword

Winter 2005/2006

DANGEROUS DOSES HAD AN IMPACT ON PROTECTING THE NATION'S drug supply beyond what I could have imagined. In the days just before the book's publication, one of the nation's largest wholesalers, Cardinal Health Inc., announced that it would close a trading division that bought discounted pharmaceuticals from other wholesalers. At the same time, New York State Attorney General Eliot Spitzer launched an investigation into the buying practices of the nation's largest pharmaceutical wholesalers.

By November 2005, the nation's three biggest pharmaceutical wholesalers, AmerisourceBergen, McKesson, and Cardinal, had pledged not to buy brand-name pharmaceuticals from the secondary market. The CVS pharmacy chain announced that it would not buy from middlemen who bought from the secondary market. Regional wholesalers made similar pledges.

States also began to change their laws. Using Florida's law as a model, the National Association of Boards of Pharmacy urged every state to overhaul its laws governing the drug supply. Nine states—Arizona, Indiana, Iowa, New Jersey, California, New Mexico, Oklahoma, Texas, and Virginia—took legislative action, with more considering it. Nevada won its battle against secondary wholesalers, reaffirming its laws and triumphing in court challenges.

To my surprise, a number of large regional and specialty wholesalers embraced the book. Companies from Morris & Dickson and Kinray to FFF Enterprises and F. Dohmen began circulating news of the book, its findings, and updates to its customers regarding the "low-life criminals" who had infiltrated the supply chain. They also offered to give a copy of the book to their customers or to patients who wanted to learn more about the counterfeiting problem.

The blood-products distributor FFF Enterprises, in Temecula, California, had for years bought directly from drug makers. In championing the book, FFF pointed out that its decision was a wise one that guaranteed the safety of its products. Others, like Morris & Dickson in Shreveport, Louisiana, changed their practices because of the book.

Skipper Allen Dickson, fifty-six, the wiry, intense, and hard-driving president of Morris & Dickson, explained that his distribution company, like almost everyone else's, purchased its pharmaceuticals directly from manufacturers but also from other wholesalers. This "trading," in which competing wholesalers bought from one another, had been a routine part of the nation's tidal flow of pharmaceuticals. It was also the vehicle that allowed dangerous drugs into the supply chain. For Dickson, the book was a wake-up call: "I did not realize that my competitors, who I was buying from, were that reckless."

Dickson and his brothers had taken over Morris & Dickson from their father, Markham Allen Dickson Sr., an ordained Episcopal minister with an engineering degree from MIT. Dickson Sr. had always taught his sons that those confronted with a problem who don't fix it should be held to a high level of disdain. So Dickson and his brothers resolved to stop buying drugs from other wholesalers. In the first months, the decision was frightening from a business standpoint. "It's a challenging position because if I'm the only one who does it, I'm dead," Dickson explained. But slowly, other wholesalers also reversed course.

The Horsemen's lives also changed dramatically. They had the satisfaction of watching reform that had resulted, in part, from their courage and candor. They became minor celebrities in the world of supply-chain integrity. In October 2005, Venema and Arias traveled to Toronto to address the Royal Canadian Mounted Police and, much to their delight, came away with framed drawings depicting police on horseback.

However, none of this happened without more heartache and struggle. By the winter of 2005, Operation Stone Cold had been effectively dismantled and four of the Horsemen—Cesar Arias, Gene Odin, John Petri, and Randy Jones—had left state employment. The most dramatic events swirled around Arias, whose bitter fight with his agency, the Bureau of Statewide Pharmaceutical Services, grew worse.

In February 2005, three months before the book appeared in stores, Arias received a notice that he had been placed on paid administrative leave, pending the outcome of an investigation. Not notified as to the charges, he was told to surrender his state equipment immediately, including the battered blue Buick. The irony could not have been greater: The day before, he had received a certificate of merit from the Florida health department for having assisted federal agents in a major Medicare fraud case.

Almost seven weeks later, the bureau charged Arias with insubordination, violating agency rules, and conduct unbecoming a public employee. Two of the five allegations dealt with his conduct after being suspended: failing to return his state car for three hours and failing to pick up a copy of the charges against him on the day they were filed—Good Friday.

The other allegations included not using a government license plate on the Buick (he used a civilian one so as not to alert those he was investigating); personal use of his government cell phone (which was permitted and for which he had paid additional charges each month); and sending e-mails critical of the bureau's management to other state employees, including Odin.

Certain the bureau had retaliated against him for his increasingly public criticism of its management, Arias hired a lawyer. He was quickly cleared of all wrongdoing and offered $26,000 plus attorney fees to resign. He did so, he said, only after his managers made it clear that in staying he would no longer be permitted to work on Operation Stone Cold. In a letter to Arias, the state's health secretary, Dr. John Agwunobi, thanked him for his seventeen years of service and noted that the health department "continues to successfully work with our partner agencies in taking the necessary steps to safeguard our citizens."

Odin resigned in protest.

The two former drug inspectors refused to go away. They embarked on a letter-writing campaign to their former bosses and other Florida officials, demanding to know whether the bureau would fulfill its stated mission to protect Floridians from bad drugs. Even Arias's mother, Maria, joined in. She wrote to Agwunobi and also to Governor Jeb Bush, saying that she was so angry over her son's treatment that she was considering dropping her decades-long affiliation with the Republican party. "My son and Dr. Odin are national heroes who were forced to resign due to the improper treatment given them and mostly because of the incompetence of their managers," her letter concluded.

She received a polite e-mail back from the state's deputy health secretary, who wrote that the health department and its bureaus would "continue the journey towards excellence. We have set ourselves on a definite course of quality improvement in all that we do."

Long the Batman and Robin of drug diversion investigators, Arias and Odin found themselves at loose ends. But John Petri, ever the organized and methodical surveillance expert, had surveyed the opportunities for the men to work together again. He formed Stonecold Investigations LLC, a licensed private-eye company. By the fall of 2005, the four Horsemen were back in business together, investigating diverters and counterfeiters for an array of private companies seeking to protect their medicine. As Odin, seventy-six, said of his new work, "I've never had this much fun with my clothes on."

Gary Venema remained at the Florida Department of Law Enforcement investigating pharmaceutical crimes. The young former prosecutor, now Stephanie Feldman Aleong, returned to Florida with her husband and took a job at Nova Southeastern University as assistant law professor and became director of the masters program in health law. She also resumed her place in the Horsemen's lives as an in-state mentor and, as always, their friend.

Throughout the years of neglect, the men had become adept at honoring each other when no one else would. On a July evening, they gathered in a suite at the Holiday Inn, along with Feldman, her husband, and two of the researchers who'd helped them work the case. Armed with gift baskets, plaques, and commemorative Horseman T-shirts, they gave speeches they'd all heard before, as several of their video cameras rolled for posterity. Full of sly humor and affection for their group, Odin declared, "I have never worked with anyone like this in my entire life, and my life has been longer than anyone else's."

In January 2006, Odin and Arias realized their final dream: to overthrow their bosses at the Bureau of Statewide Pharmaceutical Services. In November 2005, the health department's inspector general returned to the bureau—spurred by a tip from Odin. In a scathing report, it found that the bureau's program manager, Gregg Jones, had waited nineteen months to take action against a pharmaceutical wholesaler that had committed violations.

Much to the Horsemen's delight, bureau chief Jerry Hill, Jones, and another top manager all resigned. "The bureau needed to go in a different direction," said Doc Kokol, a health department spokesman. "These individuals decided they did not want to be part of the change."

In his three-page resignation letter, Hill enumerated his accomplishments, including overhauling how pharmaceutical wholesalers operate. He concluded the letter by stating, "I am most proud of the fact that I was able to attract and hire such exceptional employees."

AS STATE AND FEDERAL INVESTIGATIONS INTO DIVERSION AND counterfeiting galloped along, some thought the authorities had finally woken up. Others believed the government was on a witch hunt. Marty Bradley, the CEO of BioMed Plus, whose burgled pharmaceutical warehouse had set Operation Stone Cold in motion, fell into both categories. In March 2005, a federal grand jury indicted Bradley, his father, and six others on 288 counts of racketeering, money laundering, and other charges, alleging a conspiracy to acquire discounted medicine. The government set Bradley's bond at $10 million.

The indictment alleged that from 1999 to 2000, Bradley's company BioMed Plus had fraudulently obtained prescription drugs, laundered their origin through related companies, and diverted them to BioMed Plus for resale. Bradley and the other defendants pleaded not guilty.

Bradley embarked on an all-out effort to clear his name that involved numerous lawyers and millions in fees. "At forty I don't care if I'm penniless," he said with the zeal and optimism particular to him. "If you did something wrong, you should pay. But if you did nothing wrong, you should fight. I am strong enough, and right enough, to be able to get through it."

Even as this cloud hung over Bradley, a separate cloud extended over one of his newer employees. Neil Spence had gone to work for Bradley shortly after leaving Cardinal Health, where he had been a successful senior executive running the division NSS, which distributed lifesaving pharmaceuticals to the nation's oncology hospitals. Spence had been referred to in *Dangerous Doses* as allegedly buying medicine and taking over $15,000 in cash from Michael Carlow, sometimes in packages mailed to his house, while he was an executive at Cardinal.

Now Spence was under a methodical and slow-moving investigation spearheaded by the FDA's Office of Criminal Investigations. Investigators learned that Spence deposited the money he received from Carlow in a private bank account that he'd opened with an address similar to that of his employer, Cardinal Health, according to a person familiar with the investigation. Several former associates of Carlow received subpoenas to testify about Spence before a federal grand jury in Nashville.

A Cardinal Health spokesman, Jim Mazzola, said the company was cooperating with the FDA's OCI and the U. S. Postal Service in their ongoing investigation into Spence's activities.

Unable to make bond, Michael Carlow had remained in jail awaiting trial since his arrest in August 2003. For years he had proved his ability to rebuild from the bottom up, finding new schemes and the freedom to enact them. But this time, his situation looked difficult to escape. After posting bond for his wife and other family members; declaring bankruptcy; and selling off his two mansions, his sports cars, and his yachts, he could no longer afford to pay his private criminal defense lawyer. Within a year, he was being represented by a public defender.

He spent his mornings in jail reading the Bible with one of the wardens, according to the warden's mother, who happened to be Carlow's friend from high school.

But even as he sat in his Fort Lauderdale jail cell, Carlow's past activities would get him into further trouble. In August 2005, two years after his indictment and arrest in Operation Stone Cold, he was indicted again along with nine others, this time by the U. S. Attorney in the Western District of Missouri, for conspiring to sell $42 million in counterfeit, stolen, and illegally imported Lipitor, as well as counterfeit Celebrex.

The indictment alleged that Carlow, along with Douglas Albers (the popular Kansas City pharmacist) and his wholesale company Albers Medical Distributors; the wholesaler H.D. Smith Wholesale Drug Company of Springfield, Illinois; the drug repacker MedPro Inc. of Lexington, Nebraska; and others participated in a wide-ranging conspiracy to counterfeit, smuggle, repack, and sell Lipitor into the gray market in 2003.

In pursuing their case, Federal officials moved Carlow to a jail in Leavenworth, Kansas, where he has remained since September 2005.

Back in Miami, Jose Grillo remained in jail awaiting trial after two suicide attempts behind bars.

STILL BURNING WITH ANGER, KEVIN FAGAN CONTINUED HIS CRU-sade against bad medicine, his efforts fueling the movement toward reform. He had established himself as a corporate nightmare, writing letters to companies far and wide from Wal-Mart to Target, urging them to forego discounted pharmaceuticals. When McKesson decided in November that it would no longer buy brand-name pharmaceuticals from the secondary market, its announcement came in the form of a reply e-mail to Kevin from the company's general counsel.

"While we have no current intention of announcing this policy to the public, it has been communicated to our Board of Directors," the e-mail stated. "It has also recently been shared with those customers of ours who may have inquired about our purchasing practices."

Kevin declared, "We won."

But Kevin had other battles left to fight. On May 9, the day *Dangerous Doses* was published, his congressman Steve Israel (D-NY) held a press conference in New York City that the Horseman attended, and introduced Tim Fagan's Law, a retooled version of HR3297 that he'd announced months earlier. He then set about the work of finding cosponsors.

The bill made no compromises to the wholesale industry. It required comprehensive paper pedigrees for each drug and strong penalties against counterfeiting and diversion. "When my colleagues understand that the meds they take might pass through a strip club in Florida, they are going to want to pass my bill," he told the *Long Island Press* a month after the press conference.

There were small stirrings of political progress. On November 1, 2005, numerous witnesses including Kevin were called to testify before Congress at a Government Reform subcommittee hearing entitled *Sick Crime: Counterfeit Drugs in the United States.*

It was a day Kevin had waited for. As he walked the hall outside Representative Israel's office, he recognized the names on the doors of legislators he'd written to during his years-long quest. Few had written back. Now he would have a chance to address them personally.

Kevin testified passionately without notes, recounting the

night his son first screamed in agony from the mysterious effect of the counterfeit Epogen. He closed by asking the congressmen to "please co-sponsor this legislation, which will protect all the Tim Fagans, potentially every American citizen, from counterfeit drugs."

But numerous barriers to the bill's passage remained. The problem of domestic counterfeits had become confused with, and stalled by, the emotional debate over soaring drug prices and the demand to reimport cheaper pharmaceuticals from overseas. Several congressmen at the hearing questioned whether the alarm over domestic counterfeits wasn't simply a scare tactic by the drug makers' trade group, PhRMA, to block reimportation and preserve their monopolies and sky-high pricing.

Representative Gil Gutknecht (R-MN) even suggested to one witness that counterfeiting could be seen as an entrepreneurial act that might help lower drug prices. "If you are getting a counterfeit that is an exact copy of the name brand drug, ultimately what is the harm to the consumer?" he asked, then gave the example of a $10 pill for erectile dysfunction. "There are counterfeiters—you may call them counterfeiters, I would call them entrepreneurs—that are bringing them in and selling them for $5."

The committee chairman admonished Gutknecht that he should not encourage people to break the law by advocating counterfeiting.

Throughout first-world markets, the need for safe drugs was on a collision course with the drive for cheap drugs, as middlemen engaged in parallel trade, bringing cut-rate medicine from lower-priced markets into higher-priced ones.

In June 2005, Canadians confronted their first incident of home-grown pharmaceutical counterfeiting. Counterfeit Norvasc, a life-saving heart medicine, had been widely dispensed by a pharmacy in Hamilton County, and an alert patient had noted that the pills appeared gray, not white. The Royal Canadian Mounted Police launched an investigation into the counterfeit medicine, which is still ongoing, leading to the arrest of the pharmacist and others. But the incident punctured Canada's assumption that its own medicine was pristine. Almost simultaneously, three different counterfeit medicines, including fake Lipitor, entered England's legitimate supply. The counterfeits had resulted, in part, from parallel trade.

Experts argued that new levels of this trading within the European Union were causing counterfeits to circulate as they moved from less regulated countries like Romania and Portugal to more developed ones like England and even the Netherlands, which also discovered counterfeits in its supply.

At home, reformers viewed pedigree papers, the disclosure of a drug's previous buyers and sellers, as an essential protection against murky transactions. While the pharmaceutical wholesalers' principal trade group, the Healthcare Distribution Management Association, had begun to advocate stringent federal standards for licensing wholesalers, behind the scenes it was trying to stall or reverse new state laws that required universal pedigree papers for each drug. In October, HDMA executives met in Washington, D.C., with Carmen Catizone, the executive director of the National Association of Boards of Pharmacy, and urged him to abandon his state-level reform effort and instead back their federal legislation (which did not include pedigree requirements).

When Catizone refused, the executives claimed his group had maligned the entire wholesale industry and locked them out of negotiations over state laws. Amid shouting and accusations, Catizone explained that he had no intention of advocating weaker laws and, if anything, wanted every state to have the strongest laws possible. Calling it "private," the HDMA did not comment on this meeting.

But in a detailed written statement, the organization said that it is advocating a federal solution because state-by-state pedigree requirements had created an ineffective patchwork of regulations. The tougher uniform standards it proposes include stronger criminal penalties for counterfeiting, extensive background checks for prospective wholesalers, and FDA inspections prior to licensure.

While the HDMA recommends the use of technology to track medicine through the supply chain, it stated that paper pedigrees were "a very ineffective deterrent" to counterfeiting "because the paper record is easier to counterfeit than the drug itself." Until federal legislation is passed, the organization said it would continue to work with states and with the NABP to improve supply-chain safety.

However, Catizone believed that the wholesalers still did not want to reveal where they bought their drugs and feared that pedigree papers "would expose some things that people don't want exposed."

Wherever state laws got tough, the number of secondary wholesalers dropped and new applications plummeted, reducing some of the gateways for counterfeits to enter the drug supply. Business dried up even for more established small wholesalers. In Las Vegas, Robb Miller of Caladon Trading, who had sought to overhaul the public image of secondary wholesalers, decided to close his business in October 2005. In a good-bye letter to customers and friends, he blamed regulators who had given licenses to crooks and counterfeiters. "In order to shift the blame from their own ineptitude, [regulators] demanded extreme regulatory hurdles for everyone," he concluded, adding, "Even the major industry trade group turned its back on me and every small distributor."

But the wholesale industry still asserted itself in back-room negotiations. Two months after the congressional hearing,

Representative Israel was still seeking a Republican cosponsor. "Some of our colleagues would rather take counterfeit medicine than go against the special interests," he concluded.

IN DECEMBER, I BEGAN TO RECEIVE E-MAILS AND PHONE CALLS from ExxonMobil employees in Texas who'd become America's most recent casualties of counterfeit medicine.

Some 1,600 employees at ExxonMobil's Baytown refinery in Houston attended a company health fair in October, hoping for some protection against a threatening flu season. They did not know that Exxon Mobil had outsourced the flu shots it was providing to a local doctor, who in turn had subcontracted another doctor to buy and administer the vaccines.

The FBI began investigating after a nurse for the doctor's private company heard its employees talking about staying up all night filling syringes with water. The FBI arrested the doctor whose company had administered the shots and scrambled to recover the used syringes. Meanwhile, bewildered and frightened ExxonMobil employees got blood-screening tests to determine whether the counterfeit shots had exposed them to dread diseases like hepatitis or HIV.

This incident did not fit the exact pattern of the Epogen, Procrit, and Lipitor counterfeiting schemes. The vaccines had not moved sideways between wholesale companies and into our pharmacies. But the modus operandi was all too familiar: Someone had swapped real for fake, costly for cheap, doing so with a drug in high demand.

In an internal memo to its employees, ExxonMobil executives stated that they did not know the first doctor would contract out the shots, or that the vaccines had not come directly from the drug maker Aventis. But the subcontracting— much like an attenuated supply chain—had allowed those with shady motives direct access to patients. The effect on patients was the same. "Having my sense of security shattered has been the worst part of receiving a fake flu vaccine," one employee wrote to me. "Now I look at every pill I must take regularly and I wonder about it."

The hope of a cure remains potent medicine for any patient. Those who end up taking counterfeits never recover their confidence. Kevin Fagan likened the family's experience after Tim's liver transplant to rushing safely out of a burning building only to get mugged.

Someone asked me recently how I would define "success" when it came to the problem of counterfeit medicine. What lower-percentage chance of getting such drugs should we be aiming for? There is only one answer to that question: zero. There is no such thing as a partial solution to the problem of counterfeit medicine. State-by-state reform, while vital, does not supplant the need for a tough federal law, one that regulates wholesalers, requires pedigree papers, punishes counterfeiters and diverters, and strengthens the authority of FDA criminal investigators.

As long as anyone walking into an American pharmacy has reason to fear counterfeit medicine, this problem is not solved. Our distribution channel should be closed to dangerous medicine, which includes any whose origin cannot be guaranteed. No one with a choice would opt for medicine that has moved through a dozen hands. Every patient should ask, and have a right to know, where their medicine has been, a cause neither the Horsemen nor the Fagans would abandon anytime soon.

THE EPOGEN TRAIL TO TIMOTHY FAGAN

The chart on the following page illustrates the likely pathways that counterfeit Epogen from lots P002970 and P001091 took to reach Timothy Fagan on Long Island, based on indictments, affidavits, witness statements, invoices, and shipping and pedigree records. Solid black lines represent confirmed sales. Broken black lines represent probable Epogen sales based on documented sales of counterfeit Procrit that traveled the same path. Dotted lines indicate unknown sources for those portions of the Epogen's path. The ovals represent those who aided José Grillo in his counterfeiting scheme but were not directly involved in the sale of the Epogen.

The high-dose Epogen originated with the manufacturer Amgen Inc. Sales records show that J&M Pharmacare purchased 3,363 boxes of low-dose Epogen from Amerisource and 8,931 boxes from Cardinal between January 2001 and May 2002. Armando Rodriguez told investigators that he bought up to 200 boxes at a time of low-dose Epogen for Grillo from J&M Pharmacare. The owner of Ocean Press described for investigators how he counterfeited labels at Grillo's direction. In a statement to investigators, Silvino Morales described how Grillo came to his trailer with five hundred vials at a time for him to relabel. Witnesses told investigators they saw Grillo sell his medicine, particularly high-dose Procrit, to Nicholas Just and Paul Perito of Playpen South. Grillo himself said that he sold his medicine to Nicholas Just.

A schedule of sales maintained by Dialysist West shows that the company sold $4.4 million of the Epogen from lots P002970 and P001091 to AmerisourceBergen from September 1, 2001, to May 10, 2002. Invoices show that the medicine was shipped to Pharmabuy, a division of Amerisource in Kentucky. Pedigree papers also show that Dialysist West bought 460 boxes of lot P001091 and 812 boxes of lot P002970 from CSG Distributors; CSG had bought it from Premier Medical Group. According to a statement the owner of Premier Medical gave to investigators, all of the medicine his company sold came from Double J Consultants. The sales were brokered and delivered to him by the owner of YW Consultants.

Pedigree papers AmeRx gave to Dialysist West for 181 boxes of lot P002970 show that two boxes originated at Metro Medical and were sold to AD Pharmaceuticals and that 180 boxes originated at Medix International and were sold to Express Rx. Express Rx sold 129 boxes to Armin Medical, which in turn sold them to AmeRx. Express Rx also sold fifty-one boxes directly to AmeRx.

Mark Novosel told investigators that he bought all his medicine from Ivan Villarchao.

The owner of Coastal Medical told investigators that he bought his medicine from Paul Perito and sold it to Tradewinds Trading. Tradewinds notified Coastal by letter that the Epogen lot P002970 it had sold them could be counterfeit. Tradewinds also sent letters notifying Grapevine Trading and Rebel Distributors that the Epogen they bought from Tradewinds could be counterfeit.

Inclusion in this chart does not suggest that a company or individual participated in any wrongdoing. The text and endnotes discuss the roles of most of those listed.

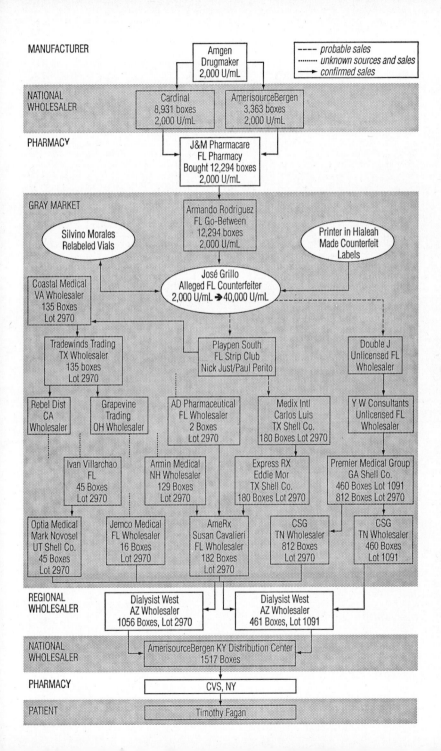

What You Can Do About Counterfeit Drugs

WATCH WHAT YOU TAKE

- Be familiar with your medicine. Examine its shape, color, and size.
- If your medicine is in pill, capsule, or tablet form, put it in the palm of your hand and examine it under a light before taking it.
- Look for altered or unsealed packaging, or changes in design.
- Make sure the packaging is pristine and has no sticky residue, which can indicate the drug was previously dispensed.

OBSERVE YOUR SYMPTOMS

- Be aware of new or unusual side effects.
- Be concerned if your medicine stops being effective.
- Be suspicious if an injectable drug stings or causes a rash.

LOOK FOR THE LATEST WARNINGS AND ANNOUNCEMENTS

- Visit MedWatch at www.fda.gov.medwatch, the FDA's Web site that lists information on drug safety, label changes, and voluntary recall announcements.
- Visit your drug manufacturer's Web site for information on current recalls.

TALK TO YOUR PHARMACIST

- He or she may have information from the manufacturer about legitimate changes in the shape, color, or taste of the product. They may know about suspicious characteristics that can help determine if a product is counterfeit.

IF YOU THINK YOUR MEDICINE IS COUNTERFEIT

- Tell your pharmacist, your doctor, and the manufacturer. Most manufacturers have phone numbers or e-mail addresses for patients with questions and concerns.

- You or your doctor should submit a report to the FDA on the MedWatch site. The form can be found at http://www.fda.gov/medwatch/report/consumer/consumer.htm.

- Once the form has been submitted, the FDA may contact you for more information. The FDA does not respond directly to each complaint of a suspected counterfeit.

- Keep a sample of your medicine as evidence, even if the manufacturer asks you to send it all back.

OTHER ADVICE

- Buy only from a reputable and licensed seller. You can check credentials through your state pharmacy board or the National Association of Boards of Pharmacy at www.nabp.net.

- If you are buying from an online pharmacy, look for the National Association of Boards of Pharmacy's Verified Internet Pharmacy Practice Site (VIPPS) seal as proof that the site has met state and federal requirements. Approved Internet pharmacies are posted at http://www.nabp.net/vipps/consumer/listall.asp.

- For more information and articles on counterfeit medicine, visit:

  The Partnership for Safe Medicines:
  www.safemedicines.org/who/

  National Consumers League: http://fraud.org/fakedrugs/

Endnotes

PART ONE

Chapter 1. A Victim of Success

Page 7 The burglary occurred January 16, 2002, as described in the offense-incident report from Miami-Dade Police Department (MDPD).

Page 10 There is no evidence that the Stone Group's representatives knew the medicine they were offering to sell had been stolen. The company's lawyer, Stanley Goodman, did not return several phone calls seeking comment about the incident.

The Stone Group is a licensed wholesaler in Boca Raton, Florida. Sales representative Sean Dana of the Stone Group contacted BioMed Plus offering to sell the drugs on January 21, as reported in a supplement to the MDPD offense-incident report of January 16, 2002. BioMed Plus's prior buying relationship with the Stone Group is documented in e-mails from Sean Dana to Marlene Caceres. The president of Bio-Med Plus, Steve Getz, negotiated with the Stone Group and the purchase price was set at $229,241, as indicated on the January 21, 2002, Stone Group invoice and reported in the MDPD report supplement.

Chapter 2. Flamingos in Missouri

Page 13 Maxine Blount was named O'Fallon Chamber of Commerce Person of the Year 2000, as noted in Senate Resolution No. 1508 on April 11, 2000.

Page 13 Maxine Blount's purchase of four vials of Procrit from the pharmacy at Schnucks in St. Peter, Missouri, is documented in the Class Action Petition, *Maxine Blount and Edward Blount, on behalf of themselves and all others similarly situated v. Amgen, Ortho Biotech, Ortho Biotech Products, Cardinal Health, Cardinal Distributions and John Doe Distributors,* filed in St. Charles County Circuit Court in Missouri on October 11, 2002.

Page 15 Ortho Biotech issued a letter on June 6, 2002, warning health-care

professionals about counterfeit Procrit vials. Another letter was issued
June 12, 2002, warning of additional counterfeit vials.

Page 16 The Courtney case has been widely documented. Articles con-
sulted or cited include: Mark Morris, Alan Balvey, Joe Lambe,
"'Staggering' Numbers Reveal Scope of Courtney's Crimes," *The
Kansas City Star,* April 20, 2002; Mark Morris, "Courtney Sentenced
to 30 Years in Prison for Diluting Medicines," *The Kansas City Star,*
December 6, 2002; and Robert Draper, "The Toxic Pharmacist," *The
New York Times Magazine,* June 8, 2003.

Chapter 3. Is Anything Okay?

Page 19 The person caught stealing is Sydney Dean Jones. For more information, see endnote page 25.

Page 21 A representative for Eckerd's did not respond to a written request
seeking comment.

Page 22 Inspector Cesar Arias first sent his commentary on Medicaid
fraud to officials in Florida's Agency for Healthcare Administration
and the Attorney General's Office on December 28, 2001.

Page 24 Theft from Jackson Memorial Hospital is widely documented.
See Gail Epstein Nieves, "Millions in Inventory Missing at Jackson,"
The Miami Herald, January 29, 2003. In Ashley Frantz's "Miramar
Man Suspected of Stealing from Hospitals," *The Miami Herald,* August 14, 2003, the vice president of the South Florida Hospital and
Healthcare Association, George Andrews, also comments, "Jackson
has between 500 and 1,000 exterior doors. It's impossible to guard all
of those entryways." Conchita Topinka, Director of Public Relations
and Community Relations, did not respond to a written request seeking comment.

Page 25 Sydney Jones's thefts from Jackson Memorial Hospital and his
arrest on September 25, 2001, are documented in an MDPD offense-
incident report and supplements and the Bureau of Statewide Pharmaceutical Services investigation report of November 13, 2001. He
became a cooperating witness and was not charged with any wrongdoing for his thefts from Jackson Memorial.

Sydney Jones's thefts from two Humana pharmacies are documented
in the Palm Beach County Sheriff's office offense report of July 3,
2001.

Page 25 The woman who bought the medicine from Sheie is Annette C.
Mantia. For more information, see endnote page 36.

Page 27 Florida Statute 499.0121 stipulates:

1. Each person who is engaged in the wholesale distribution of a prescription drug, and who is not an authorized distributor of record of such drug, must provide to each wholesale distributor of such drug, before the sale is made to such wholesale distributor, a written statement identifying each previous sale of the drug. The written statement identifying all sales of such drug must accompany the drug for each subsequent wholesale distribution of the drug to a wholesale distributor. The department shall adopt rules relating to the requirements of this written statement.

2. Each wholesale distributor of prescription drugs must maintain separate and distinct from other required records all statements that are required under subparagraph 1.

3. Each manufacturer of a prescription drug sold in this state must maintain at its corporate offices a current list of authorized distributors and must make such list available to the department upon request.

Chapter 4. The R Word

Page 35 The controlled delivery of Neupogen to Fariborz (Fred) Sheie occurred on November 15, 2001, as documented in the MDPD offense-incident report and the Bureau of Statewide Pharmacy Service (BSPS) investigative report of that day.

Page 36 Melvin Otto could not be reached for comment. A message left at his home in Davie went unreturned. Subsequently that number was disconnected and no further phone number could be found. He was not charged with any wrongdoing.

Page 36 Annette Mantia was fined $250 by the Bureau of Statewide Pharmaceutical Services, as documented in a notice of violation on April 20, 2002.

Page 36 The hand off from Melvin Otto to Mantia is documented in a Florida Department of Law Enforcement (FDLE) investigative report of November 15, 2001.

Page 39 As documented in an FDLE investigative report dated January 16, 2002, the cooperating witness [Mantia] told investigators that she'd done previous pharmaceutical deals in cash, sometimes making the exchanges in parking lots, without receipts, shipping bills or pedigree papers, all in violation of state law. She was not charged with any wrongdoing.

Pages 40–41 Sheie's addiction to narcotic painkillers is documented in his ex-wife's interview with Doug Jenkins, an investigator with the Palm Beach County medical examiner's office. Notes from this interview are included in the January 14, 2002, autopsy report of Sheie prepared by the Office of the District Medical Examiner.

Sheie hit a cyclist, as reported in the "Police Blotter," *Palm Beach Post,* March 13, 1999. He also hit another vehicle on November 16, 1999, as documented in court filings from the lawsuit, *Alfred S. Coco and Pauline Coco v. Dollar Rent a Car Systems and Fariborz Sheie,* in the Broward County Circuit Court. His arrest for driving under the influence is documented in a complaint affidavit from the Broward County Sheriff's Department dated February 16, 1999.

Chapter 5. Medicine in the Laundry Room

Page 43 The Stone Group's sales representative Sean Dana delivered the drugs to BioMed Plus on January 21, 2002. This is reported in a January 29, 2002, supplement to the MDPD offense-incident report of January 16, 2002.

Pages 44–45 Detective Paula Berris of the Miami-Dade PD was contacted by Adam Runsdorf, who said he would cooperate, as documented in an FDLE investigative report of January 21, 2002. Runsdorf described the relationship between Carlow and BTC Wholesale, as documented in the report. In interviews with investigators on January 25, 2002, various Stone Group employees also described Carlow's relationship with BTC.

Page 45 Through his lawyer John Howes, Michael Carlow declined to comment or to answer detailed written questions. He did not grant an interview or respond to a letter sent to him in July 2004 at a state jail in Fort Lauderdale, Florida.

Page 45 Carlow's arrest for armed robbery is documented in a Broward County Sheriff's Department booking record of January 17, 1973. He pleaded no contest and was sentenced to five years in prison. Carlow was arrested in September 1984 on charges of grand theft. On September 22, 1986, he pleaded guilty and was sentenced to three years probation, as documented in Alachua County Circuit Court documents relating to the case. He was arrested for selling cocaine a day later, on September 23, as documented in a rap sheet from the Alachua County Sheriff's Office.

Page 45 In June 1998, Carlow's wife, Candace Atkins, applied to the Bureau of Statewide Pharmaceutical Services for a drug wholesale permit for the company Medical Infusion Services. She listed herself as the chief operating officer. Carlow listed himself as the company's president in records filed with the Florida State Department's Division of Corporations. Candace was not listed with the division at all.

In May 2000, Carlow submitted a request to the state's pharmaceutical bureau to change the company's name to Quest Healthcare. On July 3, 2000, the Bureau of Statewide Pharmaceutical Services sent a letter to Carlow stating its intention to revoke Quest's permit due to violations of statute 499. The violations included purchases of medicine from unauthorized wholesalers, which resulted in Carlow's arrest on June 13, 2000. He pleaded no contest and was put on probation and fined, as documented in a December 1, 2000, Dade County Circuit Court order of probation.

Pages 46–47 A transcript of the investigators' interview with Thomas Atkins was obtained. The interview, as well as records of BTC's transactions with El Paso Pharmaceutical Brokers, BioMed Plus, and Omnimed, are documented in the FDLE investigative report from January 23, 2002. For more information on El Paso Pharmaceuticals, see endnotes page 72. For more information on Omnimed, see endnote page 96.

Page 49 E-mails sent between inspector Cesar Arias and his co-workers describe their frustration with the bureaucracy and lack of support from the Tallahassee office.

Page 50 The quote comes from the resignation letter of Deborah Orr, regional drug agent for the BSPS, on July 3, 2003.

Page 51 Records of sales between BTC and Omnimed are referenced in an FDLE investigative report of January 23, 2002.

Page 52 For more information on Marilyn Atkins's role as bookkeeper, see endnote page 201.

Page 53 Florida Statute 499.0121 outlines the requirements for running a prescription drug business.

Page 53 Inspectors confiscated the cannabis derivative Marinol found at the Stone Group's offices, as documented in an FDLE investigative report of January 25, 2002. In an interview with investigators on January 25, 2002, for which a transcription was obtained, the Stone Group's sales manager, Steve Gorn, said that he did not know that Marinol was a controlled substance for which his company needed, but did not have, a license from the Drug Enforcement Administration.

Page 53 The recycling of medicine is strictly prohibited by state and federal law: Medicine cannot be redispensed once it reaches an end user—a pharmacy, hospital, or patient—precisely because it cannot then be guaranteed as pure.

Pages 53–54 On January 25, 2002, investigators questioned the following
Stone Group employees: Steve Gorn, national sales manager; Brett
Jaffy, vice president of operations; Douglas Brilliant, salesman; Sean
Dana, salesman. The interviews were transcribed and documented in
the FDLE investigative report of that day. During his interview, Gorn
described how the Stone Group became a drug wholesaler. The employees questioned by state investigators described Carlow's prescription drug sales from the laundry room of his home, as documented in
transcripts of their interviews and in an FDLE investigative report.
Gorn and Brilliant also described Carlow's response to news of the
incident at BioMed Plus.

Page 54 On October 15, 2002, the Florida health department brought an
administrative complaint against the Stone Group for numerous violations of statute 499, including distributing drugs that were adulterated because they had been purchased from a residence and were not
licensed by the state. The department imposed a fine of $327,500. As
of December 2004, the complaint had yet to be resolved.

Copies of checks the Stone Group wrote to BTC were obtained.

Carlow's house was purchased in 1999 for $1.3 million, as documented in
the FDLE investigative report on January 23, 2002.

Chapter 6. The Cheshire Cat

Page 56 The history of Weston and the quote "where hyperbole meets reality," are taken from John DeGroot, "Should this city have been
built?" *Sun-Sentinel,* February 15, 1991.

Page 57 On September 20, 2001, in the case of *Medical Distribution Inc.
v. Quest Healthcare Inc. and Michael Carlow,* the United States District Court in Louisville, Kentucky, ordered Carlow to repay $98,630
plus interest to Medical Distribution Inc. The court determined that
Carlow had committed fraud and breached his contract in not repay-
ing MDI for pharmaceuticals the company had shipped.

On December 19, 2001, a United States District Court in Miami granted
MDI a break order, giving the company the right to seize Carlow's
property to satisfy the final judgment against him in the Kentucky
case.

Page 57 The Internal Revenue Service sent Carlow letters requesting payment of $15,672 and $472,591 on May 30, 2001.

Mayors Jewelers sued Carlow for a total of $72,754 including legal fees in
Mayors Jewelers v. Michael Carlow on July 31, 1998.

The Hamlet of Davie Homeowners Association sued Carlow for $2,006 in missed quarterly maintenance and fees, as documented in a notice of claim of lien for unpaid assessments on February 18, 1998.

Page 58 Information about Carlow's bankruptcy is included in a statement of financial affairs for the United States Bankruptcy Court, Southern District of Florida in Fort Lauderdale filed on July 29, 1998. James M. Meere of Debis Financial Services (Carlow's lender) confirmed that the yacht was valued at $675,000 in his affidavit to the court on October 2. The quote about his purchase of the yacht comes from a letter Carlow sent to Todd Shull of Debis Financial Services on July 16, 1997.

Page 58 Carlow's alleged front companies are described extensively in depositions taken for *Medical Distribution, Inc. v. Enhance Your Life, et al.* in the United States District Court in Louisville from the fall of 2000 to the winter of 2002.

Page 59 Seized items are detailed in the "List of Personal Property Located at the Residence of Michael Carlow," completed by the U.S. marshals present on January 20, 2002.

The 344 vials are itemized on a January 21, 2002, invoice from the Stone Group to BioMed Plus. See also endnote page 15.

Page 60 According to MDPD Records, Carlow turned himself in to the Metro-Dade warrants bureau on May 29, 1986.

Page 61 The quote comes from the letter Carlow sent to Todd Shull of Debis Financial Services on July 16, 1997.

Page 61 Carlow's Audemars Piguet watch was appraised by J.R. Dunn Jewelers of Lighthouse Point, Florida, at $110,000.

Pages 61–62 Michael Rosen and Jeffrey Schultz worked together at Local Railroad Company in Boca Raton, Florida. For a scheme the men undertook while working there, Rosen pleaded guilty to bank fraud, conspiracy, and access device fraud in 1998, and served four months in Federal prison in Miami. Jeffrey Schultz, who was the president of Local Railroad, pleaded guilty to a Federal charge of conspiracy to commit bank access fraud. He served five months in a halfway house and three months on probation, according to a deposition he gave on March 2, 2001, in the case of *Medical Distribution, Inc. v. Enhance Your Life, et al.*

During his deposition, Schultz also explained his relationship with Michael Carlow and how he had received a letter from MDI's parent company, the Louisville Public Warehouse Company, on May 18, 1999, advising him that payments on his account were "seriously delinquent in the amount of $262,500."

Page 63 Embree gave his deposition on October 26, 2001, for *Medical Distribution, Inc. v. Quest Healthcare and Michael Carlow* in the United States District Court in Miami.

Page 63 Numerous attempts to depose Carlow are described in the *ex parte* motion for break order in *Medical Distribution, Inc. v. Quest Healthcare and Michael Carlow.* On April 3, 2001, MDI entered a motion for sanctions in response to Carlow's repeated failures to appear for his deposition. He was eventually deposed on September 24, 2001, and February 12, 2002.

Page 66 Joan Bardzick sued Michael Carlow for $93,120 in returned checks. As of the final judgment on March 7, 2001, Carlow still owed Bardzick $13,120 plus legal fees, which he ultimately repaid.

Chapter 7. One Man's Trash, Another Man's Treasure

Page 68 Numerous pieces of mail for BTC and Accucare, companies that Carlow claimed were not his, were addressed to Carlow and sent to his residence in Weston.

Pages 68–71 The following were recovered in trash pulls at the Carlow residence:

1. Brochure for the Hatteras 6300, as documented in an FDLE investigative report dated February 6, 2002.
2. Letter from Lori Marvel of Two M Enterprises; two invoices to AK Distributors in Henderson, Nevada, indicating sales from BTC of $40,579.19 and $34,444.83; and invoices and shipment bills for Accucare, as documented in an FDLE investigative report dated February 9, 2002.
3. Faxes sent on February 7, 2002, from BTC Wholesale to the Arizona health department and from G&K to the Texas health department requesting license applications, and an invoice for Marinol and other drugs from Albers Medical in Kansas City to BTC, dated February 8, 2002, and a list of medicine from a sales representative at Armin Medical, sent on February 7, 2002, with Carlow's returned comments, as documented in an FDLE investigative report dated February 13, 2002.
4. Survey report for Carlow's yacht *Delicious,* a repair bill for his boat *Tenacity,* and documents related to Carlow's purchase of the Sunshine Ranches property, as documented in an FDLE investigative report of February 20, 2002.

Lori Marvel did not return a phone call seeking comment.

Page 70 Candace Carlow traded in a 2000 yellow Ferrari 360 Modena for a 2001 red Ferrari 360 Spider worth $249,000 on November 30, 2001, according to invoices from the Wide World of Cars in Spring Valley, New York. This information was also recovered in a trash pull on February 20, 2002, as documented in the FDLE investigative report of that day.

Page 71 Gary Venema sent the e-mail to Stephanie Feldman on February 6, 2002. The "W-9" he mentions is actually an IRS 1099 form from the General Star Indemnity Company to Carlow for $700,000, which is included in the FDLE investigative report of February 6, 2002.

Page 72 Díaz's contacts with Carlow and the Stone Group are documented in an FDLE investigative report of February 8, 2002.

Díaz's relationship with El Paso Pharmaceutical Brokers and his criminal history are documented in an FDLE investigative report of January 29, 2002.

Díaz was not available for comment, has no known legal representation, and as of December 2004 his whereabouts were unknown. A fugitive warrant was issued for his arrest in July 2003 for charges of racketeering and schemes to defraud issued by a Florida statewide grand jury. See also endnote pages 299–300.

Page 73 In attempting to verify the pedigree papers that the Stone Group provided to BioMed Plus, investigators learned the following:

An investigator for the Texas Department of Health, Drugs and Medical Devices Division made two attempts to inspect the offices of El Paso Pharmaceutical Brokers. No one answered the door when he knocked and his phones calls were not returned, as documented in a Texas inspection report of February 12, 2002.

In response to an e-mail inquiry from Florida state investigators, Michael Dwyer, Senior Director of Sales for Novo Nordisk, stated that Novo-Seven was never purchased by Bellco, Inc., the first company listed on the pedigree for drugs stolen from BioMed Plus, as documented in the FDLE investigative report of January 21, 2002.

Page 74 Investigators questioned Runsdorf and Gorn about Michael Carlow, Fabian Díaz, and the Stone Group's sale of blood products to BioMed Plus, as documented in the FDLE investigative report dated February 11, 2002.

Pages 75–76 Comments on Hialeah were taken from Larry Rohter, "Journal: Where Politics is Down and Dirty," *The New York Times*, November 1, 1991.

Page 76 1982 murder statistics for Dade County and Hialeah were retrieved from Florida Department of Criminal Law Enforcement, *Uniform Crime Reports, State of Florida: 1982 Annual Report*, Washington: GPO, 1983.

Page 76 From 1981 to 1991, Venema received numerous commendation letters for his actions as a police officer in Hialeah, as documented in his police department personnel file. For his actions in January 1991, he received a letter of commendation on February 9, 1990, from Rolando Bolanos, Chief of Police. Venema also received a letter of commendation on January 13, 1992, for his 1991 performance in the Organized Crime Section, even after he was reassigned to the Uniform Patrol Division on December 16, 1991.

Pages 78, 80 Mayor Raul Martinez was convicted of extortion and racketeering in March of 1991. See Larry Rohter, "Journal: Where Politics is Down and Dirty," *The New York Times,* November 1, 1991.

Page 80 Bolanos's two sons were charged with beating a twenty-three-year-old man and falsifying police reports to cover it up. See Ana Acle, Manny García, and Ajowa Nzinga Ifateyo, "Case May Grow Against Hialeah Police Chief," *The Miami Herald,* July 10, 1999.

Bolanos's suspension was reported in José Sanchez, "Hialeah Police Chief Suspended; Campaign Mail Tampering Suspected," *Sun-Sentinel,* November 5, 1993.

Page 80 The court decision was reported in Jay Weaver, "Hialeah Officers Win Bias Lawsuit; $1.4 Million Given to Non-Hispanics," *The Miami Herald,* August 8, 2000.

Pages 81–82 The investigators' search of Michael Carlow's residence is documented in an FDLE investigative report of February 15, 2002. This report describes the invoices from Joskay Corporation for "construction equipment" and documents José Benitez as Joskay's owner. A copy of the memo from Lori Marvel dated February 5, 2002, and the invoices from Joskay for "Misc. Construction Equipment," dated February 6 and 13, were obtained. A videotape of the search was also obtained.

Page 82 The formation of the task force is documented in the FDLE investigative report of February 20, 2002.

Page 83 Florida's Deferred Retirement Option Program (DROP) would have allowed Odin to collect benefits of more than $36,000 that had accrued by his retirement date.

Page 86 A presentation laying out Operation Stone Cold was given to the MDPD department chief, as documented in an MDPD report of February 26, 2002.

Chapter 8. A Cold Chain Gets Hot

Page 87 Amgen's relationship with Ortho Biotech began in 1991. See "FDA Approves Drug for AIDS-related Anemia," *Modern Health-*

care, January 14, 1991. The history of the creation and FDA approval of Epogen is explained in U.S. Congress, Office of Technology Assessment, *Recombinant Erythropoietin: Payment Options for Medicare, OTA-H-451,*Washington: GPO, May 1990.

Amgen's status as the largest biotechnology company is widely documented in publications such as *Forbes, US News & World Report,* and *Chemical & Engineering Week.*

Epoetin Alfa's U.S. sales, including that of Procrit and Epogen, were over $6 billion in 2004 according to figures supplied by IMS Health, a health-care data collection company.

Page 88 Epogen's storage information can be found on its packaging and its Web site maintained by Amgen, http://www.epogen.com. Further information on issues involved in transporting delicate drugs came from a lecture by C. Jeanne Taborsky of SciRegs at the September 2003 conference sponsored by the Institute for International Research (IIR) entitled "Supply Chain Integrity: The Industry Conference Addressing the Competency and Security of the Pharmaceutical Supply Chain."

Page 88 The 1999 outbreak and its cause were discovered and reported in Lisa A. Grohskopf, M.D., M.P.H., et al., *"Serratia liquefaciens* Bloodstream Infections from Contamination of Epoetin Alfa at a Hemodialysis Center," *New England Journal of Medicine,* May 17, 2001, Volume 344, No. 20:1491-1497.

Page 89 "Develop and project a *passion* for temperature control in your organization" is taken from a presentation and speech given by Rick Garafalo, Compliance Manager, Wyeth Vaccines, at the IIR conference in September 2003.

Pages 89-91 Descriptions of the structure of the wholesale market can be found in the following: "Profile of the Prescription Drug Wholesaling Industry," a report by the FDA released on February 12, 2001; the first interim report of the seventeenth statewide grand jury of the Florida Supreme Court, released on February 27, 2003; a series of articles in *The Washington Post* by Gilbert Gaul and Mary Pat Flaherty, October 19-23, 2003.

Page 91 In October 2003, a person associated with Cardinal showed the author a document from the Cardinal trading company listing the amounts saved by making purchases from alternative secondary distributors.

Page 92 The quote comes from John Howes, Carlow's lawyer, in a question he posed to Gary Venema during Venema's deposition in *State of Florida vs. Michael Carlow, et al.,* December 9, 2003.

Page 93 Wholesaler requirements are included in Florida Statute 499. The number of wholesalers in Florida is taken from the first interim report of the seventeenth statewide grand jury released on February 27, 2003.

Page 93 In May 2003, Brand met with the author at his office and discussed his legal representation of pharmaceutical wholesalers and his own wholesale company, Global Pharmaceutical Services.

Page 94 Evidence of Carlow's $2.5 million profit over eight months comes from G&K bank statements, Carlow's own profit reports obtained through trash pulls by Gary Venema, and a financial analysis prepared by the FDLE for the period May to December 2002.

Page 94 Henry García, Fabian Díaz, Michael Carlow, and fifteen others were identified as a racketeering cell by investigators in the course of Operation Stone Cold. A statewide grand jury charged them all with racketeering on July 18, 2003. See also endnote pages 299–300.

Henry García's lawyer, Joseph Rosenbaum, did not return phone calls seeking comment.

Page 94 The Florida pharmacist described Michael Carlow's garage as a "virtual pharmacy repacking operation," as documented in an FDLE investigative report dated July 29, 2002. The pharmacist, who is unlisted, did not respond to a request for comment left in a note for him at the guardhouse of his gated community.

Page 95 Michael McKinnon received $195,000 in prescription drugs in less than a year, as documented in the FDLE investigative report of March 22, 2002. On May 8, 2002, McKinnon was arrested by the Attorney General's Medicaid Fraud Control Unit and charged with organized fraud, grand theft, and Medicaid fraud, as documented in a report by Jack Calvar of the Medicaid Fraud Control Unit entitled "Drug Diversion in South Florida." According to the report, McKinnon cooperated and told investigators that he made $5,000 a month selling his medicine. McKinnon pleaded guilty in November 2002 to selling his medication. See Mary Pat Flaherty and Gilbert M. Gaul, "A Ring of Fraud: Medicaid Is Start Of Drug Resale Trail," *The Washington Post,* October 22, 2003.

When reached for comment at his home in Miami, McKinnon said that because of his illness, he had no recollection of his activities: "That's been a long time ago, I'm on a lot of medication and I couldn't tell you one thing."

Page 95 An indictment by a Florida statewide grand jury on July 18, 2003, charged Fabian Díaz and Michael Carlow with grand theft for their involvement in the theft of pharmaceuticals from Marty Bradley's warehouse at his company BioMed Plus on January 16, 2002. See also endnote pages 299–300.

Page 96 John Bullock signed the Texas Department of Health's drug distributor license application for JB Pharmaceuticals on December 20, 2001. Bullock recounts his introduction to Fabian Díaz and events surrounding the signing of the application, as documented in an FDLE investigative report of January 6, 2002.

Page 96 According to an FDLE investigative report of February 11, 2002, investigators inspected Omnimed's offices and found Goof Off and other products used to remove labels. On that date Michael Burman, the listed owner, also told the investigators about his meetings with Carlow and his assistant at a gas station off of I-595 to exchange medicine and money. Through his attorney, Michael Gottlieb, Michael Burman declined to comment.

Pages 96–97 Invoices and bank and shipping records show that NSS bought medicine from Carlow's companies BTC Wholesale, Quest Healthcare, Medical Infusion Services, and MedRx from 1998 through 2002. The relationship between NSS and Medical Infusion Services was initiated shortly after June 15, 1998, when Medical Infusion sent a credit application to NSS. A list of sales sent to Florida's BSPS on September 29, 2000, shows that from January 1, 1999, to June 30, 2000, NSS bought $131,000 in blood products from Medical Infusion Services. From January through June 2000, Cardinal paid $1.45 million to Quest Healthcare for blood products and other specialty pharmaceuticals purchased by NSS, according to bank records. A BTC sales ledger from August 20, 2001, to December 24, 2001, shows checks BTC received from NSS for the purchase of medicine. Shipping records from August to December 2001 show packages sent to NSS from Tom Atkins of BTC and from Mark Novosel of MedRx. A notation on BTC's January 2002 profit statement shows that the company continued to sell medicine to NSS. For more details on Spence's purchases from Carlow's companies, see endnote pages 222–223.

Page 97 Reports of Novosel's 1998 arrest appeared in *USA Today* on March 19, 1998, *The St. Petersburg Times* on September 6, 1998, and *The News-Herald* (Willoughby, Ohio) on October 17, 1997. His arrest and guilty plea are also documented in his criminal-history record from the State of Ohio Bureau of Criminal Identification and Investigation.

Page 97 According to an MDPD subject information sheet, Novosel's waterfront rental was $4,500 a month. A check receipt for $3,461.54 from Quest Healthcare made out to Mark Novosel and listed as one week's salary was included in an FDLE investigative report of May 25, 2002.

Investigators searched Novosel's home on March 28, 2002. After the search, Novosel expressed a desire to cooperate with the investigation, as documented in an FDLE investigative report of that day. His cooperation was acknowledged in his plea agreement, as documented in an FDLE investigative report of June 4, 2003.

Pages 98–99 Novosel's observation about the problem of large wholesalers buying diverted medicine is documented in several reports. See "Dirty Deals: the Drug Diversion Trade, How It Victimizes the Vulnerable and How to Stop it," The Center for Regulatory Effectiveness, Washington D.C., July 2003. See also the first interim report of the seventeenth statewide grand jury in the Supreme Court of the State of Florida, released on February 27, 2003.

Page 99 J.M. Blanco received a notice of violation on July 10, 2000, allegedly for shipping drugs into the state without a proper license. According to a compliance action, the Florida Bureau of Statewide Pharmaceutical Services waived the fine and closed the case on November 6, 2000.

Chapter 9. Stealing Time

Pages 101–102 The second attempted break-in on February 16, 2002, and the third successful break-in on May 29, 2002, at BioMed Plus are documented in MDPD offense-incident reports for those days. A copy of the security camera tape showing the third break-in in progress was made available by BioMed Plus.

Page 105 In 1985, the House Energy and Commerce Committee's subcommittee on oversight and investigations held hearings on prescription drug diversion and counterfeiting. During the hearings, Representative Gerry Sikorksi warned of the "dangers to the health and safety of unsuspecting American consumers from the operation of an illegal, subterranean diversion market." The committee released a report entitled "Dangerous Medicine" in April of 1986, expanding on the dangers of the diversion market. Examples of diversion are taken from correspondence and testimony submitted to the FDA as part of the extensive docket for the Prescription Drug Marketing Act. Documents from this docket were obtained through a Freedom of Information Act (FOIA) request to the FDA.

Page 105 From 1989 to 1992, seventeen FBI field offices investigated pharmacy fraud in Operation Gold Pill, according to an FBI training video. "They uncovered fake prescriptions, false Medicaid billings, unnecessary medical testing, and the illicit sale of prescription drugs to street dealers and corrupt pharmacists," according to the video. "Operation Gold Pill led to the arrest and conviction of over 200 pharmacists and others." CNN covered the operation on June 30, 1992. Photos of some of the drugs that were seized by Florida state inspectors were obtained.

Page 107 As described in an August 1, 1988, guidance letter issued by the FDA, the Prescription Drug Marketing Act of 1987 seeks to: "1. Ban the sale, purchase, or trade of, or the offer to sell, purchase, or trade, drug samples and drug coupons. 2. Restrict re-importation of prescription drugs to the manufacturer of the drug product or for emergency medical care. 3. Establish requirements for drug sample distribution and the storage and handling of drug samples. 4. Require a wholesale distributor of prescription drugs to be State licensed, and require FDA to establish minimum requirements for State licensing. 5. Establish requirements for wholesale distribution of prescription drugs by unauthorized distributors. 6. Prohibit, with certain exceptions, the sale, purchase, or trade of (or the offer to sell, purchase, or trade) prescription drugs that were purchased by hospitals or other health care entities, or donated or supplied at a reduced price to charities. 7. Establish criminal and civil penalties for PDMA violations."

Page 110 On August 3, 2001, Genentech, the manufacturer of Nutropin A.Q., released a warning about the counterfeit product. The case of the Michigan doctor is used as a specific example in the first interim report of the seventeenth statewide grand jury. He is not named in the report. When contacted, the family asked that their name not be used.

Page 110 In internal e-mails, state inspectors and investigators expressed their frustration that Florida's Bureau of Statewide Pharmaceutical Services issued licenses to family members of known drug traffickers.

Page 112 Max Butler shared the eulogy that his sister, Maxine Blount, composed for her own funeral.

Page 114 Cardinal's lawyers stated in their answer and affirmative defense to *Maxine Blount and Edward Blount v. Amgen, et al.* in the Circuit Court for the County of St. Charles, Missouri, that Cardinal had supplied Procrit to Shnucks pharmacy in the past, but denied that the specific vials Blount received came from Cardinal.

PART TWO

Chapter 10. "My Son Is Not a No One"

Page 123 Amgen released a warning about counterfeit Epogen on May 8, 2002.

Page 124 On October 1, 2003, lawyers for CVS Procare responded to allegations in *Timothy Fagan v. AmerisourceBergen, et al.* in United States District Court, Eastern District of New York. They denied that the company had failed to properly verify the origin of the drugs or had "purchased Epogen of unknown pedigree for re-sale to the public because it was cheaper." Todd Andrews, Director of Corporate Communications for CVS, did not return three phone calls seeking further comment.

Page 124 Amgen released a warning about Epogen lot P001091 on May 24, 2002.

Page 124 Lawyers for Amgen filed a motion to dismiss claims against the company in *Fagan v. AmerisourceBergen, et al.* Their motion, filed October 23, 2003, stated: "As a matter of law, Amgen had no legal duty to make its product impossible to alter or counterfeit, no duty to continuously monitor its product after it left Amgen's hands, and no duty to ensure that third-party criminals do not somehow misuse its products." Amgen spokesman Michael Beckerich said the manufacturer had made significant efforts to keep counterfeit drugs out of the supply chain and had worked with law enforcement to protect, detect, and prosecute illegal wholesalers. Amgen officials declined to answer any detailed questions about the company's role in investigating the counterfeits.

On August 12, 2004, lawyers for AmerisourceBergen Corporation (ABC) responded to allegations in the lawsuit, stating that "ABC admits only that it is a wholesale distributor and that it has purchased Epogen from Amgen and other lawful sources for distribution to its customers, which include pharmacies." The lawyers denied that AmerisourceBergen had purchased drugs of unknown pedigree and without prior knowledge as to how the drugs had been handled.

Page 125 For more information on the FDA's procedures for handling suspected counterfeit drugs, seen endnote page 156.

Page 126 On March 13, 2002, AmerisourceBergen received a notice of violation from the Florida Bureau of Statewide Pharmaceutical Services for purchasing and receiving drugs from a company without a Florida license, distributing a counterfeit prescription drug, failing to obtain or maintain pedigree papers, improperly storing drugs, and failing to maintain a complete audit trail.

Chapter 11. Two Streams Become One

Page 128 Through invoices obtained from ASAP Couriers, investigators learned that Mark Novosel was using Optia Medical's Utah license to ship drugs from his home in Miami, as documented in an FDLE investigative report dated March 20, 2002.

Page 129 Florida statute 499.006 defines adulterated drugs as those for which pedigree papers are "nonexistent, fraudulent, or incomplete" or those that have been "purchased, held, sold, or distributed at any time by a person not authorized under federal or state law to do so."

Page 129 The trail of drugs from Optia to Raymar to Bindley is traced through numerous sources. Several invoices from ASAP Courier document deliveries from Optia Medical to Raymar, according to an FDLE investigative report dated March 20, 2002. The wholesaler's purchases from Optia and sales to Bindley are documented in a Bureau of Statewide Pharmaceutical Services investigation report dated June 11, 2002. Employees at Raymar also gave investigators paperwork on a shipment of Panglobulin purchased from Optia on March 21, 2002, as documented in an FDLE investigative report dated March 25, 2002.

As documented in a May 22, 2002 Texas Department of Health report, investigators found 886 boxes of Procrit, worth $1.5 million, at Bindley Trading in Grapevine, Texas, and invoices from February 28 to March 20, 2002, showing that Bindley Trading bought 1,004 boxes of Procrit from Raymar Worldwide Distributors.

Page 129 The 2001 counterfeit cases have been documented in national publications. William K. Hubbard's quote on the subject is taken from Melody Peterson, "3 Fake Drugs are Found in Pharmacies," *The New York Times*, June 5, 2001.

Page 130 In September 2003, several secondary wholesalers described for state inspectors the process by which a few of the nation's largest wholesalers vetted them as vendors. In summarizing one statement about the process at AmerisourceBergen, inspectors wrote, "Once a company was approved as a vendor, transactions were done either through fax or computer. Offers were submitted to AmerisourceBergen and they would fax or e-mail a purchase order. The lowest bidder would be accepted. No human interaction was required."

In an August 8, 2004, interview, executives from AmerisourceBergen said they had significantly increased restrictions on purchasing. They said that the company buys from only twenty approved secondary whole-salers that have undergone rigorous review, including checks of criminal and other court records, affiliations, state licensing, insurance coverage, and Dunn & Bradstreet reports.

Page 130 The claim from the Big Three wholesalers that no more than 3 percent of inventory came from the secondary market appeared in Gilbert Gaul and Mary Flaherty, "U.S. Prescription Drug System Under Attack," *The Washington Post,* October 19, 2003.

Page 130 In a January 29, 2002 letter to Florida's Bureau of Statewide Pharmaceutical Services, the Health Distribution Management Association, a trade group for pharmaceutical distributors, argued that creating and relaying drug pedigrees would stifle efficiency and productivity, and force wholesalers to stop buying from the secondary market.

Page 133 According to Caremark's 2002 Annual Report, the company had $6.8 billion in net revenue.

As documented in an FDLE investigative report dated March 29, 2002, a Caremark employee contacted Marty Bradley of BioMed Plus to express concern over a shipment of Serostim purchased from First Choice Pharmaceuticals. Bradley shared the information with investigator Gary Venema, who called the employee.

Page 133 On March 22, 2002, investigators inspected Atlantic Diabetic, as documented in an FDLE investigative report of that day. According to a transcription of his interview with state investigators that was obtained, the employee of Atlantic Diabetic detailed transactions between an employee of First Choice, José Benitez, and himself. Information from the interview is also included in the FDLE investigative summary report of March 22, 2002.

Page 134 The owner of First Choice, Daniel Zarra, immediately fired his employee after learning of his buying scheme with the Atlantic Diabetic employee. Neither Zarra nor the company First Choice was accused of any wrongdoing in the course of FDLE's investigation.

In an interview, Zarra stated that his company only bought from wholesalers that were properly licensed by the Florida health department. The department had done such a poor job of monitoring the system, he said, that they had given licenses to convicted felons. He added that he was "completely for the general public being protected by all means."

Page 134 Between November 2001 and March 2002, Atlantic Diabetic recorded forty-five transactions with First Choice worth $999,891.63, as documented in the FDLE investigative summary of the relationship between First Choice, Atlantic Diabetic, and José Benitez.

Page 134 Employees of Atlantic Diabetic told investigators that they obtained all of the HIV and cancer drugs and blood products that they sold to First Choice from either José L. Benitez or his half-brother's defunct company, Brazil-US Trading. This is documented in the investigative summary report prepared by the Attorney General's Medicaid Fraud Control Unit on L&L Distributors, LLC, dated February 12, 2003. This report also notes that Brazil-US Trading's permit was suspended by the state on February 1, 2002.

Page 135 In an interview on July 23, 2004, Edward L. Hardin, executive vice president and general counsel for Caremark, explained that his company had exercised due diligence in choosing First Choice, which had a state license and insurance. He said that Caremark requested drug pedigrees from First Choice after receiving a letter from Serono, the drugmaker, warning about Serostim being sold below the wholesale acquisition cost. When First Choice declined to provide the drugs' pedigree papers, Caremark stopped buying from the company, Hardin said. When Caremark could not verify the origins of the drugs purchased from First Choice, Caremark decided to recall all products bought from the company, Hardin said.

Page 135 According to Edward Hardin, Caremark spent $2.5 million to recall all of the drugs purchased from First Choice.

Page 135 Venema's undercover purchase of one hundred boxes of Epogen took place on April 4, 2002, as documented in the FDLE investigative report of that day. Amjad Aryan, a pharmacist and owner of Robert's Drugs, called inspector Arias after being contacted by a representative of AD Pharmaceuticals regarding the sale of the Serostim and Epogen. Aryan met with the representative on April 1, 2002, to finalize the deal and a sale was set up for April 3, as documented in the FDLE investigative report dated March 29, 2002.

Sheldon Schwartz brokered the sale of the Epogen, as documented in the FDLE investigative report of that day. He would later say in an interview that he got "caught in the middle." While the deal for one hundred boxes of Epogen had seemed legitimate at the time, he said he could no longer know for sure which deals were legitimate and which ones weren't, so he got out of the industry. "I just got burned out," he said, adding, "I just couldn't trust anybody and I wasn't willing to go to jail for a couple of dollars." He was not charged with any wrongdoing.

Page 139 Brian Alan Hill and Claudio Boriminoff were identified by investigators as the men ferrying the drugs to the warehouse, as documented in an FDLE investigative report dated April 4, 2002.

Page 140 On December 20, 2000, Chantal Banatty was convicted of dealing in stolen property, according to her criminal case file in the Dade County Circuit Court. She contacted Arias on May 2, 2001, regarding the Neupogen sale, as documented in an Attorney General Medicaid Fraud Control Unit investigative note of May 2, 2001. Her license was revoked, as documented in the administrative complaint, *Department of Health v. AD Pharmaceuticals and Chantal Banatty,* of January 7, 2002.

Page 140 A chart prepared by state investigators traces the path that the one hundred boxes of Epogen traveled from Medix International in Houston to Express Rx in Dallas to L&L Distributors in Miami to Jemco Medical in Pembroke Pines to AD Pharmaceuticals in Miami.

Page 141 Ricardo Lamas's conviction and probation is cited in an investigative summary report on L&L Distributors, LLC, prepared by Jack Calvar of the Attorney General's Medicaid Fraud Control Unit on February 12, 2003.

Pages 141–142 Investigators spoke with both Lamas and Javier Rodriguez at the L&L Distributors' offices, as documented in an FDLE investigative report dated April 4, 2002.

On May 26, 2004, Javier Rodriguez was indicted on charges including racketeering and Medicaid fraud in the Broward County circuit court. He pleaded not guilty. His lawyer, Joseph Rosenbaum, did not return phone calls seeking comment. See also endnote pages 348–349.

Page 144 Craig Brand filed a class action lawsuit against the State of Florida Department of Health on behalf of United Secondary Wholesalers Inc., in the Circuit Court of Miami-Dade County on October 25, 2002.

Page 144 In an interview with the author in May 2003, Craig Brand provided an electronic business card for GPS, which listed himself as president and Ricardo Lamas as chief operating officer.

Page 144 In an interview with the author in May 2003.

Page 145 Arias's correspondence with Jon Martino is included in the investigative summary report prepared by the Attorney General's Medicaid Fraud Control Unit on L&L Distributors, LLC. Amgen declined to answer any questions relating to Jon Martino or his role in investigating the counterfeiting of Amgen's medication.

Chapter 12. The License Shrine

Page 146 On May 15, 2002, a Jemco supplier told inspectors that he and his wife had been invited on Jemco's annual trip to Mexico. Jemco also maintained records of its Mexico trips, as documented in an FDLE investigative summary of July 26, 2002.

Page 147 The Florida Department of Health cited Jemco on November 13, 2001, for selling misbranded and expired drugs, and for failing to maintain pedigrees for the drugs. Jemco was fined $3,250. Jemco's lawyer, Benjamin Metsch, contested the violations and the fine was reduced to $2,000, as noted in a revised notice of violation dated December 4, 2001.

Pages 147–150 Odin's inspection of Jemco and the inspectors' search of the unlicensed warehouse are documented in an FDLE investigative report dated April 12, 2002. The report includes a list of drugs seized, valued at almost $2 million; paperwork relating to $650,000 in business with the Stone Group; and an account of the L&L truck pulling up.

Descriptions of Jemco's offices and warehouses are based on videotapes of searches that inspectors conducted of the unlicensed warehouse on April 12 and the licensed warehouse on April 17, 2002.

Lawrence Metsch, a lawyer for Jemco, said in an interview on August 30, 2004, that the second warehouse searched by inspectors was only an auxiliary warehouse used to keep drugs that did not have to be refrigerated. He said that the state's seizure of the inventory from this warehouse had substantially impaired the company's ability to do business and forced it into receivership. He added that the state had "managed to abuse its criminal process" because "his clients had not even been indicted."

Page 153 The two companies listed on the counterfeit Epogen pedigrees were Express Rx of Dallas and Medix International of Houston. On April 4, 2002, Arias e-mailed John Gower, director of programs, drugs, and cosmetics for the Texas Department of Health, asking who owned the companies and whether they were licensed. Gower responded that Eddie Mor, the registered owner of Express Rx, had an address in Davie, Florida. On April 10, 2002, Arias contacted Debra Lemons, another investigator with the Texas Department of Health, who identified the owner of Medix International as Carlos Luis, who had an address in Miami.

Eddie Mor's lawyer, Bernie Cassidy, said that when Mor's license expired in Florida he got one in Texas. "The only thing he did wrong was not to get a full pedigree," Cassidy added. "All along he did not think he was doing anything wrong." For more information on the criminal charges brought against Mor on May 26, 2004, see endnote pages 348–349.

Page 154 Amgen released a drug warning on May 8, 2002, stating that the counterfeit lot number P002970 was in fact Epogen, but at a dose twenty times lower than indicated on the packaging.

Chapter 13. A Do-or-Die Cause

Page 156 In a written statement, the FDA explained that the agency does not always respond in person to each complaint of potentially counterfeit drugs. Typically, when counterfeit drugs are discovered, the FDA's Office of Criminal Investigations gathers an index sample of the drug and any amounts in wholesale distribution. Samples are submitted to the FDA's Forensic Chemistry Center (FCC) in Cincinnati, Ohio, for testing. In those instances where the counterfeit drugs are known to have reached the public, a Class I recall may be conducted, which instructs consumers to return drug products to the place where they were dispensed. The returned drug products are then consolidated and returned to whichever company is conducting the recall.

Page 157 The author reviewed copies of numerous e-mails Kevin Fagan sent to government officials, including the one he sent to the U.S. Department of Health and Human Services on August 25, 2002.

Page 157 A story in *Newsday* described the Fagans' situation. Although the family was not named, their lawyer, Eric Turkewitz, gave an interview. See Ridgley Ochs, "Sounding Alarm on Counterfeit Drugs," *Newsday,* June 12, 2002.

Page 157 AmerisourceBergen was number twenty-four on the Fortune 500 list for 2002; see *Fortune,* April 14, 2003.

Page 157 Fagan sent an e-mail to AmerisourceBergen's legal department on September 2, 2002, asking for an update on his son's case. When he did not hear back, he sent another message to the department on September 9. When he did not get a response to his second e-mail, he forwarded it to company chairman Robert Martini on September 10.

On a copy of the September 9 e-mail, Fagan penciled in: "Next day Zimmerman & Ms. Schulz (?) called me at work and advised I not call them or Amerisource for info. Call FDA agent in FLA. for any info!!"

Page 159 On August 4, 2003, the Fagans' lawyer, Eric Turkewitz, filed *Fagan v. AmerisourceBergen, Amgen and CVS* in the United States District Court, Eastern District of New York.

Page 160 On October 11, 2003, at the Fagans' home, Representative Steve Israel held a press conference, where he announced new legislation that he was proposing to combat the growing problem of fake pre-

scription drugs. He said the legislation would give the FDA authority to recall bad medicine, which it currently lacks, and increase
penalties for counterfeiters.

Page 161 After U.S. Senator Hillary Clinton's office intervened, FDA
headquarters made the New York District Office and New York FDA
OCI Office aware of the family's request to be personally contacted.
On September 17, 2002, an OCI Special Agent and an FDA Consumer Safety Officer visited the Fagans and obtained the evidence
they were holding, and the suspect Epogen was subsequently tested
at the FDA's FCC.

Clinton declined to be interviewed. Deputy Press Secretary Amy Bonatat-
ibus said that the senator had a blanket policy of not participating in
books. She went on to say that the issue was "obviously something the
senator's interested in, that's why she helped him."

Page 162 For information on the PDMA, see endnote page 107.

Page 162 Congressman John D. Dingell made his remarks about opponents of the PDMA in his opening statement to Congress on May 14,
1986.

Pages 162–163 Examples of industry objections to the PDMA are taken
from correspondence, testimony, and comments included in the extensive docket maintained by the FDA regarding the law and obtained through a Freedom of Information Act (FOIA) request. The
drugmaker Smith Kline & French protested the pedigree requirements in a letter dated October 3, 1988.

In a letter dated September 30, 1988, the National Association of Retail
Druggists (NARD) noted that its members, independent pharmacies,
dispensed 70 percent of all prescription drugs. In defense of the
PDMA, the letter went on to state that the members "agree that the
audit trail is essential to providing the type of evidence necessary to
convict violators and deter potential violators."

Page 164 The guidance letter issued by the FDA stated that a drug's pedigree should include "all necessary identifying information regarding
all sales in the chain of distribution of the product, starting with the
manufacturer or authorized distributor of record." The agency defined an authorized distributor as a wholesaler that could demonstrate
proof of "the existence of on-going sales by the manufacturer to the
distributor, either directly or through a jointly agreed upon interme-
diary." Two transactions in any twenty-four-month period would be
sufficient evidence of an ongoing relationship to claim authorized distributor status.

Page 165 On February 29, 2000, the federal government's Small Business

Administration sent a letter to the commissioner of the FDA siding with the wholesalers against more stringent pedigree rules.

Page 165 The final rule on pedigree papers was stayed five times as follows: May 3, 2000, until October 1, 2001; March 1, 2001, until April 1 2002; February 13, 2002, until April 1, 2003; January 31, 2003, until April 1, 2004; February 23, 2004, until December 1, 2006.

Chapter 14. A Bad Lot

Pages 166–167 The report of Hokins's inspection of the Bindley office, as well as invoices for the 1,004 boxes of Procrit that Bindley bought from Raymar, are included in a Texas Department of Health establishment detention report dated May 22, 2002.

Page 167 For more information on Bindley's merger with Cardinal, see endnote page 230.

Page 168 On May 31, 2002, Johnson & Johnson security director Alan Saleeba sent a fax to John Gower at the Texas health department confirming that Amgen had tested vials from Procrit lot P002641 and found that they were counterfeit.

Page 169 On June 4, 2002, Albert Hokins at the Texas health department sent an e-mail to Gregg Jones, pharmaceutical programs manager for the Florida health department, saying that a second lot of Procrit, P002384, had been tested and was also found to be counterfeit.

Page 170 Gregg Jones sent an e-mail to the Horsemen on May 20, 2002, describing the seizure of 1,617 boxes of Epogen from an Amerisource warehouse in Kentucky.

Page 170 Premier Medical Distributors was listed on the pedigree papers of the counterfeit Epogen, as documented in an FDLE investigative report dated May 22, 2002. In an interview with investigators, the owner of Premier, James Robert Suozzo, said that he sold the Epogen to Consumer Services Group, as documented in an FDLE investigative report dated August 6, 2002.

Page 172 Investigators learned about Medi-Plus, a company opened by a Jemco employee, through an application for a North Carolina license they found during a search of the Jemco warehouse, as documented in an FDLE investigative report dated April 17, 2002. North Carolina officials found the Epogen at Medi-Plus in late May, as documented in an e-mail sent to investigators by Gregg Jones on May 20, 2002.

On December 5, 2002, Florida investigators arrived at the offices of Medi-Plus in Yadkinville, North Carolina. There they encountered Brian Hill of Jemco, who said that he was a "consultant" for the company.

Hill's exchange with investigators is documented in an FDLE investigative report of that day.

Page 173 Peter Grasso, an investigator for the New Hampshire Board of
Pharmacy, visited the address listed for Accucare in Nashua, New
Hampshire. The dilapidated state of the house and his conversation
with the owner are documented in an FDLE investigative report
dated December 10, 2002. When Grasso returned to the house, he
found it had been fixed up, as documented in an FDLE investigative
report dated June 4, 2003.

Page 173 Venema sent the e-mail to Chris Zimmerman on May 3, 2002.

Pages 175–176 Venema sent the e-mail about Luis Perez to the other
Horsemen on May 23, 2002.

Chapter 15. Rats in the State

Page 179 The first meeting of the Ad Hoc Committee on Pedigree Papers
was held on April 30, 2002.

Page 181 The e-mail exchange amongst Gene Odin, Jerry Hill, Sandra
Stovall, and Gregg Jones of Florida's Bureau of Statewide Pharmaceutical Services, regarding Nevada's new law, occurred from October 18-23, 2001.

Page 182 Lori Bickel, associate director of regulatory affairs for the
Healthcare Distribution Management Association, sent the letter to
Jerry Hill, pharmacy services chief at Florida's health department, on
January 29, 2002. Sandra Stovall forwarded it to Robert Daniti on
January 30.

Page 182 John O. Agwunobi, secretary of the Florida health department,
sent a letter to industry representatives on February 8, 2002, announcing that he was forming an ad hoc committee charged with "present-
ing recommendations to resolve the pedigree paper dilemma that
satisfy the department's public health mission to protect the public
from misbranded and adulterated drugs, while attempting to lessen
the regulatory cost of compliance on the regulated industry."

Pages 183–185 Minutes and tapes of the first Ad Hoc Committee meeting, as well as a copy of Ricciardi's presentation for Supreme Distributors, a division of Purity Wholesale Groceries, were obtained
through a public records request to the Florida health department.

Page 186 Ricciardi's letter to the Federal Trade Commission was sent on
April 5, 2002.

Page 188 In response to written questions, Hill said that throughout his
employment with the Department of Health, "I have continually kept

supervisors abreast of all significant issues related to drug wholesale and pharmacy regulation, on an ongoing basis."

Page 189 In an interview, Feldman recalled being told by a secretary in Melanie Ann Hines's office that Hines had apparently gotten a call from "someone" important in Tallahassee. That person had gotten a call from a Washington, D.C., lawyer complaining that a young female prosecutor was out of control.

In an interview on April 10, 2004, Hines said that while she did not re-member that particular phone call, she got calls like it all the time from lobbyists for an array of interests. "I tried to keep the handling prosecutors insulated. That was my job," she said. "I answered as I was allowed under the law."

PART THREE

Chapter 16. Crazy Money

Page 196 Nicholas Just and Paul Perito's purchase of Playpen South was described by the manager of the club, Benjamin R. Ojeda, as docu-mented in an FDLE investigative report dated May 4, 2004.

Page 196 Information on Perito's medical school education and residency comes from American Medical Association records. Perito was in-volved with a home for AIDS orphans in Kenya called Nyumbani. He was commended in the home's Winter 2003 newsletter for raising $13,000 at a Miami cocktail party.

Page 197 According to the Miami Dade police report of December 29, 2002, Nicholas Just was listed as the suspect who burned the car. He was witnessed by his girlfriend's roommate threatening to burn the car.

Pages 197–198 As documented in an FDLE investigative report of May 4, 2002, Ojeda described Just's prior work history as a bouncer, his vi-olent past, and his addiction to Lortabs. An affidavit of probable cause to arrest Perito and Just, signed by investigator Gary Venema on May 25, 2004, included statements by Ojeda and an assistant manager of Playpen South, who said they saw Perito use Special K. As described in the affidavit, Perito microwaved the drug until it hardened and then he cut it up and snorted it. He then administered the drug to a dancer who became catatonic, according to the affidavit. Ojeda also stated that he saw transactions in the club's back room involving the sale of Procrit stored in a cooler.

Page 198 On May 26, 2004, Paul Perito and Nicholas Just were indicted in Broward County circuit court, on charges including racketeering, organized scheme to defraud, sale of counterfeit goods, and product tampering. See also John Dorschner, "Doctor Charged with Diluting Cancer Drug," *The Miami Herald,* May 28, 2004, and Bob LaMendola, "2 Charged in Phony Drug Scheme," *South Florida Sun-Sentinel,* May 28, 2004. For more information on Just and Perito's arrests, see endnote pages 348–349.

Through his lawyer, Jane Moskowitz, Perito categorically denied being knowingly involved in the sale of any counterfeit drugs. In a written statement, Moskowitz said Perito is a "talented, well-respected urologist who would never put a patient or anyone else in harm's way." She also noted that Perito treated AIDS patients for free and was involved with many other good works.

Just's civil lawyer, David Ryan, questioned the veracity of the statements made about his client, saying the allegations were "submitted by people who are convicted felons, real shady characters."

Just's criminal lawyer, W. Clair Lambert, later described Ojeda as a convicted felon and a vindictive ex-employee who was fired by Just and went to the authorities in retaliation. "Allegations that any drugs were brought to the club will be vigorously disputed," he said. "Any allegations of Just's drug use refer to events that happened years ago," he added. "Not to recent events."

Pages 198–199 Arias first inspected Carlos Luis's Medical Support Systems on June 14, 2002, as documented in the FDLE investigative report of that day. The Horsemen searched his home and business on July 9, 2002, as documented in an Operation Stone Cold weekly update and an FDLE investigative report of that day. Luis's lawyer, Edward O'Donnell, said that many allegations involving Luis are erroneous. He also said that drugs are often deemed adulterated on technicalities, and that he believes investigators ultimately will concede that the drugs Luis handled were not counterfeit.

Page 199 Venema's affidavit of probable cause to arrest Perito and Just stated that $943,276 in checks from Luis's company, Medical Support Systems, was deposited in Luis's account from January 10, 2001, to May 13, 2002. Perito told investigators the payment was a loan for the purchase of a health club. Perito originally told investigators that he received a smaller amount of money, $628,558, from Luis's company, as documented in an investigative summary report prepared by the Attorney General's Medicaid Fraud Control Unit on L&L Distributors, LLC, dated February 12, 2003.

Page 199 The assistant manager at Playpen South told investigators he saw Just pay a female bartender $500 to count $82,000 in cash, as documented in an FDLE investigative report dated April 22, 2004.

Pages 199–200 Information on Luis's business relationship with Just and Perito and their introduction to "Tony" was related by the assistant manager, as documented in the FDLE investigative report of April 22, 2004.

Page 200 Ojeda said he witnessed Luis's return of the Procrit and heard Just's remark about making $100,000 a week, as documented in the FDLE investigative report of May 4, 2002. Venema's affidavit also described Just and Luis studying the boxes of Procrit under a black light.

Page 200 Just and Perito's description of the drugs as "colored water" came from the assistant manager, as documented in the FDLE investigative report dated April 22, 2004.

Page 201 According to numerous records from Operation Stone Cold, Carlow operated shell companies in Florida, Texas, Maryland, New Hampshire, Missouri, and New Mexico.

Page 201 Marilyn Atkins's involvement in Carlow's companies is documented in an FDLE investigative report of December 20, 2002. An affidavit for probable cause to search her home stated that her son Thomas Atkins told investigators that she did the payroll for BTC and maintained the company's checkbook and records. Sam Whatley, the man hired to run G&K Pharma, told investigators that he was instructed to forward all phones to Marilyn Atkins's house in the evenings.

When investigators searched her home on December 23, 2002, they found paperwork for BTC, G&K Pharma, Complete Wholesale, and other Carlow-controlled companies, as documented in an FDLE investigative report of that day.

Carlow also described Atkins's role in a conversation he had with a business partner on May 23, 2003, according to a transcript of the conversation. For more information on this conversation, see endnote page 283.

Marilyn Atkins's lawyer, John C. George, did not return phone calls or respond to a written request seeking comment.

Page 202 The formation of G&K Pharma and Complete Wholesale in Odenton, Maryland, are documented in an FDLE investigative report dated April 10, 2003.

The Arizona Department of Health notified Florida's BSPS about a box of AIDS medicine that still bore a prescription label for a Fort Lauderdale patient.

Using the drug's pedigree papers, the investigators traced the medicine to JB Pharmaceuticals and Accucare, companies controlled by Carlow, as documented in an FDLE investigative report dated May 2, 2002.

Page 203 Chaille's landing of the helicopter on the police station roof on January 31, 2000, was documented in a disciplinary action report dated February 2, 2000. He received a written reprimand on February 23, 2000, six days after a disciplinary action session was held.

Chaille's declining job performance from October 25, 1999, to August 8, 2002, is documented in his employee performance evaluations, part of his Miami-Dade County personnel records dating back to 1985. Chaille submitted his resignation to the MDPD in October 2002, as documented in a letter his lawyer sent to the MDPD personnel department on October 23, 2002.

Chaille opened New Horizons Network on December 9, 2002, as documented by the state of Delaware's Division of Corporations listings. The opening of his company was also included in an MDPD field investigations weekly summary for the week of May 5, 2003.

Page 204 Mark Novosel told investigators that Carlow had asked Chaille to place Venema under surveillance, but Chaille convinced him that it would be too costly, as documented in a report on Novosel's debriefing on June 19, 2003.

Sheldon Krantz, a lawyer for Robert Chaille, made the following statement on behalf of his client: "Mr. Chaille was the owner for a brief period of time of New Horizons Network, a licensed wholesale distributor of pharmaceutical products. During the time he owned New Horizons Network, Mr. Chaille never knowingly served as a broker for adulterated or counterfeit pharmaceutical products and would never do so. In addition, he had no dealings with Michael Carlow, does not know him, and has never met him."

Chapter 17. A Special Price

Pages 207–211 According to sales records, J&M Pharmacare purchased 3,363 boxes of low-dose Epogen from Amerisource and 8,931 boxes from Cardinal between January 2001 and May 2002. The 12,294 boxes amounted to half of all Epogen sold in Florida and Georgia in that time period, according to an affidavit of October 16, 2003, in support of an arrest warrant for Maria Castro and Jesús Benitez. J&M Pharmacare's purchases of low-dose and high-dose Epogen are documented in the investigative summary report prepared by the Attorney General's Medicaid Fraud Control Unit on L&L Distributors, LLC.

Pages 208–210 Inspectors' visits to J&M and their ensuing interviews with patients and with Maria Castro, owner of J&M, are documented in the Attorney General's Medicaid Fraud Control Unit's L&L Distributors report. Details of the J&M investigation also are documented in the October 16, 2003, affidavit in support of an arrest warrant for Jesús Benitez and Maria Castro. In February 2005 Benitez pleaded guilty to the unauthorized sale of prescription drugs, received probation, and agreed to cooperate. The charges against Castro were dropped, and she agreed to cooperate.

Page 212 The information that Armando Rodriguez gave investigators about purchasing Epogen from J&M was drawn from several documents: the Attorney General's Medicaid Fraud Control Unit's L&L Distributors report; an FDLE investigative report dated August 9, 2002; and the affidavit on October 16, 2003, for the arrests of Benitez and Castro. Rodriguez was not charged with any wrongdoing. His lawyer, Norman Moscowitz, declined to comment.

Page 213 Michael N. Kilpatric, a spokesman for AmerisourceBergen, said the company has established tighter tracking systems since 2003 to help detect diversion by customers, as well as to detect counterfeits sold by vendors. An electronic system attached to sales data now tracks unusual spikes in purchase orders, he said. The company also has a more rigorous accounting system, which assigns each customer a credit limit based on anticipated purchase orders. A substantially increased order would thereby raise a red flag. The sales force has received new training on spotting evidence of diversion, he said. The company is also working to improve its tracking of medicine returned by customers, which remains a weak link in the security chain, he said. He also said that Amerisource more closely scrutinizes companies' initial orders against the medicine they return.

Page 214 Chris Zimmerman's presentation and explanation of the wholesale market is documented in minutes of the first Ad Hoc Committee meeting on April 30, 2002. Tapes of this meeting were also obtained.

Page 214 Amerisource Health and Bergen Brunswig's merger is widely documented. See Cinda Becker, "And Then There Were Three," *Modern Healthcare*, September 3, 2001.

Page 214 Bergen Brunswig issued a recall of counterfeit Serostim in early February 2001, as documented in a California Board of Pharmacy summary report for January 18 to March 8, 2001.

Pages 214–216 On June 13, 2000, the company was fined $2,000 for buying medicine from unlicensed wholesalers, as documented in a notice of violation from the Florida department of health. Nicholas W. Ghnouly, counsel for the Bergen Brunswig Corporation, sent his letter opposing the fine to the BSPS on June 7, 2001. Sandra Stovall, a compliance officer for Florida's BSPS, sent her response on June 11, 2001, and reduced the fine to $1,500.

Page 216 An invoice from an Ohio wholesaler dated August 9, 2000, shows that Amerisource saved $8 per vial on Retrovir. The cost of the Retrovir from Glaxo was obtained by an inspector at Florida's BSPS, who then sent an e-mail to colleagues (Stovall, Hill, and Arias) in the department on June 13, 2001.

On May 28, 2002, Rodney Bias, AmerisourceBergen's director of corporate security and regulatory affairs, sent a letter to Sandra Stovall at the Bureau of Statewide Pharmaceutical Services with a check for $50,000. This was payment to cover the fine that had been imposed on the company for buying drugs from unlicensed wholesalers and failing to store them properly. Bias also included a list of new policies intended to address the problems for which the company had been fined.

Page 216 Bergen Brunswig filed a lawsuit against Dialysist West in United States District Court in Arizona on August 2, 2002. In its answer and counterclaim filed on March 18, 2003, Dialysist West denied any wrongdoing and sought compensation from Bergen for the cost of the lawsuit and attorney's fees. Dialysist West then filed a third-party complaint on March 18, 2003, alleging that AmeRx, CSG, and Optia had broken their contracts with the company by providing counterfeit goods. AmeRx denied any wrongdoing and stated that it had been "tricked, defrauded and misled" by its supplier. CSG denied any wrongdoing as well, and went on to sue Optia Medical on April 21, 2003.

Page 216 According to Susan Cavalieri, Amerisource terminated her company as a vendor in the summer of 2001. Invoices sent by AmeRx to Amerisource show drugs being distributed to Amerisource's warehouses in California, Arizona, Ohio, Texas, New Jersey, Massachusetts, Missouri, Minnesota, and Kentucky.

Page 217 Pedigree records that Dialysist West gave to AmerisourceBergen in early 2002 listed AmeRx as one of the companies where the Epogen had originated.

Page 217 According to bank records obtained by investigators, AmeRx did close to $12 million in sales in 2001.

Page 218 William Walker was arrested on May 5, 2003, as documented in an MDPD complaint. He later pleaded guilty to seventeen charges, including racketeering, fraudulent use of personal identification, and violations of the health statute 499. Cavalieri's purchase of Serostim from Walker is documented in an FDA investigative report.

Page 218 Eddie Mor's lawyer, Bernie Cassidy, said that "Susan Cavalieri has made an awful lot of claims of innocence but she seems to be at the center of everything. I don't put very much stock in what she has to say about Eddie Mor."

Page 218 Investigators obtained records of sales of medicine from Eddie Mor to AmeRx during a search of Mor's home, as documented in two FDLE investigative reports dated June 14 and August 14, 2002.

Page 219 AmeRx's banking records show $800,000 in sales to Zitomer, as documented in the FDLE investigative report dated August 21, 2002. According to Cavalieri's pedigree records, the medicine she sold to Zitomer had come from companies that included AD Pharmaceuticals, Express Rx, and Medix International, as documented in an FDLE investigative report of August 4, 2002. Subsequent to the sales, these companies have been found to have distributed counterfeit and adulterated medicine.

In a search warrant served at the home of Marilyn Atkins on December 23, 2002, investigators obtained a shipping record for a package sent from G&K Pharma to Zitomer on August 17, 2002. Zitomer officials did not return phone calls or respond to a registered letter seeking comment about these purchases.

Chapter 18. The Guitar Story

Page 221 Cardinal Health sent out a press release on April 25, 2000, announcing increased revenues and earnings for the third quarter, which had ended on March 31, 2000.

Page 221 Cardinal Health announced National Specialty Services' (NSS) partnership with US Oncology in a press release dated August 9, 1999. The partnership with Novation was announced in a press release by Novation on April 24, 2000.

Page 222 A photocopy of check 5256 for $10,460—written to Neil Spence from Quest Healthcare on April 21, 2000, signed by Spence and stamped "uncollected"—was obtained from banking records that emerged as the result of a lawsuit against Carlow.

Pages 222–223 Bank and shipping records and invoices show that NSS bought medicine from Carlow's companies—BTC Wholesale, Quest Healthcare, Medical Infusion Services, and MedRx—from 1998 through 2002. The relationship between NSS and Medical Infusion Services was initiated shortly after June 15, 1998, when Medical Infusion sent a credit application to NSS. From January 1, 1999, to June 30, 2000, NSS bought $131,000 in blood products from Medical Infusion Services, according to a list of sales that NSS sent to Florida's BSPS on September 29, 2000. From January through June 2000, Cardinal paid $1.45 million to Quest Healthcare for blood products and other specialty pharmaceuticals purchased by NSS, according to bank records. A BTC sales ledger from August 20, 2001, to December 24,

2001, shows checks BTC received from NSS for the purchase of medicine. Shipping records from August to December 2001 show packages sent to NSS from Tom Atkins of BTC and from Mark Novosel of MedRx. A notation on BTC's January 2002 profit statement shows that NSS continued to buy medicine from BTC into 2002.

Page 223 For more information on Carlow's arrest in 2000, see endnote page 45.

Page 223 UPS shipping records show that MedRx sent Neil Spence three packages on September 5, 2001. Additional UPS shipping records show that BTC sent a package to Neil Spence in Gallatin on October 1, 2001. Check 639 was written from BTC to Neil Spence for "outside services" on December 13, 2001, as documented in the BTC general ledger.

Neil Spence left NSS in late 2003. He declined to comment to the author on his dealings with Michael Carlow but has effectively denied any wrongdoing. A registered letter with questions about his dealings with Carlow that was sent to his home was returned unopened. No charges have been brought against him.

Jim Mazzola, vice president of corporate communications for Cardinal Health, said that the company was launching an internal investigation into allegations relating to dealings between Neil Spence and Michael Carlow and would not comment further. He said "the company takes very seriously both the issue of criminal counterfeit activity and any allegations of wrongdoing by its employees."

Mazzola said that since counterfeit drugs surfaced in the supply chain, Cardinal has strengthened its safeguards and policies. Since January 2003, the company has significantly limited the number of secondary wholesalers with which it does business and purchases more than 98 percent of pharmaceuticals it distributes directly from the manufacturer; visits the facilities of all secondary wholesalers it buys from, rigorously reviews the companies' policies, and does background checks of business officers and their suppliers; and purchases more than one hundred drugs at high risk of being counterfeited only from their manufacturers.

In addition he said that Cardinal Health has "taken a lead role in the industry's first pilot program for radio frequency identification (RFID), which holds promise in eliminating counterfeit drugs from the supply chain." He added: "We remain diligent in protecting the safety of products in the pharmaceutical supply chain and will continue to work closely with manufacturers, pharmacy customers and industry groups to achieve this goal."

Page 224 For more information on the spreadsheet prepared by Cardinal documenting its savings from secondary wholesalers, see endnote page 91.

Page 225 A review of the purchases made by Bindley Western in 2002 comes from data analyzed by Florida's health department.

Page 226 Mary Pat Flaherty and Gilbert Gaul reported statements made by Cardinal in "U.S. Prescription Drug System Under Attack," *The Washington Post*, October 19, 2003.

In an interview on October 6, 2004, Cardinal spokesman Jim Mazzola said that by 2001 the company had made an effort to change its buying practices, but it did not fully implement the changes until the end of 2002.

Pages 227 Transcripts of statements the FBI took from subjects and targets of the Pharmoney investigation were obtained. The statements in the following pages are drawn from these transcripts.

Page 227 Joel Feldman, the magistrate judge who signed Allen's search warrants, recalled finding a stray tablet of Coumadin, a blood thinner, in his own medicine bottle; if he had taken it, it could have caused him to bleed to death during a scheduled heart surgery.

Page 227 Allen found a similar situation at Bergen Brunswig. Roger Johnson, who described himself as the company's chief secondary source buyer, pleaded guilty to wire fraud and the purchase and sale of adulterated or misbranded drugs. He told the FBI that the company encouraged the purchase of diverted medicine from the secondary market. He estimated that in 1982 alone he purchased close to $150 million in diverted pharmaceuticals for Bergen Brunswig.

Pages 227–228 Stephen Lee Asher gave statements to FBI agents on May 24, July 2, and July 3, 1985. Jack Earl Laughner gave his statement on October 17, 1985. During these meetings both men described Bindley Western's practice of buying from the secondary market, the purchase of drugs including Tagamet from Martin Thuna, and kickbacks they received from Thuna.

Asher pleaded guilty to mail fraud. Laughner pleaded guilty to mail fraud and wire fraud, as documented in the summary of criminal charges for those involved in the Pharmoney case. Thuna pleaded guilty on June 25, 1986, in United States District Court in Puerto Rico to stealing pharmaceuticals from Eastern Airlines shipping containers.

Page 229 Richard Allen testified before the House Committee on Energy and Commerce's Subcommittee on Oversight and Investigations on June 20, 1990, regarding the weakness of the PDMA and the FDA's failure to enforce it. During his testimony, he also discussed the FBI's pullout from the Pharmoney case. Allen's comment on the Tylenol poisoning cases comes from his testimony on November 25, 1986, to the Georgia State Senate's Committee on Health and Human Resources.

Page 230 Three Bindley executives from the San Dimas, California, division were convicted on charges that included mail fraud, conspiracy to commit mail fraud, and conspiracy to transport stolen goods as documented in an FDA Office of Criminal Investigations report. The FDA's investigation of Bindley-Western was widely documented. See Jeff Swiatek, "Bindley Pays $20 million for Crimes of Former Execs," *The Indianapolis Star,* August 31, 2000; and Gilbert Gaul and Mary Pat Flaherty, "U.S. Prescription Drug System Under Attack," *The Washington Post,* October 19, 2003.

The August 29, 2000, plea agreement was filed in the United States District Court of Nevada. In its plea agreement between the U.S. Attorney's Office and Bindley Western, the court stated that Bindley Western's managers at the San Dimas division "knew a substantial portion of the discount-priced product shipped to those closed-door pharmacies in Southern California and Nevada would be resold in bulk and in violation of the pharmacies' representation that they would not do so."

A June 30, 2003, FDA report noted that the agency's undercover operation corroborated that "the corporate offices of Bindley Western Drug Company had tacit knowledge of and involvement in the illicit prescription drug diversion activities." The FDA report is posted on the Internet.

William E. Bindley, Bindley Western's longtime chairman and CEO, was never charged with any wrongdoing following either investigation.

Michael McCormick, the former executive vice president and general counsel for Bindley Western, said that the government's stipulation, as well as an internal investigation by the company, made clear that the employees in San Dimas acted without the knowledge of their corporate superiors. "The FDA's internal memorandum in which it suggests that unnamed corporate officials had knowledge of and involvement" in the scheme is "unfortunate, unfair and recklessly untrue."

Page 230 John Ransom's quote is taken from an online article by Norm Heikens of *The Indianapolis Star* on March 28, 2000.

Page 230 Cardinal Health's purchase of Bindley Western is widely documented. See "Company Briefs," *The New York Times,* February 15, 2001; and *Investor's Business Daily,* February 20, 2001. Bill Bindley's subsequent gift to Purdue University is documented in *Purdue News,* September 27, 2002; and Phillip Fiorini, "Self-made Millionaire Interesting Story Behind Campaign for Purdue," *Journal and Courier,* October 6, 2002.

Page 231 Cardinal Health was twenty-third on the Fortune 500 list for 2002. See *Fortune,* April 15, 2002.

Chapter 19. "They're Going to Die Anyway"

Page 232 Grillo was indicted on July 21, 2003, for nineteen counts including organized scheme to defraud, purchase or receipt of a prescription drug from an unauthorized person, and possession with intent to distribute prescription drugs. He was indicted again on May 26, 2004, for charges of racketeering, conspiracy to commit racketeering, and organized scheme to defraud. He pleaded not guilty to both sets of charges.

Page 232 In a statement he made to investigators on August 19, 2003, Grillo said that he kept his money in a duffel bag in his closet.

Page 232 The assistant manager at Playpen South told investigators that the owner of a glass block company introduced Grillo to Perito and Just, as documented in an FDLE investigative report of April 22, 2004. On May 4, 2004, the manager of Playpen South, Benjamin Ojeda, said that the owner introduced Grillo to Just. Ojeda also said that Grillo continually changed his cell phone numbers. Statements made by Ojeda and the assistant manager are included in Venema's May 25, 2004, affidavit of probable cause to arrest Perito and Just.

Page 233 Investigators questioned Leticia Salas, Grillo's wife, on July 15, 2004, as documented in an FDLE investigative report of that day. She described Grillo's violent temper, his business meetings with Just, the cash he brought home from Playpen South, and his relationship with his father's contemporaries from Cuba. Grillo was questioned on August 19, 2003. In his proffer to investigators, Grillo stated that his father's friends from Cuba had set him up in the counterfeiting business and paid for machinery and medicine, as documented in an Operation Stone Cold update of August 24, 2003.

Rodriguez, who was not charged with any wrongdoing, was named in the indictment as the unauthorized person who sold medicine to Grillo.

Page 234 In a statement given to investigators on September 24, 2003, Silvino Cristobal Morales described his relationship with the Grillo family and how he started to work for Tony Grillo. He said that Grillo told him the medicine was legal and for weightlifters. Morales described how he relabeled the vials and how Grillo would pick up and drop off the vials in a five-gallon paint bucket. Grillo's wife also described the relationship between Morales and Grillo's family, as documented in an FDLE investigative report of July 15, 2004.

Page 235 In an affidavit in support of an arrest warrant for Maria Castro and Jesús Benitez of J&M Pharmacare on October 16, 2003, investigators stated that the wholesale acquisition cost for a box of 2,000 U/mL Epogen is $250, while a box of 40,000 U/mL Epogen is $4,700. The estimated profit of $46 million from Grillo's uplabeling is based on these figures.

Page 235 The printing errors on Grillo's counterfeit boxes are documented in the investigative summary report prepared by the Attorney General's Medicaid Fraud Control Unit on L&L Distributors, LLC, and in an FDLE investigative report dated August 26, 2002. Grillo told investigators he paid $25,000 for the printing.

Page 236 Grillo's caution is described by Armando Rodriguez, as documented in the FDLE investigative report of August 26, 2002, and in the investigative summary report prepared by the Attorney General's Medicaid Fraud Control Unit on L&L Distributors, LLC. Investigators also noted that the stoppers for both the high-dose and low-dose Epogen were the same color, as documented in an FDLE investigative report dated June 10, 2002.

Rodriguez told investigators that most of the $1.1 million he made selling Epogen to Grillo was spent on a dancer at a strip club. Rodriguez also described Grillo's reaction when investigators went to J&M and how Grillo vanished for about a month, then resurfaced and mentioned his problem with his business contact "El Viejo." His statements are documented in an FDLE investigative report of August 9, 2002.

Pages 237–239 On August 28, 2002, Rodriguez was outfitted with a recording device for his meeting with Grillo, as documented in an FDLE investigative report of that day. A transcription of their conversation was obtained.

Page 240 Grillo told investigators he received a call after his meeting with Rodriguez at La Carreta restaurant, warning him that he should avoid Rodriguez, according to the August 29, 2003, summary of Grillo's statement. He did not say who called him.

Page 241 An MDPD detective ran the vehicle identification number on Grillo's Corvette and discovered that it was stolen, as documented in an FDLE investigative report dated November 8, 2002.

Page 241 The presence of counterfeits in foreign countries is widely documented. See "Fake Drugs Threaten Both Consumers and Pharmaceutical Sector," *Jakarta Post*, June 7, 2003; "India's Cabinet Approves Death Penalty for Manufacture and Sale of Counterfeit Drugs," *Voice of America News*, December 18, 2003; Alla Startseva, "Fakes Costing Drug Companies $250M," *The St. Petersburg Times* (Russia), May 7, 2002. The World Health Organization has analyzed reports on counterfeit drugs from member countries and created a policy paper on how to combat the problem internationally. In 2003, the Pharmaceutical Security Institute, a research organization funded by drugmakers, prepared a fifty-five-page annual internal report for its members that documents counterfeiting in foreign and domestic markets. The author obtained a copy of the document.

An FDA report released on February 18, 2004, entitled "Combating Counterfeit Drugs" listed the number of counterfeit cases as six in 1997, four in 1998, six in 1999, six in 2000, twenty in 2001, twenty-two in 2002, and twenty-two in 2003.

According to the FDA's OCI, the number of cases that included but were not limited to counterfeiting were even higher: nine in 1997, five in 1998, eleven in 1999, six in 2000, twenty-one in 2001, twenty-seven in 2002, thirty in 2003, and fifty-four in 2004.

Six hundred thousand patients may have received counterfeit Lipitor, John Theriault, vice president of global security for Pfizer, said in his testimony on April 5, 2004, to the Department of Health and Human Services Drug Importation Task Force.

Page 242 Randy Jones was approached by the 280-pound bodyguard, as documented in an Operation Stone Cold report dated February 10, 2003.

On July 29, 2002, James Suozzo told investigators about his heroin addiction and his education level. Suozzo's arrest record is documented in the investigative summary report prepared by the Attorney General's Medicaid Fraud Control Unit on L&L Distributors.

Page 243 FDA investigators arrested Eddy Gorrin on February 27, 2003, as documented in a federal court filing and an FDLE investigative report dated February 28, 2003. On June 11, 2003, Gorrin pleaded guilty to selling counterfeit pharmaceuticals. On September 4, 2003, he was sentenced to thirty-seven months' imprisonment, as documented in court records for the case. See also Larry Lebowitz, "Three Miami-Area Men Admit to Sale of Tainted Fake Drugs," *The Miami Herald*, June 12, 2003.

Page 243 At least fifty of the 1,458 wholesalers licensed by Florida were under suspicion for counterfeiting or diversion activities, according to a February 2003 report by the Florida Legislature's Office of Program Policy Analysis and Government Accountability entitled "Counterfeit and Diverted Drugs Threaten Public Health and Waste State Dollars."

Page 243 On September 17, 2002, at the request of prosecutor Penezic, investigators began to list the targets for their investigation. The first list included thirty-three people, with suggested penalties ranging from a $50,000 fine and house arrest to a $5 million fine and twenty-five years in prison.

Page 244 The number of subpoenas comes from a log maintained by investigators. The seizures of $14 million in adulterated medicine and cash are documented in an Operation Stone Cold summary from late 2002. The figure of $3 million in assets returned to the state was included in an outline created by investigators in late November 2002.

Page 244 Venema sent his e-mail encouraging the other investigators to keep pursuing the case on October 10, 2002.

Page 245 Venema's comments about the sluggish pace of the health department are taken from an FDLE status report dated June 19, 2002.

Pages 245–246 Venema recounted his whispered conversation with Penezic in an e-mail to the investigators on August 1, 2002.

Page 247 Carlos Alvarez, director of the MDPD, issued a department-wide memo on June 18, 2002, stating that all cases involving suspicious pharmaceutical products would now be directed to the Unlicensed Practitioner Unit.

Pages 248–250 A briefing document prepared by the FDLE for the meeting with Governor Jeb Bush on December 17, 2002, explained how drugs were diverted and sold back into the legitimate market and how this affected Medicaid spending in Florida. The briefing included the recommendation that the governor empanel a grand jury and strengthen statute 499.

Chapter 20. They Know We Know

Page 253 The Ad Hoc Committee on Pedigree Papers released its recommendations on October 31, 2002. The committee recommended that wholesalers be required to post a $100,000 bond as a guarantee against violations, renew their permits annually, and designate an employee responsible for all receipt of medicine and warehouse activity. The committee suggested giving BSPS more authority, increasing the agency's staff and budget, and paying for the additional costs with higher licensing fees. It also recommended increasing penalties for certain offenses by making them felonies.

Page 254 The *Orlando Business Journal* article quoting Ed Homan appeared on April 14, 2003.

Page 255 The release of Arias's personal information on February 17, 2003, was documented as a security breach in a management review of BSPS prepared by the health department's inspector general.

Pages 255–256 In December 2004, Hill responded in writing to questions regarding criticism of the bureau's performance and of his management. He said that the bureau had been working constantly to improve the satisfaction and productivity of its employees and to protect Florida's citizens from counterfeit drugs.

Page 256 As documented in an FDLE investigative report dated September 6, 2002, Bradley called Venema after speaking with another wholesaler who told him that the police were asking questions about Bradley.

Bradley's former associate pleaded guilty to racketeering and Medicaid fraud charges on April 16, 2001. He received seven years' probation and was ordered to pay almost $3.5 million in restitution, as documented in a judgment from the Leon County Circuit Court.

Page 257 Odin and Arias seized a total of $445,477.50 in adulterated Lipitor and Celebrex from a wholesaler in Jensen Beach, Florida, as documented in an FDLE investigative report dated April 11, 2003, and in a BSPS investigation report dated April 12, 2003. A significant portion of the medicine was in Arias's Buick that day.

Page 258 Shuchu Hung's daughter saw the advisory about the counterfeit Procrit lot P007645 in the *Chinese World Journal*, March 12, 2003.

Pages 259–262 Through his lawyer Christopher Lyons, David Ebanks declined to comment.

Ebanks's first arrest is documented in an MDPD report dated April 9, 2003, and in an FDLE investigative report of the same day. His second arrest, on June 1, 2003, is documented in an FDLE investigative report of that day.

Investigators learned that Ebanks had been secretly using the identity of a patient at a drug rehabilitation clinic where he worked, as documented in FDLE investigative reports dated February 24, March 24, April 3, and May 2, 2003.

Page 263 Bonnie Basham, a lobbyist for the Healthcare Distribution Management Association, did not return several phone calls seeking comment.

E-mails between health department officials and wholesalers' lobbyists as they worked to revise proposed legislation were obtained through a public records request to the Florida health department.

Page 265 The Prescription Drug Protection Act (PDPA) was signed by Governor Jeb Bush on June 13, 2003. The PDPA imposed stricter regulations on wholesalers and increased penalties for violations, including a life sentence for any wholesaler whose medicine, if counterfeit, caused the death of a patient.

Chapter 21. Inspector Arias Goes to Washington

Page 267 The FDA released an alert on May 23, 2003, that Albers Medical Distributors was voluntarily recalling three lots of Lipitor. In testimony before the Department of Health and Human Services' Drug Importation Task Force on April 5, 2004, John Theriault, vice president of global security for Pfizer, recounted the recall of eighteen million tablets of Lipitor the previous spring. He said that previously the American people had believed that "counterfeits were distributed only by illicit brokers or unapproved pharmacies that proliferate on the Internet," and that developed countries like the United States were immune to the dangers of counterfeit drugs, a myth that was destroyed with the Lipitor recall.

In a July 14, 2003, letter from Representative Billy Tauzin, chairman of the Committee on Energy and Commerce's Subcommittee on Oversight and Investigations, to FDA Commissioner Mark McClellan, Tauzin recounted the conference call with Pfizer on June 9, in which Pfizer linked the counterfeit Lipitor to a Miami wholesale company.

Page 268 On June 5, 2003, the leaders of the House Committee on Energy and Commerce's Subcommittee on Oversight and Investigations sent a letter to FDA commissioner Mark McClellan. The letter described the backlogged state of the Miami mail facility, sought an explanation for the release of 1,233 unapproved boxes of generic Viagra, and requested a list of drugs shipped to, held at, and released from the facility since June 7, 2001. The information about the Miami facility was also contained in a memo sent by Representative Billy Tauzin to members of the oversight committee announcing the hearing, "A System Overwhelmed: The Avalanche of Imported, Counterfeit and Unapproved Drugs in the U.S.," on June 24, 2003.

Page 269 Evidence of Carlow's growing business and profits comes from Carlow's financial records, which were pulled from his trash by investigators and analyzed by the FDLE.

Pages 269–270 A trash pull on April 5, 2003, uncovered paperwork between Carlow's accounting firm and the IRS regarding $290,000 Carlow owed the IRS, as documented in an FDLE investigative report of that day. The note to the Bahamas wholesaler was discovered on March 22, 2003, as documented in an FDLE investigative report of that day.

Page 270 Albers's inclusion on the list for the 2001 Ernst & Young Entrepreneur of the Year was publicized in *The Business Journal of Kansas City*, May 25, 2001. Albers Medical Distributors has been a rotation practice site for pharmacy students since 1977, as stated on the company's Web site.

Doug Albers's lawyer, Cathy Dean, declined to answer any questions regarding her client or the ongoing investigation of his company.

During an interview with FDA investigators on March 12, 2003, an FDA agent asked Albers "why he gets so many counterfeit drugs coming through his company," as documented in an OCI interview memorandum of that date. Albers responded that he always checks whether a company is licensed to sell pharmaceuticals before making a purchase, but that he stopped buying drugs from companies in Florida because of the counterfeit drug problem, according to the memorandum. When asked by the FDA agent whether the price of the drug would indicate it's a counterfeit, Albers stated that he would not buy a drug that was 50 percent below the wholesale cost because it might be counterfeit.

Page 270 The growth in Albers's annual sales from $1 million to $47 million resulted from his expanding his community pharmacy into pharmaceutical distribution, as reported in *The Business Journal of Kansas City*, June 29, 2001.

Page 271 The Missouri pharmacy board, in a complaint filed on April 9, 2003, with the state's administrative hearing commission, cited numerous instances when inspectors found drugs at Albers that the company had bought from unlicensed, out-of-state distributors. In six visits from May 2000 to June 2002, inspectors found drugs purchased from up to forty-four unlicensed distributors. In its answer to the board on June 17, 2003, Albers denied all complaints against it.

Page 271 Documents sent from Albers to G&K specified that products were returned because they were damaged, short-dated, or had the wrong national drug code.

Page 271 Venema joined the FDA in Missouri to conduct a search warrant at Albers Medical on March 27, 2003. According to the FDLE report, investigators found records indicating that Albers had done business with Carlow's companies—BTC, Accucare, and G&K Pharma—and other companies under investigation in Florida. Photographs from the search show shipping records, invoices, and the proposal for a partnership with Cardinal.

Page 271 The jump in Carlow's sales to Albers from $3 million to $11 million a month is noted in an FDLE investigative report dated April 2, 2003.

Page 272 The repackaging company outside of Chicago, Local Repack in Richton Park, Illinois, recalled 7,189 bottles of counterfeit Lipitor in August 2003. Phil & Kathy's Inc, the parent company of Local Repack, sued the FDA in mid-August demanding the return of $2.5

million in medicine the FDA seized the month before. On April 13, 2004, Phil & Kathy's Inc signed a consent decree in the United States District Court for the Northern District of Illinois agreeing to comply with all FDA regulations. See David Schwab, "Another Firm Recalls Counterfeit Lipitor," *The Newark Star-Ledger,* August 8, 2003, and Susan Todd and David Schwab, "Lipitor Suit Shows How Medicine Enters the U.S.," *The Newark Star Ledger,* August 28, 2003.

Page 272 The FDA first announced that Albers was recalling three lots of Lipitor on May 23, 2003. On June 3, 2003, the FDA announced another recall involving more counterfeit Lipitor. Pfizer issued a statement that day, saying that it had reported its suspicions about the counterfeit medicine immediately to the FDA after receiving complaints from pharmacists and physicians about the bitter taste of the pills. The company also announced that it had no relationship with Med-Pro or Albers, the companies allegedly involved in the distribution of the counterfeit Lipitor.

Page 272 In response to the recalls, Rite Aid asked customers to stop taking the Lipitor dispensed between April 3 and May 23, 2003, as a precaution.

On October 1, 2003, a class action lawsuit on behalf of patients who received counterfeit Lipitor was filed against Rite Aid, Albers, and others in the superior court of New Jersey.

Lipitor is the top-selling medicine in the United States, with $6.8 billion in sales and 68 million prescriptions written in 2003, as reported by IMS Health, a health-care data collection company.

Page 273 In the mid-1980s Cruz was convicted of trafficking in cocaine in both state and federal courts, as documented in a criminal complaint filed on December 5, 2003, by FDA Special Agent Stephen M. Holt in *United States v. Julio Cesar Cruz* in Western Missouri District Court.

On November 5, 2003, an associate of Julio Cruz told investigators that he, Cruz, and two other men had come up with the Lipitor counterfeiting scheme while in federal prison.

Chapter 22. The Ultimate Box Case

Page 283 A transcript of Steven "Doc" Ivester's conversation with Carlow at the Quarterdeck Restaurant on May 23, 2003, was obtained. Ivester first approached investigators with information about Carlow in mid-May 2003, as documented in an FDLE investigative report of May 21, 2003.

On August 11, 2003, during Carlow's bail hearing in *State of Florida v. Michael Carlow and Candace Carlow* in Broward County, Ivester testified that Carlow had an affair with his girlfriend in May 2003, which led to his decision to talk to investigators.

On September 8, 2003, during Carlow's bail hearing, Venema testified about Carlow's involvement with an apparel company, Clarity Industries.

Page 284 In a trash pull on April 5, 2003, investigators found information on Carlow's investment in Navigator PC on a sheet of stockholder information for the company, as documented in an FDLE investigative report of that day.

Page 285 During Carlow's bail hearing on August 11, 2003, Jean McIntyre testified that "Carlow was going to open up a pharmaceutical company called World Pharma. I was going to operate it with him" once the investigation was resolved. She also admitted to having an affair with Carlow during the past year, saying their relationship had been "professional for many years. It has been personal for the last year."

Page 285 The big explosion to which Carlow is most likely referring happened little more than a month before his conversation with Ivester at the Quarterdeck. On March 27, 2003, FDA agents searched Albers Medical in Kansas City, Missouri. Albers Medical was one of Carlow's biggest clients, as documented in an FDLE investigative report dated April 2, 2003.

Page 287 The *Sun-Sentinel* also ran two articles by Bob LaMendola and Sally Kestin, "Fake Drugs Find Their Way into Rx supply," May 25, 2003, and "Former Convicts Try a Safer Venture: Pharmaceuticals," May 26, 2003.

Page 287 During Carlow's bail hearing on August 11, 2003, Ivester recounted Carlow's statement that investigators did not have anything on him since they had given whatever they had to the *Sun-Sentinel*.

Page 287 A copy of Carlow's "Offshore Wealth Preservation Planning Business Structure Diagram" was obtained by investigators, as documented in the FDLE investigative report of May 21, 2003. Ivester also testified that he found the document in a folder in the Navigator PC office.

Page 287 Investigators based their estimate—that $54 million worth of medicine passed through Carlow's many shell companies—on his own financial records.

Page 289 Bob Sparks, a spokesman for the Statewide Prosecutor's Office, said it has been the policy of the office to offer cases to the local state attorneys. "Many times they do decline and ask us to go ahead, and this was one of those times," Sparks said.

Page 291 The editorial that Arias was reading from appeared in the *Sun-Sentinel* on May 28, 2003, and was entitled "Crackdown Needed Now." In part the editorial blamed the health department for Florida's booming trade in adulterated medicine, claiming that the department was understaffed and had failed to enforce existing laws. "Someone has been asleep at the switch," the editorial stated.

Page 294 The inspector general's management review report for the period of March 18 to July 21, 2003, completed on September 3, 2003, stated that the "perceived negative work environment in the Bureau over the past few years has resulted in the loss of experienced staff, high turnover in some positions, and increased workload for some employees. Additionally, a lack of information flow back and forth and lack of trust in senior and mid-level leadership has contributed to employee dissatisfaction."

Six BSPS inspectors sent a letter to the deputy secretary of health, Annie Neasman, on September 11, 2003, which was obtained. They wrote: "We believe that the management of BSPS and their DO NOTHING POLICY bears much of the responsibility for not taking a leadership role to stem these illegal activities and PUT THE PUBLIC AT RISK to receiving RX drugs which were tainted in some manner." They added, "the field staff respectfully submits that it is extremely difficult, if not impossible, to work with the current management team."

PART FOUR

Chapter 23. The Rosetta Stone

Pages 299–300 From July 14 to July 17, 2003, the seventeenth statewide grand jury heard evidence of alleged racketeering by Michael Carlow and seventeen co-conspirators. The grand jury returned indictments on charges of racketeering and conspiracy to commit racketeering against Michael Carlow, Candace Carlow, Thomas Atkins, Marilyn Atkins, Henry García, Fabian Díaz, Joel De La Osa, David Ebanks, José Benitez, Dariel Tabares, Michael Burman, Ivan Villarchao, Lazaro Villarchao, Joseph Villaneuva, Arturo Godinez, Tom Martino, and Julio Cesar Cruz. Some suspects also were indicted on charges of grand theft, dealing in stolen property, sale or delivery of a controlled substance, and violations of Florida's health law, statute 499. Another person, Gisela Gonzalez, was charged with organized scheme to defraud. Mark Novosel was indicted separately on charges of organized scheme to defraud on June 6, 2003.

Through her lawyer, Christopher Pole, Candace Carlow declined to comment. Thomas Atkins's lawyer, Kevin Kulik, did not return phone calls seeking comment.

Grillo was charged with nineteen counts including organized scheme to defraud, purchase or receipt of a prescription drug from an unauthorized person, and possession with intent to distribute prescription drug. He pleaded not guilty. José Grillo's lawyer, José Quinon, declined to comment on his client's case.

The arrests took place on July 21, 2003. Bail amounts were documented in an FDLE investigative report of July 17, 2003. All of the suspects pleaded not guilty. The indictments and arrests were widely covered in the *Washington Post,* the *Miami Herald,* the *Sun-Sentinel,* and the *Record* (Bergen County, New Jersey). State Attorney General Charlie Crist and FDLE Assistant Special Agent in Charge Michael Mann also were interviewed on National Public Radio's *All Things Considered* on July 22.

Neither Michael Carlow nor José Grillo responded to letters sent to them in jail requesting interviews. Michael Carlow's lawyer, John R. Howes, said that his client did not want to discuss the charges or any other matters. Howes wrote in an e-mail that Carlow "wants these frivolous charges disposed of so that he can get on with his life."

Page 302 For information on Cruz's earlier conviction on cocaine charges, see endnote page 273.

Roy Kahn, Julio Cruz's lawyer, said that his client had no relationship with Michael Carlow except that Albers Medical Distributors in Kansas City had funneled money to Cruz through Carlow's companies to pay Cruz for the foreign Lipitor that he was bringing into the country.

Kahn further said that while Cruz had pleaded not guilty, he was cooperating with federal authorities as the main witness in an ongoing investigation into Albers. For information on Albers's response, see endnote page 270.

Pages 308–310 On July 23, 2003, investigators interviewed Gina Catapano, as documented in an FDLE investigative report of that day. She said she had known Carlow for eight years. She began doing secretarial work for him in early 2002 and created invoices for drugs on G&K Pharma and Accucare letterhead. Catapano stated that she did not know that Carlow might have sold medicine illegally until she saw a newspaper article linking him to pharmaceutical crimes. When she confronted him, he denied the allegations, Catapano told investigators. Investigators' efforts to retrieve Catapano's files is also documented in an FDLE investigative report of that day.

Catapano was never charged with any wrongdoing.

Page 310 Michael Carlow's bail hearing took place on July 28, 2003, at the Broward County courthouse in Fort Lauderdale. A second hearing to determine the Carlows' assets took place on September 8, 2003.

Page 311 A number of lawsuits followed after the Carlows offered the mortgage on their ranch house as collateral in order to make bail. The bail bond firm's underwriters, the American Bankers Insurance Company of Florida, sued the Carlows on October 22, 2003, in Broward County Circuit Court to foreclose the mortgage deed on their property. A subsequent lawsuit in January 2004 filed by Marvin and Wanda Perkins against the Carlows, the Bail Bond Firm LLC, and the American Bankers Insurance Company laid claim to proceeds from the foreclosed mortgage.

Chapter 24. A Wink and a Nod in Las Vegas

Page 312 The first hearing for *Nevada State Board of Pharmacy v. Dutchess Business Services and Legend Pharmaceuticals* was held October 14-16, 2003. The hearing resumed January 14-15, 2004. Sixteen witnesses testified.

Page 313 The National Pharmaceutical Wholesale Information Group's Pharmaceutical Regulatory Compliance Seminar took place on October 16, 2003, at the Monte Carlo Resort & Casino in Las Vegas, Nevada. Topics of discussion at the wholesalers' meeting included "The Florida Experience: What It Means For You" and "Diversion: What It Is and What It Isn't." A notice informed participants that the FDA and the HDMA had withdrawn from the conference.

Pages 318 The Pharmaceutical Security Institute, a research organization funded by drugmakers, documented counterfeiting in foreign and domestic markets in a fifty-five-page annual internal report for its members, the "PSI 2003 Situation Report." The author obtained a copy of the report.

Page 319 Pfizer stated in a December 11, 2003, press release that it would cut off business to wholesalers that sold its drugs to unlicensed wholesalers. Eli Lilly announced that it had halted sales to five wholesalers because they were caught buying a counterfeit Lilly drug from secondary wholesalers. See Jeff Swiatek, "Lilly Cuts Off Five Wholesalers; Indianapolis Company Cites Purchases of Counterfeit Drugs," *The Indianapolis Star,* December 12, 2003.

Page 320 For more information about Nevada's tightened regulations and the resulting drop in wholesale companies, see Katherine Eban,

"Pharmacy Fakes," *Self,* March 2003; and Mary Pat Flaherty and Gilbert Gaul, "Nevada Gets Tough, With Mixed Results," *The Washington Post,* October 22, 2003.

Page 320 Four Nevada wholesalers filed a lawsuit against regulators Louis Ling and Keith Macdonald in United States District Court in Nevada on March 13, 2003.

Page 321 The notice of intended action and accusation in *Nevada State Board of Pharmacy v. Dutchess Business Services and Legend Pharmaceuticals* was filed on August 21, 2003.

Page 322 On March 21, 2003, the United States District Court in Nevada granted the wholesalers a temporary restraining order against Ling and Macdonald that prohibited the regulators from disclosing further information about the wholesalers to those in the industry. The court granted the wholesalers a second temporary restraining order against Ling and Macdonald on the day of the hearing that barred them from participating in it.

Page 322 Paul DeBree made his son-in-law Benjamin Ross a principal in his company Las Vegas Pharmaceutical Distributors (LVPD), as described in the notice of intended action and accusation filed in *Nevada State Board of Pharmacy v. Dutchess Business Services and Legend Pharmaceuticals.* From November 1996 to April 1998, pharmacies illegally resold deeply discounted drugs to LVPD. In turn, the company sold over $4 million of the medicine to Bindley Western. Ross pleaded guilty to conspiracy on July 5, 2003, in the Federal District Court in Nevada.

Page 325 On January 31, 2001, Robert Lynn and Kelly Burke—two AIDS patients who received counterfeit Serostim—filed a class action lawsuit against Serono, McKesson, Chronimed Holdings, and Procare Pharmacy of Berkeley in San Diego County Superior Court. The California lawsuit stated that Lynn and Burke had injected a counterfeit drug and had been deprived of the beneficial effects of the Serostim. The case was settled in the spring of 2002, although none of the defendants admitted liability. See Katherine Eban, "Pharmacy Fakes," *Self,* March 2003.

Page 325 On October 22, 2003, Jesús Benitez and Maria Castro of J&M Pharmacare were arrested and charged with organized scheme to defraud and unauthorized sale of drugs, as documented in an MDPD arrest warrant.

Page 326 In his plea agreement for *United States v. Per Loyning* in the Miami-Dade Circuit Court, Bill Walker, also known as Per Loyning, pleaded guilty to seventeen charges including racketeering, fraudulent use of personal identification, and violations of the health statute 499.

Page 328 On February 5, 2004, the Nevada pharmacy board fined Dutchess Business Services $1 million and Legend Pharmaceuticals $371,000 and revoked the licenses of both companies. The board found that the companies purchased and resold adulterated and misbranded drugs, created and provided false pedigrees, and did business with unlicensed wholesalers.

Chapter 25. The Education of Kevin Fagan

Page 331 Eli Lilly notified health-care professionals on May 4, 2002, that Zyprexa tablets had been removed and replaced with tablets stamped "aspirin." The FDA sent out an alert about counterfeit polypropylene mesh used in hernia repairs on December 19, 2003. A wholesaler in Tennessee recalled adulterated Risperdal in May 2004, according to an FDA enforcement report from August 25, 2004. A secondary wholesaler in California recalled two lots of counterfeit Kaletra, according to an FDA enforcement report on November 26, 2003.

Page 331 On September 9, 2004, in an undercover operation titled "Operation Pill Collector," the FBI broke up a drug ring that allegedly trafficked in diverted medicine valued at $56 million. Federal and state agents in New Jersey, New York, Tennessee, and Georgia arrested seventeen suspects. See Jason George, "FBI Says Arrests Disrupted Trade in Stolen or Expired Medicine," *The New York Times,* September 10, 2004 and John P. Martin and Susan Todd, "Feds Crack Prescription Drug Network," *The Star-Ledger,* September 9, 2004.

Page 331 On December 4, 2003, the man accused of counterfeiting Lipitor, Julio Cruz, told investigators of warehouses in both Costa Rica and Florida that contained huge amounts of counterfeit, adulterated, and authentic medicine that was to be intermingled and shipped into the American market.

Pages 332–333 On February 17, 2004, the FDA stayed the final rule on pedigree papers for the fifth time until December 1, 2006.

Page 333 After counterfeits surfaced, middlemen began suing each other to assign blame and recoup lost money. The lawsuits began with a complaint filed in *Bergen Brunswig v. Dialysist West* in United States District Court in Arizona on August 2, 2002. Third-party complaints soon involved four other companies: AmeRx Pharmaceutical run by Susan Cavalieri; CSG Distributors, run by Joe Parisi, who was arrested in September 2004 during Operation Pill Collector; Optia Medical, a company whose license Mark Novosel used; and Premier Medical, run by James Suozzo, who cooperated with Florida investigators.

On August 19, 2003, Grillo spoke briefly with investigators about his counterfeiting operation, but did not fully cooperate.

As documented in an FDLE investigative report dated August 9, 2002, Armando Rodriguez told investigators he sold Grillo up to four hundred boxes a week of low-dose Epogen that he obtained from J&M Pharmacare. Sales records from Amerisource and Cardinal show that J&M Pharmacare purchased 3,363 boxes of low-dose Epogen from Amerisource and 8,931 boxes from Cardinal between January 2001 and May 2002.

Pages 333–334 In a statement given to investigators on September 24, 2003, Silvino Cristobal Morales described how he relabeled the vials of medicine for Grillo.

Page 334 On June 11, 2003, the Pharmaceutical Research and Manufacturers of America, the drugmakers' principal trade group, urged the FDA to implement pedigree regulations immediately.

Page 334 On October 2, 2003, the FDA released the Counterfeit Drug Task Force Interim Report to the public. The report stated that "for an electronic pedigree to become universally adopted, industry or national standards would have to be developed and implemented."

Page 335 The conference entitled "Supply Chain Integrity: The Industry Conference Addressing the Competency and Security of the Pharmaceutical Supply Chain" was held September 24–26, 2003, and sponsored by the Institute for International Research.

Page 336 According to an August 24, 2003, Stone Cold update, Grillo told investigators he sold three hundred boxes of high-dose Epogen to Nicholas Just and Paul Perito at Playpen South.

Pedigree papers for medicine that AmeRx sold to Dialysist West show that Epogen from lot P002970 originated at Carlos Luis's company, Medix International, and was sold to Eddie Mor's company, Express Rx.

According to an FDLE investigative report on Eddie Mor, he bought $2.1 million worth of drugs from Luis from February 14 to April 30, 2002.

Investigators obtained records of sales of medicine from Eddie Mor to AmeRx during a search of Mor's home, as documented in an FDLE investigative report of June 14, 2002, and an FDLE investigative report dated August 14, 2002.

Information on Luis's business relationship with Just and Perito and their introduction to "Tony" was related by the assistant manager, as documented in the FDLE investigative report of April 22, 2004.

Pages 336–337 On February 18, 2004, the FDA released its final report on drug counterfeiting titled "Combating Counterfeit Drugs." The report said the FDA had decided to stay paper pedigree requirements once again because track-and-trace technology could be implemented by 2007.

The author obtained a copy of the twenty-five-page draft document, "Counterfeit Drug Task Force Interim Report," dated September 5, 2003. When asked about the draft report, FDA officials declined to comment on its contents. Paul Rudolf, Senior Advisor for Medical and Health Policy, said "any report goes through drafts." William Hubbard, Senior Associate Commissioner for Policy, Planning and Legislation, explained: "We tend to write hundreds of pages that no one will read. The final report just has our recommendations and a discussion of our recommendations."

Pages 338–339 Information on Ricciardi and Purity Wholesale's political contributions were compiled from public campaign records.

The voluntary distribution integrity pledges signed by members of the wholesalers' trade groups went into effect on July 1, 2003. The associations pledged that member companies would notify the FDA and drugmakers of any suspicious products within five days of receiving them.

The list of members of the Pharmaceutical Distributors Association was obtained through a Freedom of Information Act request to the FDA. The member companies are Associated Medical Distributors, Inc.; Columbia Medical Distributors; AK Medical Supply Co., Inc.; Chicago Medical Equipment and Supply; J M Corporation; LAL Consultants Group; JAM Pharmaceutical; High Country Medical; Grand Canyon Medical Enterprises; Expert-Med, Inc.; Drugmax, Inc.; DIT Healthcare Distribution, Inc.; Advance Medical Sales; MC Distributors; MedSource Direct; Michigan RX Brokerage, LLC; National Pharmaceutical, Ltd.; PDI Enterprises, Inc.; Parke Medical Supply; Priority Pharmaceuticals; Purity Wholesale Grocers, Inc.; R & S Sales, LLC; Rx Drug Services; Rebel Distributors Corp.; Resource Healthcare, Inc.; South Pointe Wholesale, Inc.; and Wise Choice Health Care.

Page 339 On October 15, 2003, QK Healthcare sued Valley Drug, accusing the company of selling counterfeit Lipitor. Valley Drug is a division of Drugmax, a PDA member. While Valley Drug said it had gotten the pills directly from Pfizer, Drugmax president Bill Lagamba said the Lipitor had been purchased from Alliance Wholesale in Chicago. See Jeff Harrington, "Drug Seller Sued Over Distributing Phony Pills," *St. Petersburg Times*, October 16, 2003.

Page 339 The California Attorney General subpoenaed records from

Advanced Medical Sales in an investigation into price gouging of flu vaccines. See "Local Flu Vaccine Distributor Named in Price Probe," *San Luis Obispo Tribune*, October 26, 2004.

Page 339 Banking and pedigree records show that four PDA members— R&S Sales, Rebel Distributors, PDI, and Grand Canyon Medical— did business with Dutchess Business Services, according to the notice of intended action and accusation in *Nevada State Board of Pharmacy vs. Dutchess Business Services and Legend Pharmaceuticals* filed on August 21, 2003.

Page 340 According to an investigative summary prepared by the FDLE, Suozzo said that his company, Premier Medical Group, had handled only high-dose Epogen and Procrit. All of it had come from an unlicensed middleman, YW Consultants, which had gotten it from another unlicensed middleman, Double J Consultants. He said no pedigree papers had existed for the medicine, and that he sold it all to CSG Distributors in Knoxville, Tennessee. A schedule of Epogen sales prepared by Dialysist West show that the company had bought Epogen from CSG Distributors.

Page 341 On August 4, 2004, the Securities and Exchange Commission announced that Bristol-Myers Squibb agreed to pay $150 million for perpetuating a "fraudulent earnings management scheme" by selling large amounts of products to its largest wholesalers ahead of demand to boost revenue. Bristol-Myers Squibb agreed to the payment without admitting or denying the allegations.

Page 342 GlaxoSmithKline's decision to sell different-colored tablets as an anti-diversion tactic is noted in Lew Kontnik's "2003 Year End Pharmaceutical AntiCounterfeiting Insight Report and Forecast." Kontnik is a pharmaceutical anti-counterfeiting specialist.

Pamela Williamson-Joyce testified about Serono's new pricing strategy and sales policy for Serostim in *Nevada State Board of Pharmacy v. Dutchess Business Services and Legend Pharmaceuticals*. See also Katherine Eban, "Pharmacy Fakes," *Self,* March 2003.

Page 342 On January 22, 2003, Ortho Biotech sent out a notice about new tamper-proof packaging for Procrit and warned that wholesalers who bought the company's products from any other source would be terminated as authorized distributors. For more on speculative buying, see Stephanie Saul's "Making a Fortune by Wagering that Drug Prices Tend to Rise," *New York Times*, January 26, 2005.

Page 343 The FDA's public meeting of its anti-counterfeit drug initiative was held on October 15, 2003. Attorney Eric Turkewitz was one of more than seventy people who spoke that day.

Page 345 On July 29, 2004, Judge Sandra Feuerstein granted Amgen's motion to dismiss the company from the lawsuit, but denied similar petitions by Amerisource and CVS, as stated in the opinion and order in *Timothy Fagan v. Amerisource Bergen, Amgen, CVS Corporation and Procare Pharmacy* filed in United States District Court, Eastern District of New York.

Page 345 See Ilisa Bernstein, Pharm. D., J.D., and Paul Rudolf, M.D., J.D., "Counterfeit Drugs," *The New England Journal of Medicine*, April 1, 2004.

Page 346 Figures for sales and prescriptions were taken from top-line industry data, as reported by IMS Health, a health-care data collection company.

Page 346 See Gardiner Harris, "Tiny Antennas to Keep Tabs on U.S. Drugs," *The New York Times*, November 15, 2004.

Epilogue

Pages 348–349 On May 26, 2004, the Seventeenth Statewide Grand Jury returned indictments against Carlos Luis, Javier Rodriguez, Eddie Mor, Edward Safile, Adalberto Hernandez, Paul Perito, Nicholas Just, and José Antonio Grillo. The amended indictment included charges of racketeering, conspiracy to commit racketeering, organized scheme to defraud, and Medicaid fraud. They all pleaded not guilty.

Just and Perito were released on bail, and Perito's license to practice medicine was reinstated while he awaited trial. Through their lawyers, both men vigorously denied being involved in the sale of counterfeit drugs.

Rodriguez's lawyers did not return calls seeking comment. Luis's lawyer, Edward O'Donnell, said the government "jumped the gun" on its investigation. For statements by Mor's lawyer, see endnote page 153.

Page 350 Investigators documented the medicine and cash they confiscated in a chart, "Operation Stone Cold, Seizures to Date." Figures for the statewide IVIG Medicaid payments come from a report investigators submitted to the state health department in summer 2004.

Page 351 On August 5, 2004, Candace Carlow filed for Chapter 11 bankruptcy in the U.S. Bankruptcy Court, Southern District of Florida.

Page 351 Grillo's wife met with investigators in July 2004 and said that when her husband returned from a night at Playpen South, he often had a wad of cash of $25,000. Though he took her to meetings, she told investigators she rarely witnessed his transactions.

Pages 352–353 In financial documents filed with the SEC, Cardinal disclosed that both the SEC and the U.S. Attorney for the Southern

District of New York were investigating its classification of revenue in the company's pharmaceutical distribution business. On July 28, 2004, the Associated Press reported the two investigations and the departure of the company's chief financial officer. The quote comes from a statement released by the chief financial officer at the time.

Page 353 On November 21, 2003, a federal grand jury in the Western District of Missouri indicted Diana S. Coelyn, an executive at the wholesaler H.D. Smith in Springfield, Illinois, and Christopher Wayne Lamoreaux, the president and chief executive officer of NuCare Pharmaceuticals Inc., a drug repackager in Anaheim, California.

In one indictment, Coelyn was charged with mail and wire fraud for allegedly directing Albers to pay "secret commissions to her" in excess of $400,000 after she arranged for H.D. Smith to buy Lipitor and other medicine. Lamoreaux was charged with mail fraud. Investigators alleged that he directed Albers to pay $115,000 in secret commissions to a consulting firm he owned. See Dan Margolies, "Charges Brought in Lipitor Scheme," *Kansas City Star*, November 25, 2003.

On July 15, 2004, Lamoreux was found guilty of two counts of mail fraud. He was sentenced to twenty-one months in federal prison and three years of supervised probation, according to the FDA and the United States Attorney's office for the Western District of Missouri.

On November 19, 2003, after being sued by Pfizer Ireland Pharmaceuticals and others, Albers Medical Distributors Inc. filed a third-party complaint in U.S. District Court accusing various companies of breach of contract, unjust enrichment, negligence, fraud, and negligent misrepresentation. The defendants included OTS Sales, Inc.; Paul Krieger and Diana Coelyn; Carlow's companies G&K Pharma LLC and Complete Wholesale Distribution; Julio Cruz's wholesale company Pharma Medical LLC; and Robert Chaille's drug wholesale company, New Horizons Network.

Doug Albers's attorney, Cathy Dean, told *The Kansas City Star* in an article on November 20, 2003, "Obviously, Albers was misled and victimized by these defendants." On April 29, 2004, Albers withdrew its application to renew its permit with Florida's Bureau of Statewide Pharmaceutical Services.

Page 353 Investigators reported that they seized drugs valued at $15.2 million from a Miami warehouse on December 2, 2003. Investigators also uncovered receipts that Julio Cruz, using the alias Rick García, kept for pill-making equipment that he purchased from Kirby Lester Inc. For more on the scheme, see Susan Todd, "How Fake Lipitor Was Sold," *The Star-Ledger* (Newark), December 20, 2003.

Julio Cruz entered into a plea agreement on January 22, 2005, with the U.S. Attorney for the Western District of Missouri, in which he pleaded guilty to three counts of conspiracy to sell, and selling, counterfeit Lipitor.

Page 353 The Florida Department of Law Enforcement nominated Operation Stone Cold for a Davis Productivity Award in November 2003. In its application, the department estimated that the agents involved had put in 9,165 hours and had "literally saved an unknown number of lives of Florida citizens by seizing $17,834,471 in counterfeit Rx drugs."

Page 354 The author obtained copies of the resignation letters of drug inspectors Jim Gerber and Deborah A. Orr, e-mail correspondence of a third inspector who was forced to resign, and the letter the six inspectors wrote to Annie Neasman, Deputy Secretary of Health, on September 11, 2003.

Page 355 The reports by the inspector general for the Florida health department include the management review report for March 18, 2003–July 21, 2003 and the pharmaceutical bureau's internal audit, inspections, investigations, and monitoring report for July 1, 2001–November 30, 2003.

Page 355 In May 2004, the Tallahassee office of BSPS advertised for the job that Cesar Arias held as the manager of the Miami field office. The job was described as "work planning and managing, or coordinating the major operations of an agency. Duties and responsibilities include managing the daily operations and planning the use of resources."

Page 356 On November 12, 2004, the Florida Department of Health issued a press release announcing Jerry Hill's appointment to the National Drug Advisory Coalition.

Afterword

Additional interviews for the afterword not listed in the "Interviews of Note" section include:

FFF ENTERPRISES
Patrick M. Schmidt, CEO

INDIANA BOARD OF PHARMACY
Joshua Bolin, former Director

MORRIS & DICKSON
Markham (Skipper) Allen Dickson Jr., President
Markham Allen Dickson Sr., former President

NATIONAL ASSOCIATION OF BOARDS OF PHARMACY
Eleni Anagnostiadis, Professional Affairs Director

Page 358 The decisions by Cardinal Health Inc., AmerisourceBergen Corporation, and McKesson Corporation to stop buying brand-name pharmaceuticals from the secondary market have been widely documented. On September 22, 2005, Amerisource issued a news release announcing its decision to stop buying pharmaceuticals from the secondary market. Cardinal announced its decision internally to its employees in May 2005. McKesson informed Kevin Fagan of its decision in an e-mail on November 11, 2005.

For more information, see: Heather Won Tesoriero, "Cardinal Health Ends Drug Trading," *The Wall Street Journal*, May 6, 2005; Melissa Davis, "AmerisourceBergen Dims Gray Market for Drugs, TheStreet.com, September 26, 2005; Heather Won Tesoriero, "McKesson Takes Step on Curbing Fake Drugs," *The Wall Street Journal*, November 14, 2005.

The CVS pharmacy chain announced in a corporate statement on May 23, 2005, that it would no longer buy pharmaceuticals from wholesalers who traded drugs in the secondary market. For more on this decision, see Julie Appleby, "CVS gets Pickier on Which Drug Suppliers to Use," *USA Today*, May 24, 2005.

Page 359 On June 21, 2005, the CEO of Kinray, Stewart Rahr, sent the first in a number of reports to his customers entitled "Dangerous Criminals/Poison Doses, Part I." The report states, "The government's ongoing investigation into the counterfeit Lipitor scheme continues to reveal the identities and roles of those involved. They are a collection of low-life criminals."

Pages 360–361 Interdepartmental memos show that on February 9, 2005, Cesar Arias was notified in writing by the Bureau of Statewide Pharmaceutical Services that he was being placed on administrative leave subject to the outcome of an investigation. The letter did not state the reason for his suspension. But it requested that he turn in all state equipment and not use his office. The letter also stated, "You shall not initiate any contact with departmental staff unless requested to do so as part of this investigation."

On March 28, 2005, a letter he received from the bureau's chief, Jerry Hill, stated that the health department, the bureau's parent agency, was considering dismissing him for charges including: insubordination; violation of law or agency rules; and conduct unbecoming a public employee.

Arias retained a lawyer and threatened to sue the bureau for retaliating against him as a whistleblower. The bureau quickly dropped its dismissal proceedings. On May 3, a settlement agreement stipulated that

Arias would resign immediately, and be paid $26,000, plus $23,000 in attorney fees, to settle the matter.

Pages 362–363 The investigative report by the inspector general, HIG 05-169, covered the review period of September 14, 2005, to October 31, 2005, and was completed on November 7, 2005.

On March 23, 2005, a federal grand jury in Savannah, Georgia, indicted Marty Bradley, his father, six of their associates, Bradley's corporation Bio-Med Plus, and an associated Georgia company, Interland Associates, on 288 counts of racketeering activities.

The sweeping indictment alleged that the defendants had engaged in a multistate scheme to fraudulently acquire blood products and defraud Medicaid. All those indicted pleaded not guilty and prepared for a criminal trial in Savannah, Georgia, scheduled for February 2006.

Page 364 Neil Spence did not respond to several requests for comment, made by phone, through e-mail, and by certified letter.

Page 365 On August 21, 2005, a federal grand jury in Kansas City, Missouri, indicted Douglas Albers, Michael Carlow, and nine others, as well as three businesses—Albers Medical Distributors, H.D. Smith Wholesale Drug Company, and MedPro Inc.—for allegedly conspiring to sell counterfeit, illegally imported, and misbranded Lipitor. The indictment also states that some of the individuals and corporations additionally trafficked in other illegally imported and diverted drugs, including Coumadin, Celebrex, and Bextra.

Pages 370–371 Events surrounding the counterfeit flu vaccine dispensed at an ExxonMobil health fair on October 21, 2005, are described in several documents: an affidavit filed by Kevin Lammons, special agent with the Federal Bureau of Investigation, in the U.S. Southern District of Texas court; a fourteen-count federal indictment by the U.S. Attorney's Office in the Southern District of Texas against Dr. Iyad Abu El Hawa and Martha Denise Gonzales, a private doctor and his colleague, who were involved in acquiring and administering the vaccines; and an internal memo by ExxonMobil executives to its employees on November 4, 2005, entitled "Update #9—Flu Vaccinations."

Glossary

Adulterated: A drug that has been compromised through improper storage, handling, or transport, or is of unknown or improperly documented origin.

ADR: An authorized distributor of record is a wholesaler that purchases drugs directly from manufacturers in a regular and ongoing relationship.

Bad Medicine: A general term for medicine that is adulterated, counterfeit, diverted, or otherwise cannot be guaranteed as safe.

Big Three: The nation's three biggest drug wholesalers are Cardinal Health Inc., AmerisourceBergen Corp., and McKesson Corp. They sell 90 percent of the medicine that moves through the wholesale market.

BSPS: The Bureau of Statewide Pharmaceutical Services, a division of Florida's Department of Health, regulates drug, medical device, and cosmetic manufacturers, prescription drug wholesalers, and oxygen and ether retailers.

Closed-door pharmacy: A pharmacy that receives deeply discounted drugs from manufacturers on the condition that it dispenses them only to patients in medical institutions such as nursing homes.

Clotting factor: A human protein in the blood that interacts with platelets to help blood clot.

Cold Chain: A protected path of temperature control and careful handling designed to ensure the integrity of a drug as it moves from manufacturer to patient.

Counterfeit: A drug that has been deliberately and fraudulently altered or mislabeled and may contain no active ingredient, a diluted active ingredient, or the wrong active ingredient.

Diverted: Drugs that middlemen often obtain through fraud or misrepresentation outside of intended distribution channels. The drugs are generally sold below market rate and accompanied by incorrect pedigree papers, obscuring their origin and making their purity impossible to guarantee.

Epoetin alfa: A synthetic version of a human protein, erythropoietin, that boosts red blood production for patients suffering from kidney failure and those who have undergone organ transplants or cancer treatments. It was developed by Amgen Inc., and is sold under the brand names Epogen and Procrit.

FDA: The U.S. Food and Drug Administration is the federal agency that regulates the safety of the nation's food and medicine.

FDA OCI: The FDA's Office of Criminal Investigations enforces laws created to ensure the safety of America's food and medicine.

FDLE: The Florida Department of Law Enforcement has statewide jurisdiction and is the state's lead police agency.

Gray Market: The loosely regulated buying and selling of discounted drugs between wholesalers.

HDMA: The Healthcare Distribution Management Association is the leading trade group for drug wholesalers, including the Big Three.

Immune globulin intravenous (IGIV): A solution made from human blood plasma that contains antibodies to help protect the body from disease when a patient's own immune system is weak. It is sold under the brand names Gamimune, Gammagard, Iveegam, and Panglobulin.

Lipitor: Helps to lower cholesterol. Made by Pfizer Inc., this is the world's best selling drug.

Lot number: A unique number assigned to each production run of a given drug. Particularly useful in the event of a recall.

MFCU: The Medicaid Fraud Control Unit within the Florida Attorney General's Office investigates fraudulent Medicaid billings.

Neupogen: A synthetic version of a natural substance that increases white blood cell growth for patients undergoing cancer treatments or bone marrow transplants. It is manufactured by Amgen.

NovoSeven: A synthetic protein that acts as a blood-clotting agent for patients suffering from hemophilia. It is manufactured by Novo Nordisk Inc.

PDA: The Pharmaceutical Distributors Association is a trade group representing secondary wholesalers.

PDMA: The federal Prescription Drug Marketing Act of 1987 aimed to end the diversion of medicine by restricting the resale of pharmaceuticals by hospitals, pharmacies, and other end users and prohibiting the sale of drug samples.

PDPA: Florida's Prescription Drug Protection Act became law in June 2003. It imposed new restrictions on drug wholesalers and stricter criminal penalties for those who traffic in adulterated medicine.

Pedigree paper: A document listing each previous buyer and seller of a drug. It is intended to function as an audit trail and proof of a drug's origin.

Retrovir: Used to treat infections caused by HIV and manufactured by GlaxoSmithKline.

Secondary wholesaler: A local or regional wholesaler that typically stocks a limited variety of medicine and buys from other wholesalers rather than from manufacturers.

Somatropin: A synthetic version of human growth hormone used to treat dwarfism and kidney disease in children, as well as AIDS wasting syndrome in adults. It is sold under the brand names Nutropin and Serostim.

U/mL: Units per milliliter is used to denote the concentration of medicine in a milliliter of suspension liquid.

Uplabeled: A drug that has been fraudulently relabeled to appear stronger than it actually is and is sold at a higher cost.

WAC: The wholesale acquisition cost is the initial listed drug price set by the manufacturers.

Zyprexa: Used to treat psychosis in patients suffering from schizophrenia and bipolar disorders. It is manufactured by Eli Lilly and Co.

Interviews
of Note

FLORIDA

Government

DEPARTMENT OF HEALTH

John O. Agwunobi, Secretary of Health

Annie Neasman, Deputy Secretary

Robert Daniti, General Counsel

Lindsay Hodges, Information Officer

Jerry Hill, Chief, Bureau of Statewide Pharmaceutical Services

Gregg Jones, Pharmaceutical Program Manager, Bureau of Statewide Pharmaceutical Services

Sandra Stovall, Compliance Officer, Bureau of Statewide Pharmaceutical Services

Cesar Arias, Drug Inspector Supervisor, Bureau of Statewide Pharmaceutical Services

Robert Loudis, former Drug Inspector, Bureau of Statewide Pharmaceutical Services

Joseph Nycz, former Drug Inspector, Bureau of Statewide Pharmaceutical Services

Gene Odin, Drug Inspector, Bureau of Statewide Pharmaceutical Services

Deborah Orr, former Drug Inspector, Bureau of Statewide Pharmaceutical Services

ATTORNEY GENERAL'S OFFICE

Melanie Ann Hines, former Statewide Prosecutor

Stephanie Feldman, former Assistant Statewide Prosecutor

Oscar Gelpi, Assistant Statewide Prosecutor

Robert Penezic, former Assistant Statewide Prosecutor

Jack Calvar, Senior Investigator, Medicaid Fraud Control
Unit, Miami Bureau

FLORIDA DEPARTMENT OF LAW ENFORCEMENT

Tim Moore, former Commissioner

Amos Rojas, Jr., Regional Director

Michael Mann, Assistant Special Agent in Charge

Gary Venema, Special Agent

Paige Patterson-Hughes, Public Information Officer

MIAMI-DADE POLICE DEPARTMENT

Alan Mandelbloom, former Captain

John Petri, Sergeant, Career Criminal Unit

Randy Jones, former Detective, Career Criminal Unit

Steve Zimmerman, Detective, Career Criminal Unit

Richard Trujillo, Detective, Police Operations Bureau

BROWARD COUNTY SHERIFF'S OFFICE

Lisa McElhaney, Sergeant

AGENCY FOR HEALTH CARE ADMINISTRATION

Rhonda Medows, former Secretary

SENATE COMMITTEE ON HEALTH, AGING
AND LONG TERM CARE

Barry Munro, Attorney

Gail Tedder, Staff Assistant

STATE SENATE

Senator Walter "Skip" Campbell Jr. (D-Tamarac)

Senator Durrell Peadon (R-Crestview)

Roger Robinson, Assistant, Senator Peadon

HOUSE OF REPRESENTATIVES

Representative Ed Homan (R-Temple Terrace)

FEDERAL

Food & Drug Administration

OFFICE OF THE COMMISSIONER

William Hubbard, Associate Commissioner, Office of Policy and Planning

OFFICE OF EXTERNAL RELATIONS

Larry Bachorik, Assistant Commissioner for Public Affairs, Office of Public Affairs

Jason Brodsky, Public Affairs Specialist, Office of Public Affairs

OFFICE OF POLICY AND PLANNING

Ilisa Bernstein, Pharm. D, J.D., Senior Science Policy Advisor

Paul Rudolf, M.D., J.D., Senior Advisor for Medical and Health Policy

OFFICE OF REGULATORY AFFAIRS

John M. Taylor III, Esq., Associate Commissioner for Regulatory Affairs

Benjamin England, former Regulatory Counsel to the Associate Commissioner for Regulatory Affairs

OFFICE OF CRIMINAL INVESTIGATIONS

Terrell L. Vermillion, Director

James Dahl, Assistant Director, Special Programs

Doug Fabel, former Special Agent in Charge, Miami Field Office

Tom Keeley, former Special Agent in Charge, San Diego Field Office

Kent Walker, former Special Agent in Charge, Miami Field Office

U.S. House of Representatives

Representative Steven Israel (D-NY)

HOUSE ENERGY AND COMMERCE COMMITTEE
HOUSE ENERGY AND COMMERCE COMMITTEE
    Chris Knauer, Investigator
    David Nelson, Economist

U.S. Customs and Border Protection

    Vincent Hauck, Chief Inspector
    Ralph Pisani, Assistant Chief Inspector
    Janet Rapaport, Director, Office of Public Affairs

Department of Justice

    Jim Sheehan, Associate United States Attorney, Eastern District, Pennsylvania

    Gale McKenzie, Assistant United States Attorney, Northern District of Georgia

U.S. District Court

    Joel Feldman, Magistrate Judge, Northern District of Georgia

Internal Revenue Service

    Janice Weller, Case Agent

MANUFACTURERS

Amgen

    Michael Beckerich, Associate Director, External Communications

Eli Lilly

    Ed Sagebiel, Corporate Communications Manager

Ortho Biotech

    Doug Arbesfeld, Vice President of Public Relations
    Elizabeth Hansen, Packaging Specialist
    Mark Wolfe, Director of Public Affairs

Pfizer

> Bryant Haskins, Director of Corporate Media Relations

Purdue Pharma

> J. Aaron Graham, Vice President, Corporate Security

Serono

> Carolyn Castel, Public Relations Director

TAP Pharmaceuticals

> Barbara Tolbert, Manager, Customer Service/Trade
> Relations

PATIENTS

Counterfeit Epogen

> Timothy Fagan, patient
>
> Jeanne Fagan, mother
>
> Kevin Fagan, father
>
> Lewis Teperman, M.D. Director of Transplantation, New
> York University Medical Center
>
> Eric Turkewitz, Attorney for the Fagans
>
> Gary Lindeman, Pharmacy Manager, Dickinson Memorial
> Hospital

Counterfeit Lipitor

> Pat Rocco, Attorney, Shalov Stone & Bonner
>
> Thomas Sobol, Attorney, Hagens Berman

Counterfeit Nutropin AQ

> Richard Cherlin, M.D.
>
> Annabelle Foo, Pharmacy Manager, Bay Area IV Therapy

Counterfeit Procrit

> Ed Blount, husband of patient, Maxine Blount
>
> Max Butler, brother of Maxine Blount

Patti Silvey, friend of Maxine Blount

Don Downing, Attorney, Stinson Morrison Hecker, represents the Blounts

Gretchen Garrison, Attorney, Stinson Morrison Hecker, represents the Blounts

Bach Ardalan, M.D., treated patient who received adulterated drug

Shuchu Hung, patient

Shyou Hung, daughter

Counterfeit Serostim

Kelly Burke, patient, California

Rick Roberts, patient, California

STATE INVESTIGATORS/REGULATORS

Georgia Drugs and Narcotics Agency

Richard Allen, Deputy Director

Nevada Board of Pharmacy

Louis Ling, General Counsel

Keith Macdonald, Executive Secretary

New Jersey

Michael Mordaga, Chief of Detectives, Bergen County Prosecutor's Office

Texas Department of Health

John Gower, Director, Drugs, Devices & Cosmetics

Albert Hokins Jr., former Investigator, Drugs, Devices & Cosmetics

NATIONAL ORGANIZATIONS

Center for Regulatory Effectiveness

Bruce Levinson, Director, Pharmaceutical Policy Project

Lew Kontnik Associates

Lewis Kontnik, Principal

National Association of Boards of Pharmacy

Carmen Catizone, Executive Director

National Association of Chain Drugstores

Steve Perlowski, Vice President, Industry Affairs

Pharmaceutical Security Institute

Tom Kubic, Executive Director
Peter York, Intelligence Analyst

Product Surety Center

Myles Culbertson, Director

SciRegs Consulting

C. Jeanne Taborsky, Consultant

US Pharmacopeia

Claudia Okeke, Associate Director
Eric Sheinin, Vice President, Standards Development
Sherrie L. Borden, Director of Public Relations

EXPERTS

SRD (Systems Research and Development)

John Bliss, Chief Privacy Officer

Business Research Services

Bruce Siecker, President & Treasurer

deKieffer & Horgan

Donald deKieffer, Partner

Global Options

Neil Livingstone, CEO

Goldman Sachs

> Christopher McFadden, Analyst

University of Pennsylvania Law School

> Paul Robinson, Professor

University of Texas at Austin

> Marv Shepherd, Director, Center for Pharmacoeconomic
> Studies

Z+ Partners

> Andrew Zolli, Founder

NATIONAL DISTRIBUTORS

AmerisourceBergen

> Michael Kilpatric, Vice President, Corporate & Investor
> Relations
>
> Chris Zimmerman, Senior Director, Security and
> Regulatory Affairs
>
> Larry Lonergan, Vice President, Distribution Center
> Manager, New Jersey

Cardinal Health

> Jim Mazzola, Vice President, Corporate Communications
>
> Tom Blaylock, former Business Unit Manager, National
> Specialty Services (NSS)

Bindley Western Industries

> Jack Earl Laughner, former President, Bindley Western
> Drug Company
>
> David Dyck, former Manager, San Dimas, California
> division, Bindley Western Drug Company

> FORMER LAWYERS
>
> Michael McCormick, General Counsel, Bindley Western
> Drug Company
>
> Larry Mackey, Partner, Barnes & Thornburgh LLP

REGIONAL AND SECONDARY DISTRIBUTORS

Florida

AD PHARMACEUTICALS
Sheldon Schwartz, former broker

AMERX PHARMACEUTICAL
Susan Cavalieri, former Principal

BIOMED PLUS
Marty Bradley, CEO & Founding Partner
Steve Getz, Sales Manager
Rene Perez, Purchasing Manager
Guy Bailey, Jr., Attorney, Bailey & Dawes

FIRST CHOICE PHARMACEUTICAL WHOLESALERS
Daniel Zarra, President

RAYMAR HEALTHCARE
Ray Maranges, Principal

California

REBEL DISTRIBUTORS
Ron Ressner, President & CEO

Kentucky

MDI
Alejandro Hoyos, Associate Attorney, Holland & Knight
Dennis Murrell, Attorney, Middleton Reutlinger

R&S SALES
Steve Shirley, President

Nevada

CALADON HEALTH SOLUTIONS
Robb Miller, President

DUTCHESS BUSINESS SERVICES
Paul DeBree, former President

LEGEND PHARMACEUTICAL, INC.
Lance Packer, former President

TRADE GROUPS

Healthcare Distribution Management Association (HDMA)

Ron Streck, former President & CEO
Jon Borschow, Board Officer
Anita Ducca, Director, Regulatory Affairs
John Howells, Director of e-Business
Amanda Forster, Director of Public Relations and
Communications

Pharmaceutical Distributors Association (PDA)

Bruce Krichmar, Legislative/Regulatory Affairs Director
Ross McSwain, Attorney, Blank, Meenan & Smith
Anthony Young, Attorney, Kleinfeld, Kaplan & Becker

PHARMACY SERVICES

Caremark, Inc.

Edward L. Hardin, Executive Vice President and General
Counsel
Gerard Carney, Spokesperson, Gavin Anderson & Company

Robert's Drugs

Aiman Aryan, Vice President of Pharmacy Operations

MICHAEL CARLOW'S WORLD

Business Associates

COLEMAN INSTITUTE
Norman Embree, former Principal
Hazel Hoylman, former Community Outreach
Coordinator

ENHANCE YOUR LIFE HEALTH CENTER, INC.
Jeffrey Schultz, former consultant

G&K PHARMA
    Sam Whatley, former salesperson

JB PHARMACEUTICALS
    John Bullock, listed former owner

MEDRX
    Mark Novosel, former administrator

NAVIGATOR PC
    Steven "Doc" Ivester, President & CEO

QUEST HEALTHCARE
    Michael Rosen, former administrator

Bail Bondsmen

THE BAIL BOND FIRM
    Rick Arenas, Bail Bondsman

INTERNATIONAL FIDELITY SERVICES
    Edward Sheppard, Bail Bondsman

NATIONAL SURETY SERVICES
    Russell Sabish, Bail Bondsman

High School Friends

    Sharon Demark
    Sarah Ellis

Lawyers

    Craig Brand, Attorney for Michael Carlow
    John Howes, Attorney for Michael Carlow

ASSORTED FLORIDA

    Joan Bardzick, President, A1 Check Cashing
    Wendy Blum, Territory Business Specialist, BMS Virology
    Sydney Dean Jones, pharmacy technician
    Benjamin R. Ojeda, former Manager, Playpen South

Acknowledgments

Many people helped to shape this book, encouraging and supporting me throughout the long months of reporting and writing.

The book first began as an article in *Self* magazine. I am grateful to the entire team there including news director Sara Austin for her terrific editing and news judgment; executive editor Dana Points; Meg P. D'Incecco, then the public relations director; senior editor for research Patricia J. Singer, who not only fact-checked the piece assiduously but pointed the way to new information; and editor-in-chief Lucy S. Danziger, who makes the magazine wonderful to write for and to read.

This book would not exist without the immense talent and experience of my agent, Liz Darhansoff, and my editor, Jane Isay. Liz, in addition to representing me, I thank you for your patience, loyalty, and friendship. They mean the world to me. Jane, you envisioned this book, brought it forth in two-and-a-half backbreaking years, and journeyed with me through every reporting adventure and every draft. Though you are my first and only book editor, I can't imagine there's a better one anywhere.

The Harcourt team has been a joy to work with. Jenna Johnson's quiet competence and unfailing good judgment made the process speedier and smoother than any of us expected. David Hough brought his tremendous experience, talent, and good cheer to every page. Jennifer Gilmore and

Michelle Blankenship worked in wonderful ways to get out the word. Donna Wares and Lisa Wolff helped to shape and polish the manuscript. And many thanks to Harcourt's legal eagles Susan Amster and Ed Klagsbrun, a terrifying lawyer and prince of a man.

I was lucky to have help from three top research assistants. The book benefited immeasurably from the work of Felicia Mello, who brought her able reporting, razor-sharp analytic skills, and tremendous brainpower to the project. Among her many contributions, she developed and maintained a database of facts and figures that proved crucial to developing the book's narrative.

Molly Eger stepped into a pressure cooker of a deadline, fearlessly organizing and analyzing thousands of pages of documents, a monumental effort that would have frightened off a less skilled information manager. In South Florida, Crystal Mattingly delved into the tangle of court and police records scattered across many databases in many different courthouses and, time and again, extracted needed documents.

Several grants allowed me to expand the scope of this book. I am grateful to the Alfred P. Sloan Foundation for its generous support, and particularly to Doron Weber, who runs the foundation's Public Understanding of Science and Technology program, for believing in this book.

The Fund for Investigative Journalism also gave essential support. I am grateful to John C. Hyde, David Burnham, and the Fund's other board members for their generous grant and am honored to have the backing of an organization that has helped so many of the independent journalists I admire.

A special thank-you must go to my sources (they know who they are), who have stuck by me for years, provided vital information, and believe, as I do, that the best disinfectant is sunlight.

I owe a debt of gratitude to Wowa Badian, who gave me a home away from home in South Florida. His remarkable generosity, and the warm hospitality of Jana and Friedrich Katz and Joey and Vera Badian, made this book possible.

I owe particular thanks to those stalwart friends who read the manuscript in progress. Jennifer Gonnerman, an exceptional journalist, gave generously of her time and talent. The changes she recommended improved the book significantly. Philip Friedman, a master novelist, read right down to the wire and gave great advice on story structure. My *Self* editor and valued friend, Sara Austin, gave me detailed suggestions that brought flow and coherence to rough passages. My cousin Dr. Jessica Kandel read with a keen eye and saved me from several missteps. My father, Michael O. Finkelstein, was a painstaking proofreader.

Thanks must also go to friends who brainstormed on words, ideas, matters of design, and gave other important advice including Ellen Avenoso, Maureen N. McLane, Lisa Waltuch, Tina Bennett, Kate Robin, Andrea Stern, Bonnie Schwartz, Vivian Berger, and Rachel Simmons.

I am grateful for the encouragement of my friends and former colleagues at the *New York Times,* particularly Walt Bogdanich and Chris Drew, investigative reporters at the top of their game. I thank Kurt Eichenwald for his sage advice.

Friends including Tracy Straus, Brad Hoylman, and Maryam Mohit, to name just a few, were understanding in the face of my virtual disappearance for over a year. The daily companionship of Lindy Friedman, our late afternoon coffees, and the company of Joel, Esme, and Elsie also kept me going.

Thank you to my entire family for their unflagging support in the book, including my brother, Matthew Dalton, a keen journalist and elegant writer; my sister, Claire Finkelstein;

brother-in-law, Leo Katz; and my niece, Tessie, a source of joy to us all. I can never thank my parents, Elinor Fuchs and Michael Finkelstein, enough for their love, their belief in me, and the example they set through their own excellent writing. It has made all the difference.

And finally, thank you to my beloved husband and best friend, Ken Levenson. He tolerated my long absences, encouraged me at every step, and has been my equal partner through every draft and every dream. He has made me happier than I ever imagined I could be. Our Newfoundland, Lola, has offered the kind of patient, supportive, and quiet company that no author should be without.

Index

Page numbers in parentheses refer to the page numbers at the front of each endnote.